RNA Genetics

Volume III
Variability of RNA Genomes

Editors

Esteban Domingo
Scientist
Instituto de Biologia Molecular
Facultad de Ciencias
Universidad Autonoma de Madrid
Canto Blanco, Madrid, Spain

John J. Holland
Professor
Department of Biology
University of California, San Diego
La Jolla, California

Paul Ahlquist
Associate Professor
Institute for Molecular Virology
and
Department of Plant Pathology
University of Wisconsin-Madison
Madison, Wisconsin

CRC Press
Taylor & Francis Group
Boca Raton London New York

CRC Press is an imprint of the
Taylor & Francis Group, an **informa** business

CRC Press
Taylor & Francis Group
6000 Broken Sound Parkway NW, Suite 300
Boca Raton, FL 33487-2742

First issued in paperback 2020

© 1988 by Taylor & Francis Group, LLC
CRC Press is an imprint of Taylor & Francis Group, an Informa business

No claim to original U.S. Government works

ISBN-13: 978-1-315-89734-9 (hbk)
ISBN-13: 978-0-367-65745-1 (pbk)

Library of Congress Cataloging-in-Publication Data

RNA genetics.

   Includes biblioigraphies and index.
   Contents: v. 1. RNA-directed virus replication --
v. 2. Retroviruses, viroids, and RNA recombination --
v. 3. Variability of RNA genomes.
   1. Viruses, RNA. 2. Viral genetics. I. Domingo,
Esteban. II. Holland, John J. III. Ahlquist, Paul.
[DNLM: 1. RNA--genetics. QU 58 R6273]
QR395.R57   1988   574.87'3283   84-22432
ISBN 0-8493-6666-6 (vol. 1)
ISBN 0-8493-6667-4 (vol. 2)
ISBN 0-8493-6668-2 (vol. 3)

A Library of Congress record exists under LC control number: 87022432

**Visit the Taylor & Francis Web site at
http://www.taylorandfrancis.com**

**and the CRC Press Web site at
http://www.crcpress.com**

THE EDITORS

Esteban Domingo was born in Barcelona, Spain (1943), where he attended the University of the city, receiving a B.Sc. in Chemistry (1985) and a Ph.D. in Biochemistry (1969). Subsequently he worked as postdoctoral fellow with Robert C. Warner at the University of California, Irvine (1970 to 1973) and with Charles Weissmann at the University of Zürich (1974 to 1976). Work with Weissmann on phage Qβ led to the measurements of mutation rates and the demostration of extensive genetic heterogeneity in phage populations.

He is presently staff scientist of the Consejo Superior de Investigaciones Científicas at the Virology Unit of the Centro de Biología Molecular and Professor of Molecular Biology at the Universidad Autónoma in Madrid. For the last 10 years his group has contributed to research on the genetics of foot-and-mouth diseases virus (FMDV). They have characterized the extreme genetic heterogeneity of this virus and have established cell lines persistently infected with FMDV. Dr. Domingo has been adivser to international organizations on FMDV and problems of research in biology and medicine. His current interests include the relevance of FMDV variability to viral pathogenesis and the design of new vaccines, subjects on which his group collaborates with several laboratories in Europe and America.

John J. Holland, Ph.D., is Professor of Biology at the University of California at San Diego, La Jolla, California. He received his Ph.D. in Microbiology at U.C.L.A., did postdoctoral work, and was Assistant Professor in the Department of Microbiology, University of Minnesota. He was Assistant Professor and Associate Professor of Microbiology at the University of Washington in Seattle and Professor and Chairman of the Department of Molecular Biology and Biochemistry at the University of California at Irvine before moving to the University of California at San Diego, and spent a sabbatical year as a visiting scientist at the University of Geneva, Switzerland. He has published numerous papers in the field of virology, and presently, he and his colleagues do research on rapid evolution of RNA viruses, virus-immunocyte interactions, defective interfering particles, persistent infections, and related areas.

Paul Ahlquist, Ph.D., is Associate Professor of Molecular Virology and Plant Pathology at the University of Wisconsin, Madison. He is a member of the American Society for Virology and currently serves on the executive committee of the International Committee on Taxonomy of Viruses. His research interests include virus genome structure, organization and evolution, RNA replication and gene expression mechanisms, and viral gene functions.

CONTRIBUTORS

Paul Ahlquist, Ph.D.
Associate Professor
Institute for Molecular Virology and
  Department of Plant Pathology
University of Wisconsin-Madison
Madison, Wisconsin

Christof K. Biebricher, Ph.D.
Research Biochemist
Max-Planck Institut für
Biophysikalische Chemie
Göttingen, West Germany

John F. Bol, Ph.D.
Head of Plant Virus Research
Department of Biochemistry University
  of Leiden
Leiden, Netherlands

Candace Whitmer Collmer, Ph.D.
Curator Department of Virology
American Type Culture Collection
Rockville, Maryland

Esteban Domingo, Ph.D.
Scientist
Instituto de Biologia Molecular
Consejo Superior de Investigaciones
  Cientificas
Canto Blanco, Madrid, Spain

Manfred Eigen, Ph.D.
Professor and Director
Max-Planck Institut für
Biophysikalische Chemie
Göttingen, West Germany

Bernard N. Fields, M.D.
Professor and Chairman
Department of Microbiology and
  Molecular Genetics
Harvard Medical School
Boston, Massachusetts

Roy French, Ph.D.
Assistant Professor
Department of Plant Pathology
University of Nebraska
Lincoln, Nebraska

John J. Holland, Ph.D.
Professor
Department of Biology
University of California at San Diego
La Jolla, California

Alice S. Huang, Ph.D.
Professor of Microbiology and
  Molecular Genetics
Harvard Medical School
Director, Laboratories of Infectious
  Diseases
Children's Hospital
Boston, Massachusetts

J. M. Kaper, Ph.D.
Research Chemist
Microbiology and Plant Pathology
  Laboratory
Plant Protection Institute
Agricultural Research Service
U.S. Department of Agriculture
Beltsville, Maryland

Paul K. Keese, Ph.D.
Research School of Biological Sciences
Australian National University
Canberra, A.C.T., Australia

Michael B. A. Oldstone, M.D.
Member
Department of Immunology
Scripps Clinic and Research Foundation
La Jolla, California

Peter Palese, Ph.D.
Professor
Department of Microbiology
Mount Sinai School of Medicine
New York, New York

John J. Skehel, Ph.D.
Division of Virology
NIMR
London, England

Frances I. Smith, Ph.D.
Assistant Professor
Department of Microbiology
Mount Sinai School of Medicine
New York, New York

P. J. Southern, Ph.D.
Associate Member
Department of Immunology
Scripps Clinic and Research Foundation
La Jolla, California

Robert H. Symons, Ph.D.
Department of Biochemistry
University of Adelaide
Adelaide, South Australia, Australia

L. van Vloten-Doting, Ph.D.
Director
Research Institute Ital
Wageningen, Netherlands

Jane E. Visvader, Ph. D.
Molecular Biology and Virology
 Laboratory
The Salk Institute
San Diego, California

Elizabeth A. Wenske, Ph.D.
Research Microbiologist
New Technology Research
E. I. duPont de Nemours and Company,
 Inc.
North Billerica, Massachusetts

D. C. Wiley, Ph.D.
Department of Biochemistry and
 Molecular Biology
Harvard University
Cambridge, Massachusetts

Flossie Wong-Staal, Ph.D.
Chief
Section on Molecular Genetics of
 Hematopoietic Cells
Laboratory of Tumor Cell Biology
National Cancer Institute
Bethesda, Maryland

RNA GENETICS

Volume I
RNA-DIRECTED VIRUS REPLICATION

Volume II
RETROVIRUSES, VIROIDS, AND RNA RECOMBINATION

TABLE OF CONTENTS

GENETIC HETEROGENEITY OF RNA GENOMES

GENE REASSORTMENT AND EVOLUTION IN SEGMENTED RNA VIRUSES

ROLE OF GENOME VARIATION IN DISEASE

Genetic Heterogeneity of RNA Genomes

Chapter 1

HIGH ERROR RATES, POPULATION EQUILIBRIUM, AND EVOLUTION OF RNA REPLICATION SYSTEMS

Esteban Domingo and John J. Holland

TABLE OF CONTENTS

I. INTRODUCTION: EARLY EVIDENCE OF GENETIC VARIABILITY OF RNA VIRUSES

It was often observed in the initial studies on mutagenesis of RNA viruses that wild-type revertants arose and contaminated clonal mutant preparations. In addition, it was noted that the so called "wild-type" virus included significant proportions of mutant genomes. Granoff[1,2] identified several spontaneous plaque-type (small, turbid) mutants of Newcastle disease virus (NDV) which occurred at variable frequencies that occasionally reached 1.5×10^{-1}. Pringle[3] characterized many *ts* mutants of vesicular stomatitis virus (VSV) Indiana serotype that had been induced by various mutagenic agents. The yield of revertant phenotypes in such preparations was quite high and variable. For one mutant it reached 0.19 (ratio of plaques of mutant and wild type plated at the restrictive temperature). For others, it was <0.001, either because they were affected in more *stable* genomic positions (Section VII) or because their *ts* phenotype was the result of multiple point substitutions or of deletions. When analyzed for its ability to synthesize RNA at the restrictive temperature, one mutant was shown to contain 10^{-3} revertants. One "subclone" derived from this mutant did not synthesize viral RNA in the infected cells, while another "subclone" of the same preparation did so normally.[4] Genetic instability was found with other VSV mutants.[5,6] In those studies the frequency of spontaneous *ts* phenotype in VSV was 1% and 2% with nonpermissive temperatures of 39.0 and 39.8°C, respectively. The early evidence of genetic variability of rhabdoviruses has been reviewed.[7-9]

From the onset, influenza viruses became the prototype of "variable" virus, although it is now clear that the genetic variation of most RNA viruses is generally comparable to that of orthomyxoviruses (Sections III to VI). During the early 1940s, F.M. Burnet studied the rapid phenotypic variations represented by the O (original) to D (derivative) transition in agglutinating ability of influenza virus.[10] Burnet stated that passage of a natural isolate of a virus in some laboratory host until it becomes less pathogenic is a process that "must necessarily be something of an exercise in population genetics".[11] He recognized the need to define the distribution of genotypes in the initial and final virus populations, the frequency of mutation, and the selective effect of the environment, as well as the difficulty in obtaining genetically homogeneous populations of viruses.

For reovirus, the frequency of spontaneous *ts* mutants was estimated to be 0.3%.[12] Among picornaviruses, variation in the behavior of individual mutants made very difficult the early genetic studies on RNA recombination. To keep mutants free of revertants, passages of viral stocks were kept to a minimum; even with such precautions, some mutants could not be used because of their genetic instability (review in Cooper,[13] and references therein).

Lesion-type mutants of plant viruses were recognized early and Kunkel showed that a porportion as high as one out of 200 lesions were induced by variant tobacco mosaic virus (TMV).[14] Further evidence of high mutability of TMV was provided by Mundry and Gierer.[15] As documented in other chapters and as reviewed by van Vloten-Doting,[16] plant viruses show extreme adaptability to new environments, and adaptation has often been shown to be concomitant with selection of mutant genomes.

Extreme variability was also clearly documented in initial studies with the RNA bacteriophages. Valentine et al.[17] found that a stock of bacteriophage Qβ contained 8% *ts* mutants. Furthermore, wild-type revertants frequently arose on the plaques formed by *ts* mutants. They concluded that replication errors constantly "replenished the mutant pool" and that roughly "one base in 3×10^4 was misread." Plaque morphology mutants, antiserum-resistant, amber, *ts*, and azure mutants of the RNA phages were described, and they included revertants at frequencies of 10^{-3}, except for several *ts* mutants that did not revert and were assumed to contain either multiple point substitutions or deletions (reviewed by Horiuchi[18]).

A fruitful line of research was opened by Spiegelman and colleagues who developed an in vitro system for the replication of purified Qβ RNA. They amplified RNA molecules in a "serial transfer experiment": a series of tubes contained the standard reaction mixture with Qβ replicase, but no template; the first tube was seeded with Qβ RNA and incubated to allow synthesis of progeny RNA; then an aliquot was transferred to the second tube and the process of synthesis and amplification of Qβ RNA could continue for extended periods of time.[19] This experiment in its exact format, or in modified versions, has been useful in the study of the evolution of RNA replicated in vitro and of viruses passaged in cell culture. The results of Spiegelman's group represented the first amplification of an infectious genome in the test tube and paved the way for numerous interesting experiments on evolution of self-replicating RNA molecules in vitro.[20] Phage RNA replication is reviewed by Biebricher, C. and Eigen, M. in Chapter 1 of Volume I.

The first evidence of a nucleotide sequence variation within "one virus" was obtained by Wachter and Fiers,[21] who showed two different 5' terminal sequences in RNA from cloned phage Qβ. Weissman et al.[22] considered that a possible complication in nucleotide sequence determination of phage RNA was that "an apparently phenotypically homogeneous phage stock might contain multiple variants at various sites on the RNA" (see Section V). Phage RNA sequencing, however, progressed quickly and by the mid 1970s, Fiers and colleagues had elucidated the complete nucleotide sequence of phage MS2,[23-25] a historic achievement in biology. Work originated in Spiegelman's, Fiers', and Weissmann's groups has had many important consequences for our understanding of RNA genome organization, replication, and variability, as documented in several other chapters of these volumes.

Most of the nucleotide sequences of viral RNAs during the 1960s and eary 1970s were elucidated by time-consuming enzymatic methods, since molecular cloning of cDNA and rapid nucleotide sequencing techniques were not yet available. This precluded the analysis of even a few of the genomes that constitute viral populations and, thus, the early suggestions of considerable variability of RNA viruses relied mostly on indirect genetic evidence. More recently, as methods of sampling of viral nucleic acid sequences have become available (Section II), the results have increasingly supported the concept of extreme variability of RNA genomes. As we document in Sections III to VII, the RNA systems that have been carefully analyzed have proven genetically heterogeneous in that they consist of complex mixtures of variant genomes. Upon RNA replication, new variants constantly arise and the proportion of each of them increases or decreases in competition with the other variants of the population. The latter is in a dynamic equilibrium that may be defined by an "average" or "consensus" sequence.[26,27] The composition of this consensus sequence is dependent, in the most critical way, on the environment imposed upon the system. A substantiation of this view and of the view that such genetic organization for RNA viruses has important biological implications, requires reviewing a number of analyses that have been carried out in recent years on several RNA genomes. We will first define the types of RNA genome analyses to be considered, the methods available to carry them out, and the results of nucleotide sequencing. Finally, we will propose a *population equilibrium model* for RNA genomes, and discuss its implications.

II. RNA GENOME POPULATIONS AND METHODS FOR THEIR ANALYSIS

For any RNA virus or nonviral RNA genetic element, a comparison of genomic nucleotide sequences may involve the following types of viral populations: (1) independent isolates obtained from different hosts, locations and times; (2) isolates that are somehow related epidemiologically, such as belonging to the same disease outbreak or to a defined geographical area; or isolates from the same infected individual obtained at different times or from different sites; (3) individual genomes within a single isolate. We define isolate here as the set of viruses obtained from an intact host organism or from one subset of tissues or cells from that host; and (4) finally, one may consider the analysis of genomes within clonal populations, such as viruses derived from a plaque produced on a cell monolayer. Obviously, since current methods do not allow direct sequencing of an individual molecule, each genome must undergo amplification (by replication in some host cell) before analysis, in order to provide the amount of material required. This will usually be necessary for analyses (1) and (2) and is clearly unavoidable for sequencing "individual" genomes or clonal populations as in (3) and (4). Additional viral populations of interest are provided by viral clones amplified in a directed fashion under controlled environmental conditions, for example, by serial passage in different host cells or host organisms. The resulting populations can then be subject to the analysis (3) or (4).

Several *direct* sequencing or *indirect* phenotypic methods are available at present for the analysis of viral populations.

A. Direct Methods

They involve comparison of RNAs, either by sampling of nucleotide sequences (oligonucleotide fingerprinting) or by determination of nucleotide sequences (RNA or cDNA sequencing or cDNA synthesis and molecular cloning and sequencing).

1. Oligonucleotide Fingerprinting[28]

The RNA is digested with a base-specific ribonuclease (generally T_1) and then subjected to two-dimensional gel electrophoresis. The position of the oligonucleotides on the gel is critically dependent on their size and charge. Thus, point mutations, insertions, or deletions in each oligonucleotide can be readily and reproducibly detected.[27-30] This sampling method has two limitations: (1) only 5 to 12% of the sequences (those in large, well-resolved oligonucleotides) are amenable to screening when the RNA has a complexity of 4 to 30 kilobases, the range for entire RNA genomes; however, a higher proportion of sequences can be analyzed by fingerprinting subgenomic-size RNAs selected by hybridization to cloned cDNA segments copied from viral genomes,[31,32] (2) little direct information on the location and nature of the mutations in the viral genome is obtained. Nevertheless, T_1-oligonucleotide fingerprinting has contributed decisively to our understanding of RNA genome variability.

2. RNA Sequencing

Viral RNA can be sequenced directly by enzymatic extension of an oligonucleotide primer using reverse transcriptase. Two approaches have been used: the sequence may be derived by including dideoxynucleoside triphosphates as chain terminators in the reaction mixture[33] or, alternatively, the cDNA copy may be sequenced by the chemical degradation method of Maxam and Gilbert.[34] Restriction enzyme fragments or synthetic oligonucleotides (generally 10 to 20 residues long) are used as primers.

3. cDNA Synthesis, Molecular Cloning, and DNA Sequencing

A double-stranded cDNA copy of the viral RNA is synthesized using reverse transcriptase and DNA polymerase I, and then cloned into viral or plasmid vectors, prior to amplification and sequencing. Some vectors, such as phage M13 derivatives, have been designed for direct DNA sequencing.[35,36] Alternatively, the cloned cDNA may be fragmented with restriction enzymes and sequenced.[34] Methods 1. (oligonucleotide fingerprinting) and 2. (RNA sequencing) yield the *average* or *consensus* RNA sequence of the population under study (see Section VII), while method 3. (sequencing of cloned cDNA) yields the sequence of one (or several) reverse transcription copies of individual molecule(s) constituting the population. These, of course, include any base substitutions, additions, or deletions which occur during reverse transcription.[37,38]

Single-base substitutions in RNA molecules have also been detected in RNA-RNA or

RNA-DNA duplexes by cleavage with ribonuclease and electrophoretic analysis of the fragments.[39,40] Also, single point mismatches were detected in hybrids by electrophoresis in denaturing gels.[41,42] For viruses that replicate via a DNA intermediate, restriction enzyme cleavage analysis of intracellular viral DNA has been used for the quantitation of viral genome heterogeneity.

A discussion of recombinant DNA techniques, as applied to the study of RNA virus gene function expression and regulation, can be found in Chapter 3 of this volume.

B. Indirect Methods

Indirect methods do not involve an analysis of the viral genome, but instead, some measurable phenotype is used to infer a genetic difference at the locus encoding a protein involved in that phenotype. Viral protein analyses (electrophoretic mobility, isoelectric focusing, peptide fingerprinting, and amino acid sequencing) belong to this group. Also, quantitation of the proportions of spontaneous *ts*, plaque morphology, etc. mutants, as in the examples of Section I, or the frequency of infectious viral particles resistant to monoclonal antibodies (MAbs) are *indirect* methods of viral population analyses. Next we review the information that *direct* and some of the *indirect* methods have provided for quantitation of heterogeneity among natural viral isolates.

III. GENETIC HETEROGENEITY OF NATURAL POPULATIONS OF RNA VIRUSES

RNA viruses are ubiquitous in nature. They constitute the most abundant group of cellular parasites. It has been estimated that more than 70% of all known viruses of differentiated organisms are RNA viruses.[43,44] They are diverse in virion structure, genome organization, and in replication strategy. Some of them are endemic to certain areas of the world. Others exist in many natural reservoirs without causing disease. On occasions, and for reasons often not understood, a virus suddenly causes an outbreak of acute disease and spreads rapidly among susceptible hosts. Some produce only one illness with a well-defined symptomatology. Others cause two or more distinguishable conditions, and then the question arises whether "one" or "several" viruses are involved. Some viral diseases, such as poliomyelitis, were already described in ancient civilizations, and others, such as epidemic keratoconjunctivitis and acquired immune deficiency syndrome (AIDS) have recently appeared as "new" diseases.

There is considerable information currently being published on oligonucleotide maps and nucleotide sequences of viral RNAs, and on antigenic characterization of viral proteins using MAbs. This short review can focus only on a few investigations dealing mainly with genetic heterogeneity and variability of RNA genomes. The reader will find reviews for specific groups of viruses, for other RNA genetic elements, and on the biological relevance of variant RNAs in other chapters of this volume.

The early serological analysis of cross reactivity among different viruses using polyclonal antisera and the molecular studies using nucleic acid and hybridization suggested variation among viruses that had been classified as belonging to the same group. Genomic RNA analyses by T_1-oligonucleotide fingerprinting and nucleotide sequencing have amply confirmed those early suggestions.

A. Negative Strand RNA Viruses

Genomic RNAs of different isolates of VSV (all viruses belonging to one rhabdovirus serotype) produced T_1-oligonucleotide fingerprints that revealed either a close relatedness or considerable divergence (10 to 20 oligonucleotide differences when examining 10% of the genomic sequences).[45] This study showed that, as expected, distinct fingerprints do not imply lack of a close serological relationship, as exemplified by the VSV New Jersey isolates. The results indicated also that divergence was more pronounced when viruses had an independent natural origin than when they differed only in passage history in the laboratory, although obviously the number of genome doublings that had occurred in each situation was not known.

The homologies between the L, N, M, and G genes of VSV NJ and the corresponding ones of the IND serotype are in the range of 50 to 80%.[46,47] The NS gene was 40.6% homologous and the deduced amino acid sequence 32%, with a likely conservation of phosphorylation sites in the protein.[48] Hydropathicity plots suggest, however, overall structural conservation of the encoded proteins.[49] Within VSV-IND, the M-protein gene of the Glasgow and Orsay strains differ in eight nucleotides and both differ from the San Juan strain in 13 nucleotides that result in 6 amino acid substitutions.[50] (see also Volume I, Chapter 7). The NS gene of vesiculoviruses is extremely variable.[50a]

The morbilliviruses that cause acute measles, acute measles encephalitis, or subacute sclerosing panencephalitis (SSPE) constitute a related set of genomes.[51-53] RNA of viral isolates from clinical cases of these diseases showed 15% or less, common, large T_1-oligonucleotides, in spite of all isolates being virtually identical by classical serology.[53] No consistent differences in the oligonucleotide maps were seen that could serve to distinguish measles from SSPE viruses. The transition from a classical measles infection into a different clinical syndrome, such as SSPE, is not well understood.[51,52] Studies of the virus that persists in the brain of SPPE patients, of SSPE tissue culture cell lines, and of cells persistently infected with measles virus (MV) suggest a defect in the matrix (M) protein synthesis or in its stability so that mature virions cannot be produced.[51,52,54-57] In viral RNA extracted from the brain of a SSPE patient, the monocistronic MV M m-RNA was entirely, or to a large extent, replaced by a bicistronic phosphoprotein-matrix (P-M) mRNA,[58] an abnormal messenger spanning two adjacent genes on the MV genome. Furthermore, direct nucleotide sequence analysis around the P-M intercistronic region revealed several point mutations[59] that distinguished that segment from the corresponding one of the Edmonston MV. One of the point mutations led to a stop codon at triplet 12 of the M reading frame. In this revealing study, two overlapping cDNA clones obtained from viral RNA of the same brain differed by 1% of nucleotides, and the sequenced segments of either clone differed from the corresponding region of Edmonston MV RNA in 2.8 to 3.7% of nucleotides.[59] These results suggest a rapid evolution of MV RNA during persistence in the brain and, although the number of different viral RNA molecules sequenced is still limited, this and previous analyses[53] strongly suggest that viral RNAs in SSPE consist of related, nonidentical nucleotide sequences. Thus, one possibility is that upon infection by MV, variant genomes with different tropisms and altered replication cycles find their way into neuronal cells and, in conjuction with the host immune response, induce the new pathology in rare individuals. This possibility is very relevant to concepts of population equilibrium and variation during replication of RNA genomes that are discussed in Section VII.

Another morbillivirus, canine distemper virus (CDV), can also cause a neurological disease in dogs. MV and CDV show antigenic cross-reactivity and nucleotide sequence homology.[60,61] A segment of the nucleocapsid protein (NP) of the two viruses shows 88% amino acid homology. Two short amino acid segments are also found in the nucleocapsid of Sendai virus, another paramyxovirus.[62] Conserved amino acid sequences between the NPs of human parainfluenza virus type 3, Sendai virus, MV, and CDV have been identified.[63,64] The M protein-coding regions of MV and CDV are 67% homologous. At the amino acid level, the homology is 76%, and both M proteins are 32 to 35% homologous with the M protein from Sendai virus.[65] Little homology exists among the fusion proteins of different paramyxoviruses.[66]

T_1-oligonucleotide fingerprinting indicated that avian paramyxoviruses isolated in different countries between 1975 and 1978 were each genetically unique.[67] A lower divergence (minimum 4% nucleotide differences) was scored among isolates from one prefecture in Japan during 1980.[68]

The 3' terminal nucleotide sequences of L and S RNAs of the arenaviruses lymphocytic choriomeningitis virus (LCMV), Tacaribe and Pichinde are conserved.[69] The comparison of the first 150 nucleotides at the 3' end of S RNA from the LCMV and LCMV-WE strains revealed 82% homology. LCM-WE and Pichinde S RNAs are 54% homologous, and the corresponding N protein products have 51% homologous amino acids.[70,71] The comparisons suggest maintenance of the folding of the protein. The homology for the glycoproteins encoded at the 5' half of S RNA of those viruses is 39%.[72] Genetic and phenotypic variations among LCMV strains have been recognized.[73]

The comparison of the nucleotide sequences of the M RNA from two bunyaviruses, snowshoe hare virus (SSHV) and bunyamwera virus (BV) shows overall 55% homology, but some segments are 80% homologous.[74,75] The respective protein products (encoded in the complementary RNA, see Volume I, Chapter 9) show 43% amino acid sequence homology and a similar hydropathy profile. Amino acid variations occur not only between SSHV and BV, but also between those viruses and La Crosse virus and its variant L74 (Figure 8 in Lees et al.).[75]

Heterogeneity within one virus, some amino acid sequence conservation among different groups of viruses, and predictions of conservation of secondary structure of viral proteins in spite of limited sequence homology are observations often encountered with different systems.

B. Positive Strand RNA Viruses

Picornaviruses, the simplest animal RNA viruses, constitute an important group of human and animal pathogens that has been the subject of many revealing studies. The nucleotide sequences of the three prototype polioviruses of serotype 1, 2, and 3 indicate that their homology extends to 71% of the genomic nucleotides with 80% of the substitutions being silent; this comparison suggests that the three serotypes derive from a common ancestor.[76] That each "serotype" is represented by multiple related sequences was first demonstrated clearly by Kew et al.,[77] who showed that poliovirus serotype 2 and 3 vaccine strains can undergo more than 100 mutations during replication in only one or two individual humans. Analysis of clinical isolates of poliovirus type 1 suggested that about 100 base changes were fixed in the viral genome during 13 months of epidemic transmission.[78] As expected, genetic and antigenic heterogeneity was detected among 17 poliovirus type 3 isolates (1939 to 1958).[79] Stanway et al.[80,81] found that attenuation of poliovirus type 3 was brought about by a maximum of ten point mutations, three of them involving amino acid substitutions. Evans et al.[82] showed that a single nucleotide substitution at position 472 (from U in the type 2 Sabin vaccine to C in mutants) occurred regularly and rapidly during replication in the intestine of vaccinated persons. Also, Jameson et al.[83] sequenced the point mutations in antigenic variants of the Sabin type 1 vaccine which appeared in normal infants several days postvaccination. Administration of plaque-purified W-2 strain of poliovirus type 2 to mice resulted in recovery of viruses with mutations in up to 1.2% of genomic nucleotides from paralyzed mice and up to 0.3% from nonparalyzed mice.[84] Thus, in the latter study, a set of quite different genomes were observed following infection by a plaque-purified virus. Minor et al.[85] immunized a 4-month-old child with a Sabin live vaccine containing the three serotypes of poliovirus, then characterized type 3 viruses shed during the next 73 days of fecal excretion. A variety of point mutations and recombinations with the other two serotypes were observed in mutant strains which appeared during in vivo replication. However, antigenic variation was observed at minor antigenic sites, but not at the major immunodominant site. An antigenically unusual type 3 poliovirus caused a limited outbreak of poliomyelitis in Finland between August 1984 and January 1985.[86] Analysis with MAbs suggested variations in the major immunodominant site involved in virus neutralization, a site previously shown to be quite conserved among unrelated type 3 viruses.[86,87] The viruses isolated during this outbreak in Finland were genetically and antigenically heterogeneous.[86,87a] Omata et al.[88] used recombinational analysis to map the attenuation phenotype of poliovirus type 1 Sabin vaccine. They observed that the loci influencing attenuation were spread out over several areas of the vaccine virus genome, indicating that a number of different mutations were involved. Also, none of the in vitro phenotypic markers alone could serve as a good indicator of attenuation or neurovirulence. Finally, Kohara et al.[89] have shown that infectious cDNA clones[90] of poliovirus Sabin strains may be used as a stable repository of inocula for vaccine production. This could lower the risk of reversion to virulence upon repeated passage of vaccine repository virus.

Foot-and-mouth disease virus (FMDV), the causative agent of the most important animal viral disease world-wide shows a remarkable antigenic diversity, with seven serological types and more than 65 subtypes having been recorded by classical serological analyses.[91-93] The genome organization of FMDV has been studied by molecular cloning and expression of viral gene products in recombinant plasmids.[94-99a] Considerable interest exists in capsid protein VP1, since early[100-103] and more recent evidence[104-108] indicates that it contains a major immunological site involved in virus neutralization, and thus either VP1 or synthetic peptides with the amino acid sequence of its relevant epitopes could be the basis for new synthetic vaccines.[108] The entire coding region of several strains[95-98] and the segment encoding the VP1 of more than 40 FMDV strains[110-120a] and isolates have been sequenced. These analyses have shown extensive genetic polymorphism among strains of different serotype[110-115] or of the same serotype.[115-120] RNA from FMDVs isolated within single outbreaks of disease differ in 0.7 to 4.0% of nucleotide positions.[119-122] Genetic heterogeneity has been found among viruses isolated the same day from one geographical location,[119] and distinct RNAs have been identified within single FMDV isolates.[116,121] The rate of fixation of mutations in the viral genome during an outbreak of FMD was evaluated by nucleotide sequencing of the VP1 gene and by T_1-oligonucleotide fingerprinting of genomic RNA segments selected by hybridization to cloned cDNA probes representing defined regions of the viral genome.[119] Values ranged from $<0.04 \times 10^{-2}$ to 4.5×10^{-2} substitutions per nucleotide per year, depending on the time period and genomic segment considered. Structural protein-coding segments evolve at rates up to sixfold higher than nonstructural protein-coding regions.[119,123,124] Thus, natural isolates from localized outbreaks of FMD consist of a continuum of related nonidentical genomes.[119,121] An even more rapid rate of genetic and antigenic variation occurred during persistent FMDV infections of cattle following infection with plaque-purified FMDV.[124a]

Takeda et al.[125] have demonstrated the continuous rapid evolution of the genome of enterovirus 70, a newly appeared human disease agent that causes an acute hemorrhagic conjunctivitis first recognized in Ghana in 1969. RNA from four coxsackie A10 viruses isolated in Japan in 1978 had 85 to 93% indistinguishable T_1 oligonucleotides, but the isolates from 1981 to 1982 shared only 17 to 34% of their large oligonucleotides.[126] Variations were also shown among six independent isolates of Drosophila C virus, an insect picornavirus, and the T_1-oligonucleotide maps suggested that some isolates were mixtures of more than one virus.[127] Hepatitis A virus (HAV) was considered rather homogeneous because independent isolates proved serologically indistinguishable. However, the interesting study of Siegl and colleagues[128,129] demonstrated that the genomic RNA of independent isolates of HAV differs in 1 to 6% of genomic positions, as estimated by oligonucleotide fingerprinting, and the value is in agreement with that deduced from nucleotide sequence comparisons.[130-132] Siegl and colleagues found that different HAV isolates varied in phenotypic properties such as host cell specificity, optimal replication temperature, etc., an observation commonly seen with natural isolates of other viruses such as FMDV[91,120a,133] or HAV.[134] Phenotypic heterogeneity is an important consequence of genetic heterogeneity of RNA viruses and it is discussed in Section VII.

Partial or complete genomic nucleotide sequences have now been determined for over 40 picornaviruses, including, in addition to those mentioned above, rhinoviruses,[135-138] echovirus 9,[138] encephalomyocarditis virus,[139] and coxsackievirus B3.[140] The predicted amino acid sequences for the four capsid proteins VP1, VP2, VP3, and VP4 have been compared by Palmenberg[141] after dividing the picornaviruses into the following four groups: group one: FMDVs; group two: EMC, Mengo, and Theiler's viruses; group three: HAVs; group four: polio, coxsackie, and rhinoviruses.[142] Values for homology between amino acids were based on the criteria of Staden.[143] Consensus sequences were defined for each of the four groups individually, and then for all viruses considered. In the latter comparison, a consensus amino acid could be assigned to about one third of the residues.[141] An examination of the compiled sequences allows several conclusions that have also been drawn from analyses of other groups of viruses, some included in previous paragraphs. In comparing two viruses, the number of amino acid substitutions varied enormously depending upon on the particular segment examined within one protein and upon the viruses under consideration. For example, the overall number of substitutions for all capsid proteins between Rhinovirus 2 and 14 is about 20-fold higher than between two isolates of HAV. Within the FMDVs the variations are such that no consensus amino acid could be assigned to certain positions, notably those known to be major antigenic determinants in VP1. In spite of variations in primary sequence, the viral proteins and the virions from different picornavirus groups have a very similar spatial organization. This is suggested by predictions of higher order structure of the viral polyprotein or of individual proteins.[131,138] This has been elegantly established for complete virions by solving the structure of several picornaviruses by X-ray diffraction.[144-147] Crystals of other picornaviruses, are presently under analysis by X-ray crystallographic methods.[148] It is to be expected that as more three-dimensional structures for picornaviruses are elucidated, it will become possible to unravel the chemical subtleties needed to preserve shape, while permitting the fixation of variant amino acids.

A similar picture emerges from studies with more complex positive strand RNA viruses, namely, alphaviruses, flaviviruses, and coronaviruses. One of the first systematic analyses of a natural population of an RNA virus was carried out by Trent and colleagues on the alphavirus Venezuelan equine encaphalomyelitis virus (EEV).[149] The oligonucleotide maps of RNA from EEVs representing different subgroups revealed variations that, assuming a random representation of the mutations among the T_1-oligonucleotides, must affect a few hundred genomic positions. Even closely related isolates (such as TC-83 and PTF-39 which differed in two oligonucleotides) probably deviate in 10 to 20 genomic positions. Similar conclusions were reached in a study of isolates of western equine encephalomyelitis virus.[150] Genetic and antigenic heterogeneity was also found among isolates of Sindbis virus.[151] Partial or complete nucleotide sequences of different alphaviruses have been determined,[152-158] and they are discussed by Strauss and Strauss in Chapter 4 of Volume I. Comparison of the structural glycoproteins of eight alphaviruses suggest conservation of the three-dimensional conformations in spite of variation in the primary sequence.[159] Moreover, conservation of certain segments among nonstructural proteins of Sindbis virus and of plant viruses has been recognized.[160]

Flaviviruses are an important group of pathogens that include more than 60 members. Complete nucleotide sequences have been obtained for two strains of yellow fever virus (YFV), the 17D vaccine strain[161] and its parental virulent strain,[162] and for West Nile virus[163,164] (and references therein). Partial sequences are also known for Murray Valley encephalitis virus (MVEV), St. Louis encephalitis virus, Japanese encephalitis virus, and two serotypes of Dengue virus. These sequences and their implications for the evolution of flaviviruses are reviewed in Chapter 4 of Volume I. As is the case for other viruses, the extent of homology varies greatly depending upon the protein considered and upon the domain within a protein. For example, overall amino acid sequence homology between YFV 17D[161] and MVEV, 1951 Australian isolate[165] is 42%; residues 92 to 123 of protein E, the viral hemagglutinin are 79% homologous, whereas for residues 125 to 183 the homology is only 12%. Since E protein possesses the major antigenic determinants, Dalgarno et al.[165] have suggested that group-, type-, and strain-specific epitopes might correlate with domains of decreasing conservation (see also Section IV). Again, the predicted hydrophobicity of the flavivirus polyproteins is similar even in regions of limited amino acid homology. However, despite likely structural conservation, the analysis of isolates of the flavivirus St. Louis encephalitis virus (SLEV) by T_1-oligonucleotide fingerprinting revealed a distribution of nonidentical genomes.[166] Trent and colleagues divided 57 SLEVs into several geographical varieties and showed that viruses within one variety constituted a related set of genomes that were distinct from the set from another variety. They concluded that "evolution of SLEV within a geographical area is a continuing process." The same concept applies to other flaviviruses such as Dengue virus[167] and YFV.[168]

Although less extensive, the studies with coronaviruses provide evidence for a notably heterogeneous group of viruses. Among them, murine hepatitis viruses (MHVs), in spite of sharing similarities in structure and replication strategy, cause diverse diseases such as hepatitis, encephalomyelitis, and gastroenteritis.[169] The replication of coronaviruses is reviewed by Lai in Chapter 6 of Volume I. The genomes of several MHV virus strains analyzed by Lai and Stohlman[170] differed by multiple mutations, and each genome had a number of unique, large T_1-oligonucleotides.[170,171] Two variants from the neurotropic strain JHM were isolated after serial passage of the virus in suckling mouse brain. They were both genetically and phenotypically distinct.[172] The nucleotide sequences of the nucleocapsid gene of two different coronaviruses, MHV JHM and avian infectious bronchitis virus (IBV), are very divergent, except for short stretches of homology, while the two closely related Beaudette and Massachusetts M41, strains of IBV show variations in 4% of residues, plus a 184-base insertion present in the 3' extracistronic region of the Beaudette strain and absent in M41. The coding region shows 5% amino acid variation between the two viruses.[173]

In addition to quite high variability due to point mutations, coronaviruses exhibit a remarkably high rate of recombination. This is apparently a result of their discontinuous and nonprocessive mode of replication which generates free segmented RNA intermediates. This can allow a high rate of RNA recombination by a copy choice mechanism.[174] RNA recombination is reviewed by King in Chapter 7 of Volume II.

Thus, analysis of genomes from natural isolates of viruses allow the following general conclusions to be drawn. For each virus group there are multitudes of deviant genomes around one particular sequence that we may call the "prototype" virus. Different "prototypes", however, exist for viruses corresponding to particular geographical isolates. As we document in Section V, a "prototype" virus usually consists also of multiple nonidentical genomes for which an "average" or "consensus" sequence represents the most appropriate description. Viruses that differ considerably in the primary sequence of their proteins maintain, however, similar folding of individual proteins and similar particle shape. Nevertheless, mutants of plant viruses with altered architecture have been described,[16] and predictions of different secondary structure are computed for highly variable regions such as segments of the major envelope protein of HTLV-III. Stretches of primary sequence homology have been identified among quite different viruses, in particular, in the viral RNA replicases.[160,175-178] The significance of those observations is discussed by Zimmern in Chapter 10 of Volume II. The differences in amino acid sequences revealed by the analyses summarized above are expected to be reflected in antigenic variations of the corresponding viral proteins. Next we summarize results of analysis of RNA viruses using MAbs.

IV. ANTIGENIC HETEROGENEITY OF NATURAL POPULATIONS OF RNA VIRUSES

Most groups of viruses have been analyzed with regard to their reactivity with MAbs directed to several structural and nonstructural proteins. This allows comparisons among groups, strains, and isolates. A considerable number of studies of RNA viruses with MAbs have been previously reviewed.[9,179,180] We present some recent examples that support two previous conclusions drawn from studies with different systems: (1) viral isolates that were either antigenically indistinguishable or strongly cross-reactive using polyclonal sera can often be divided in two or more antigenic groups when MAbs are used for their comparison; and (2) some MAbs react with several different viruses within one family, while others react only with representatives of one group, subgroup, or even one set of isolates.

The division of respiratory syncytial virus (RSV) into two subtypes has been recently proposed on the basis of the analysis of the reactivity of nine strains and isolates with a panel of 31 MAbs.[181] The complex genetic relationship among morbilliviruses (Section III. A) finds its counterpart at the protein level with regard to their reactivity with MAbs: virus group specific, virus group cross-reactive, virus type specific as well as a number of intertypic epitopes have been defined.[182] Each four strains of mumps virus also showed unique epitopes,[183] and epitope variability is remarkable among strains of Newcastle disease virus.[184] Of four groups of MAbs prepared against human parainfluenza 3 virus (PI3), only some of group 1 (defined for their high hemagglutination inhibition, neuraminidase inhibition, and plaque neutralization titers) cross-reacted with bovine PI3, indicating that the human and bovine viruses share at least one epitope and differ in many others.[185] Similar conclusions have been reached in the analysis of arenaviruses.[186,187]

Among MHV strains, strong conservation of epitopes has been found for the internal N protein and for the membrane glycoprotein E1, while variations occur within the peplomer glycoprotein E2, the target for neutralization antibodies.[188,189] One determinant A(E2), however, was found on all strains. This stable epitope appears to be determined by the primary amino acid sequence, and thus it is likely that it corresponds to one of the constant protein domains present in most viral proteins, as documented in Section II. In fact, the frequency of occurrence of variants resistant to that MAb was very low.[189] These studies[188,189] also illustrate the difficulties often found when trying to correlate a particular antigenic structure with a phenotype such as viral pathogenicity (see Section VII). Analyses of transmissible gastroenteritis virus (TGEV), another coronavirus, revealed antigenic conservation of sites on the N and E1 proteins when nine TGE strains were compared.[190] Variations were seen with MAbs directed against the peplomer protein E2. However, the frequency of variants resistant to MAbs directed against protein E2 from TGE was lower than 10^{-9} (no resistant virus could be isolated).[191,192] This extremely low frequency probably reflects the strong biological pressure which can act to select against variation at certain sites.

Using MAbs, Prabhakar et al.[193,194] found high frequencies of antigenic variants among clinical isolates of coxsackie B4 virus. They used a sensitive microneutralization assay to detect epitope negative viral variants present at low frequency. They concluded that changes in one or more epitopes are detected at frequencies exceeding 10^{-2}. Antigenic variants were also detected within plaque-purified populations (see Section V). In this revealing study, both the disappearance and appearance of reactive epitopes upon virus passage was demonstrated.[194] Some epitopes appeared to be stable, but unique to viruses derived from one human isolate.[195] It is likely that complex patterns of reactivity with MAbs will also be found with other picornaviruses.[196,197] The epitopes of coxsackie B4 virus were divided into highly, moderately, and poorly conserved. This distinction is being currently substantiated for other RNA viruses such as flaviviruses[198] and morbilliviruses.[199] *Constant* and *variable* epitopes may be defined on the basis of their occurrence in independent viral isolates, or of the frequency within a clonal preparation of the virus of infectious particles which do not react with the relevant MAb. The presently available evidence suggests that the two definitions will not necessarily lead to the same classification of epitope variability. Presumably, this reflects the different selective pressures acting on a virus during replication in natural environments and during growth of viral clones in cell cultures. In the next section we review studies which examine clonal populations of viruses with regard to their genetic and antigenic heterogeneity.

V. GENETIC AND ANTIGENIC HETEROGENEITY OF CLONAL POPULATIONS OF RNA VIRUSES

The evidence presented in Sections III and IV leaves little doubt that natural populations of RNA viruses consist of multitudes of nonidentical genomes. In this section we review evidence that clonal populations of RNA viruses are also genetically heterogeneous. We term clonal populations those that were derived from a single genome which was amplified by any procedure (compare Section II). Clonal populations of viruses occur in nature when a single successful virion colonizes a susceptible host, probably a fairly frequent event. Replication in a multicellular organism provides multiple physical and biological environments, which could allow for differential growth of variant genomes such as the selection of multiple genomes upon infection of mice with plaque-purified poliovirus[84] (Section III.B). In the laboratory, plaque-purified virus propagated in cell cultures gives clonal preparations. Very different selective pressures are provided by propagation in lytic infections or by establishing persistent infections. We review here analyses of viruses which underwent a minimum of amplifications in their host cells, and of others subject to more extensive replication, but in all cases the infections were initiated by a single genome.

Evidence that phage Qβ consists of complex mixtures of variant genomes was obtained by the group of Weissmann in Zürich. The phage, after its isolation in Kyoto in 1961[200] was passaged for many years in the laboratories of Watanabe (Tokyo) Spiegelman (Urbana) and Weissmann (New York and Zürich, 1965 to 1974). In 1974, Shapira and Billeter in Zürich cloned phage Qβ from a multiply passaged phage stock. The T_1-oligonucleotide fingerprint of this clone termed "AS" lacked one large T_1 oligonucleotide present in the parental viral stock. At this time, Weissmann and colleagues were testing the infectivity of RNA mutant molecules generated by in vitro site-directed mutagenesis (Section VI.A), by transfection of *Escherichia coli* spheroplasts, and fingerprinting of phage RNA derived from individual plaques. T_1-oligonucleotide fingerprints from some RNA progeny deviated from that of clone "AS" and of its parental Qβ populations.[201] A systematic analysis by T_1-oligonucleotide fingerprinting of RNA from more than 160 phage clones (equivalent to the screening of about 70,000 nucleotides in search for mutations) revealed that 15% of them showed a fingerprint that deviated from that of the parental populations. Assuming that the mutations were randomly distributed among the infectious molecules, the results suggested that each viable genome differed in one to two positions from the "average" or "consensus" sequence of the parental population.[27] The study also showed that serial passage of the cloned Qβ resulted in the generation of variant genomes while the "average" T_1-oligonucleotide map of the RNA remained unchanged during the 50 serial infections involved in the experiment. Billeter sequenced and mapped on the Qβ genome all large T_1 oligonucleotides resolved on the fingerprint.[202] This permitted the identification and precise positioning on the Qβ genome of several mutations found on variant genomes. Thirteen mutations affected regions encoding structural phage proteins, a few were extracistronic and only one was located within the replicase gene, reflecting a bias against variation of nonstructural proteins seen in several viruses of eukaryotes (Section III and Chapter 10 of Volume II). Four of the variants were tested in growth competition experiments with uncloned phage

Qβ. Interestingly, they grew at a rate relative to the wild type of about 0.87.[27] Thus, variant genomes arise in the course of replication of phage Qβ that have a selective disadvantage when grown in competition with the other genomes that constitute the "wild-type" population (Section VII).

Vesicular stomatitis virus (VSV) may be propagated in cell culture in lytic or noncytocidal persistent infections. The genetic analyses of plaque-purified VSV and VSV serially passaged under different culture conditions has revealed either rapid genome evolution or quite strong conservation of genome sequences, depending upon passage conditions (reviewed by Holland et al.).[9] When a plaque-derived clone of VSV was employed to initiate prolonged persistent infection of BHK-21 cells in culture, virus derived from this carrier culture has undergone continuous rapid evolution over all areas of the genome. Likewise, serial undiluted passages of acute infection of BHK-21 cells by cloned virus produced extensive and rapid sequence evolution. In contrast, serial dilute passages of the same cloned VSV showed very little sequence evolution, despite hundreds of passages of acute infection in BHK-21 cells. Obviously, under the latter conditions the equilibrium pools during each acute passage are dominated by the most fit original consensus sequence (or closely related sequences) used to initiate the passages. In the former cases, constant selective pressures exerted by defective interfering (DI) particles repeatedly destabilize equilibrium pools and lead to the continuing emergence of new DI particle-resistant virus mutants (S di$^-$ mutants) and of new DI particles which coevolved.[203-205] This is direct evidence that a fit variant can maintain its consensus sequence dominance for extended periods, whereas it may be rapidly displaced whenever selective forces favor the emergence of any of the multitude of other variants which are always present in a quasispecies RNA virus population.[206] The quasispecies or extremely heterogeneous nature of RNA genomes and its biological implications are discussed in Sections VII and VIII. The theoretical basis of the quasispecies concept is treated by Eigen and Biebricher in Chapter 12 of this volume.

Heterogeneous populations of infectious FMDV were generated upon serial passage of plaque-purified FMDV C_1 or O_1 in cell cultures.[207] The T_1-oligonucleotide maps of 34 individual clones derived from the multiply passaged populations indicated that each infectious genome differed in about four to eight mutations from the "average" or "consensus" nucleotide sequence that defines the complete population. This estimate of heterogeneity agrees well with the values for phage Qβ[27] and VSV.[208] In contrast to the results with phage Qβ, passaging of FMDV led to rapid selection of a different "average" nucleotide sequence. This observation was interpreted by Sobrino et al.[207] to be the result of adaptation of the FMDV natural isolate to the new environment provided by cell culture conditions. It must be emphasized, however, that during serial passage of phage Qβ, new genetic variations are as likely to occur in the constant environment provided by *E. coli* in laboratory conditions, as suggested by the similar level of genetic heterogeneity attained with the two viral systems. The critical difference is, however, that mutations able to endow the virus with a measurable selective advantage will occur rarely in the case of phage Qβ, since the virus has been "testing" the same or a very similar environment for many generations (compare the origin and history of passages for Qβ and FMDV in Domingo et al.).[27,121] In favor of that interpretation is the experimental observation that an increase of the temperature of propagation of Qβ in *E. coli* from 37.0 to 41.5°C led to RNA variation readily selected and detected on T_1-oligonucleotide maps. This represented a shift in the "average" sequence of the phage genome.[209] It is also noteworthy that at least some of the FMDV variants selected in cell culture were phenotypically different from the parental virus, as shown by a threefold increase in the yield of infectious particles per cell.[207] Also, FMDVs with altered host range, *ts*, etc. have been isolated from BHK-21 cells persistently infected with FMDV, in spite of the fixation of a limited number of mutations in the viral genome.[32,209a] Thus, when selective presures are very strong, such as those exerted by DI particles of VSV,[9,203-205] the viral genome undergoes extensive and rapid genetic variation. When no strong selective pressures are present, viruses can undergo a slower, but also continuous, evolution, such as in the passages of phage Qβ[27,209] or of FMDV in cell cultures.[32,207,209a] In the latter situation, viruses better adapted to their environment than the parental ones were also selected among the pool of continuously arising variants.[207,209a] Also, the appearance of antigenic (or other) variants upon passage of FMDV in cell cultures[91,210] is best explained by selection or drift acting on a quasispecies (extremely heterogeneous population), rather than acting on mixtures composed of a limited number of variant viruses (Section VII and Chapter 12 of this volume).

For several systems, sequencing of a number of independent clones of cDNA copied by reverse transcription of viral RNA revealed numerous mutations indicative of heterogeneity.[211-213] Because misincorporations during reverse transcription[38] are probably about as frequent as for RNA replication (Section VI), it is difficult to ascertain whether mutations exist in template RNA molecules or are introduced during cDNA synthesis. Of course, when the same mutation is found two or more times on independent cDNA clones, there is little doubt that the variation occurred at the RNA level.[37,208,212]

Analyses of RNA molecules replicated in vitro or of clonal populations of RNA viruses and other genetic elements provide an increasing wealth of evidence that RNA genome populations are extremely heterogeneous.[9,206,214-216] Additional examples are discussed in several chapters of this volume. It is particularly noteworthy that the sequence heterogeneity of minivariant Qβ RNAs amplified after cloning was very rapidly restored (Chapter 12 of this volume, Section VII.D).

The concept that genetic elements which replicate with a limited fidelity will lead to a set of related, nonidentical elements was initially proposed on a theoretical basis by Eigen and colleagues.[217,218] They coined the term ''quasispecies'' to define the set of related elements, in heterogeneous population such as those constituting nucleic acid clones (Chapter 12 of this volume). Domingo et al.[206] have reviewed the evidence for, and the relevance of the quasispecies (extremely heterogeneous) nature of many viral RNA populations, since it provides the raw material for adaptability, rapid evolution, and phenotypic flexibility of RNA genomes (Section VII).

Extensive heterogeneity at the nucleotide level predicts that unless all or most mutations are silent (that is, that they do not result in an amino acid substitution), heterogeneity will also be found in the antigenicity of RNA viruses. Indeed, MAbs have afforded an independent indirect method to quantitate variants in clonal viral populations. Infectious particles which do not react with a MAb are generally found at frequencies higher than 10^{-6} and up to 10^{-2} [180,193,194,219-225] In the study of Diamond et al.[224] it was shown that variant polioviruses selected for their resistance to MAbs directed against two neutralization epitopes of capsid protein VP1 included one or two mutations located away from the critical determinant, and even on a different structural protein. This illustrates the profound phenotypic effect that single amino acid substitutions may have on a virion (see Section VII). Also, several independent mutations or groups of mutations may yield a very similar phenotype, such as absence of binding of virions to an antibody molecule. Diamond et al.[224] also reported that despite plaque purification, three isolates resistant to one MAb appeared to be mixtures of either two single mutant genomes or of a double mutant together with wild type.

In contrast, as indicated in Section IV, for some viruses, variants resistant to certain MAb occur at low frequency.[191,192,226] It must be emphasized that the observed frequency of occurrence of an antigenic variant in a virus population need not correlate with the mutation rate at the nucleotides that encode the relevant amino acids. In addition to mutation rates, such frequency is also the result of the relative fitness of the resulting variants and the parental genomes (Section VII). Few direct measurements of mutation frequencies and mutation rates have been obtained for RNA genomes. Next we review the limited number of methods employed to date and their results.

VI. MUTATION FREQUENCIES, MUTATION RATES, AND VARIABILITY OF RNA GENOMES

The genetic heterogeneity of clonal populations of RNA viruses (Section V) must be the result of variant genomes generated at any stage of the life cycle of the virus and of the varying replication rates of those viruses relative to that of the parental genomes. One critical parameter in the quantitation of RNA variability is, therefore, the rate at which genomic alterations (point mutations, additions, deletions, rearrangements, etc.) occur in RNA viruses. Few measurements have been carried out on the frequency of occurrence of point mutations, and they have concerned only one or few nucleotides of the entire genomes studied. The stage is set, however, for additional measurements, and those obtained so far suggest extremely high mutation rates. Next, we summarize the results and compare them with those obtained with larger DNA genomes.

A. Reversion Rate of an Extracistronic Mutant of Phage Qβ

Elegant experiments by Flavell, Weissmann, and colleagues in Zürich on in vitro site-directed mutagenesis of Qβ RNA permitted for the first time the preparation of viral RNAs with mutations at defined, preselected sites on the genome.[227-230] Several nucleotides were modified by this procedure, in particular, from the 3' extracistronic region, a segment that plays an important function in RNA replication and which is conserved among related RNA phages.[231-233] An interesting demonstration of the power of this methodology was the synthesis in vitro of an infectious extracistronic mutant RNA with an A → G transition at position 40 from the 3' end (A_{-40} → G).[26] The mutant RNA replicated less efficiently than its wild-type counterpart both in vitro and in vivo. Upon infection of *E. coli*, the plaque-purified mutant virus reverted spontaneously to the wild-type sequence (G_{-40} → A), an event that was quantitated by fingerprinting RNA from the mixed viral populations. By following the reversion of independent mutant clones, and by growth competition experiments between wild type and mutant phage, Batschelet et al.[234] estimated that the transition G_{-40} → A occurred at a rate of about 10^{-4} per genome doubling. In these experiments, the relative growth rate of mutant A_{-40} → G to wild-type virus was 0.25. This value was entered into the calculation, so that the calculated mutation rate at this site was probably quite accurate.

B. Direct Measurement of Mutation Frequencies

Recently, a direct method that can be applied to many different sites on any RNA genome has been developed by Steinhauer and Holland.[208] This method utilizes the absolute specificity of certain ribonucleases to produce oligonucleotides of defined size and sequence from viral genome RNA and from RNA transcripts. "Error oligonucleotides" are generated whenever the base at the consensus cleavage site is substituted by a base which is not a cleavage substrate. For example, mutation of a G residue at any chosen position to a U, C, or A residue will prevent T_1 ribonuclease cleavage at that position and a larger "fusion" oligonucleotide will be formed in place of the two smaller expected consensus T_1 oligonucleotides. These error oligonucleotides can be isolated and quantitated relative to the "correct" consensus oligonucleotides to provide an estimate of the base substitution frequency at that position. The "error oligonucleotides" can also be sequenced to demonstrate the base substitutions directly. Sequential high resolution one- and two-dimensional acrylamide gel separations are required to purify consensus and "error" oligonucleotides. This is best done after annealing of synthetic complementary oligonucleotides to the chosen segment of viral or transcript RNA. This allows purification of the DNA-RNA hybrid to greatly enrich for the oligonucleotide segments being examined. Using this approach, a base substitution frequency between 10^{-3} and 10^{-4} has been estimated for a number of VSV genome G positions.[208,235] Although it is tedious and time consuming, this method can be applied to numerous sites on any sequenced viral (or other) RNA molecule.[235]

C. Other Measurements of Mutation Frequencies

Palese and colleagues have carefully measured rates of influenza virus mutation and evolution.[236,237] They showed that the rate of evolution of the NS gene is quite constant in nature at about 2×10^{-3} base substitutions per site per year with most substitutions being maintained for long periods.[236] They also calculated a viable mutation rate for the NS gene of influenza A virus and of the VP1 gene of poliovirus.[237] By repetitive sequencing of gene segments of many individual plaques, they calculated the rate of viable, random, unselected mutations to be 1.5×10^{-5} for the segment of influenza virus NS protein, and less than 2.1×10^{-6} mutations per nucleotide per infectious cycle for the poliovirus VP1 segment. They estimated five infectious cycles for production of the original plaques, and it was shown that all seven NS gene variants obtained represented neutral mutations as regards kinetics of growth in MDCK cells.[237] Clearly this very high rate of viable neutral mutations verifies an even higher rate of total mutations (i.e., of total polymerase errors) and provides a basis for the extremely rapid molecular clock of the influenza virus NS gene. The lower viable mutation rate for the poliovirus VP1 gene could be due to lower polymerase error or to more stringent selective forces operating on that gene segment (or both). Heterogeneity in VP1 of poliovirus, however, has been observed in preparations of plaque-purified poliovirus (Section V).[224] Palese discusses evolution of influenza viruses in Chapter 6. It will be extremely informative to compare viable mutation rates and total polymerase rates for a number of RNA (and rapidly evolving DNA) viruses. It was shown recently that amber mutants of poliovirus reverted at a frequency of 2.5×10^{-6} by an A → C transversion. Assuming that the other two base changes possible at this A site occur at similar frequency, this site would mutate at a rate of about 10^{-5}. However, different types of substitutions may be site and enzyme dependent. Avian myeloblastosis virus reverse transcriptase, for example, preferentially catalyzes transversion mutations (see Section IV, D and F).[238b]

From the reversion frequency of a *ts* mutant of Sindbis virus, Durbin and Stollar[238] have calculated a point mutation frequency of $<5 \times 10^{-7}$. If this value is due only to a single mutation at the sequenced site, then this is evidence for extensive differences in mutation rates at different sites in RNA viruses. However, the necessity for complementing reversion mutation(s) at one or more distant sites cannot yet be excluded.

The availability of a number of in vitro systems for transcription and/or replication of RNA from animal viruses (Volume I, Chapters 1, 2, 7, and 8) should allow estimates of fidelity of copying of RNA templates by enzyme preparations. The fidelity of purified poliovirus RNA replicase was measured by the rate of misincorporation of noncomplementary substrates on synthetic homopolymeric RNA templates. The ratio of noncomplementary to complementary ribonucleotides incorporated was 1 to 5×10^{-3}. This high error frequency varied with changes in the reaction conditions and it was higher when the elongation rate increased.[239] It is not possible at present to assess the relevance of those measurements to RNA replication in vivo (see also Section VI.E).

D. Mutation Rates of Retroviruses

Zarling and Temin[240] showed that the spontaneous mutation of chicken sarcoma virus Bratislava 77 type II (a virus with low efficiency of infection of duck cells) to virus type III (high efficiency of infection of duck cells) occurred at a high rate, estimated as one mutational event in about 50 cumulative cell replications. Darlix and Spahr[38] sequenced T_1-ribonuclease-resistant fragments (structured RNAs) of Rous sarcoma virus RNA Pr-C and variants of Pr-B (LA23), and found a very high incidence of spontaneous mutations (620 base changes, including insertions and deletions, in about 10,000 nucleotides sequenced which covered 3900 nucleotides of the viral genome). It is now clear that retroviruses share with other RNA genomes a very high rate of mutation and the capacity of very rapid evolution.[240-242] These changes can include rapid antigenic variation of animal retroviruses, such as visna virus[243] and canine infectious anemia viruses,[244] and human retroviruses.[245-248] This rapid evolution of the viruses associated with AIDS will pose important problems for its control. It is very likely that all genomes which utilize RNA as their genome, or to replicate their DNA genome (as for caulimoviruses, hepadnaviruses, and retrotransposons) will exhibit high rates of mutation and of evolution during periods of extensive replication.

E. RNA Recombination and Reassortment in the Generation of Variability

It is obvious the RNA genome recombination and reassortment can greatly add to the potential variability of rapidly mutating RNA genomes, not only by generating new variants, but by bringing together mutations in one segment with sequences (and mutations) in another. This should frequently allow an otherwise unfit mutation to survive and even to dominate in a competitive quasispecies population. It is interesting in this regard that in a child fed all three Sabin vaccine serotypes, the majority of novel antigenic variants shed during the next 52 days represented type 2-type 3 intertypic recombinants carrying minor antigenic mutations in structural protein genes distant from the sites of recombination (which were in nonstructural genes).[85] Additional examples of generation of poliovirus recombinants from wild and vaccine strains are reviewed in Chapter 7 of Volume II, Section II.E.

Rotaviruses in chronically infected, immunodeficient children were found to contain genomes with atypical profiles: normal RNA segment 11 was missing and additional bands of dsRNA consisting of concatemers of segment-specific sequences were observed migrating slower than the segments they were derived from.[249] Upon serial passage at high multiplicity of infection in cell cultures, virus from different plaques of one isolate of bovine rotavirus (BRV) evolved to yield viable viruses with unique RNA segment patterns. Genome segment 5 was lost and it was replaced by new bands, shown to be concatemers of sequences related to the missing segment 5.[250] Probably a deletion to yield a defective segment 5, followed by duplications giving concatemers of different lengths, led to the formation of the new genome segments. Interestingly, viruses with the rearranged genomes grew in the absence of standard virus, and in mixed infections they overgrew the standard virus, in serial high-multiplicity infections, but not at low multiplicity of infection.[250] The BVRs with genome rearrangement produced abnormal mRNAs and abnormal protein profiles in infected cells; in one case a novel protein was observed.[250] Rotaviruses with rearranged genome segments do reassort, and rearranged dsRNA can replace normal RNA segments structurally and functionally,[251] and can also be replaced by normal RNA segments.[252] Analysis of over 500 individual plaques of one recombinational mixture (between a human rotavirus with genome rearrangements and bovine rotavirus) showed extensive heterogeneity of genotypes within the viral population,[253] and host cell factors appeared to exert selective effects which influenced the frequencies with which certain reassortants were isolated. Viruses with novel bands were generated upon serial passage at high multiplicity of infection of a six-time, plaque-purified parental virus with genome rearrangements.[254] Thus, distributions of different genomes and extremely heterogeneous populations can come about by rearrangements and reassortment of dsRNA genomes in support of the quasispecies structure for rotaviruses.[255] It will be extremely interesting to study at the nucleotide sequence level all those novel rotavirus genomic segments. Variability in animal reoviruses is treated in Chapter 5 and deletion mutants of dsRNAs of plants and fungi in Chapter 8 of this volume.

RNA recombination provides one mechanism for large mutational jumps as such those required to connect distant points of the "sequence space" (Chapter 12 of this volume, Section III) of an RNA. Even for those RNA viruses in which recombination is rare, or as of yet undocumented, it could play a major role in long-term evolution. In addition, the sequence homologies observed among quite different RNA viruses (see Section III.B) might be generated and maintained (at least in part) by RNA recombination. In view of the extreme variability (including recombination and reassortment capacity) of RNA genomes, it is not surprising that animal genomes devote a considerable number of gene equivalents to immune cell differentiation and immunoglobulin and T cell receptor recombination; nor is it surprising that hypermutation mechanisms operating on certain immunoglobulin gene segments can generate codon mutation rates approaching 10^{-4} per cell generation.[256]

It would be very desirable to extend measurements of mutation frequencies and mutation rates to different sites of various systems. The availability of infectious RNA from cloned cDNA copies of the genome of several bacterial, animal, and plant RNA viruses should allow the preparation of additional well-characterized mutants for such measurements.

F. RNA and DNA Mutation Frequencies

The mutation rates for RNA genomes are much higher than the corresponding values that have been estimated for the DNA of their hosts. The latter are in the range of 10^{-7} to 10^{-11} per nucleotide per replication.[257-260] What is the molecular basis of the wide difference in fidelity of copying between replicating host DNA and viral RNA? For thermodynamic reasons, all template-copying processes have a limited accuracy. When a replication mechanism involving base-base interactions was first proposed for DNA, the suggestion was made that rare tautomeric forms of the usual bases could induce misincorporations during DNA synthesis.[261] Pathways for the generation of transitions and transversions in DNA were proposed by Topal and Fresco.[262] Pu-Py mispairs with one of the two bases in an unfavored tautomeric form — that is, G or T in *enol* form and A or C in *imino* form — can account for the occurrence of base transitions. Pu-Pu mispairs with the free nucleotide substrate in the *syn* configuration will result in transversions. Quantitation of the proportion of rare tautomers in purine and pyrimidines by spectral analyses yield values in the range of 10^{-4} to 10^{-5}.[263-266] Early work was quoted by Topal and Fresco.[262] Saenger[267] has recently reviewed this subject. Unusual tautomers may occur in the template strand and in the free nucleotide substrate both at the incorporation site on the growing chain and at the subsequent proofreading step. The combined probabilities that a rare tautomer is first misincorporated and then that the misincorporated base will survive the proofreading step may thus approximate estimates of mutation rate of DNA deduced from genetic studies. Not only tautomeric shifts, but also the *syn-anti* rotation of the nucleotide substrate, depurination, protonations, and deprotonations of bases are likely to influence misincorporation rates during both DNA and RNA synthesis. Information for DNA has been obtained in fidelity assays using either synthetic polynucleotides or viral DNAs as templates. Very different misincorporation rates have been scored depending upon the nature of the template and the enzyme, and of reaction conditions (presence of added factors, metal ions, imbalances in the relative concentrations of the nucleotide substrates, etc.).[258,260] In comparative analyses, natural DNA and alternating homopolymers were copied with a greater accuracy than single-base homopolymers. In the latter case, proofreading activities appear to be suppressed and misincorporation rates approach those found for RNA genomes. Studies of the melting of double-stranded homopolymeric chains with defined mispairs indicated that the tautomer that predominated at a mispaired base was influenced by the surrounding nucleotides, probably through selection of the tautomeric form which best stacked with neighboring bases on the chain.[268] That the position of a nucleotide in the template chain has subtle effects on its biochemical behavior was suggested by the different incorporation of A or G directed by N^4-hydroxyCMP at two positions of the Qβ minus strand. At position 15 from the 5′ end of the minus strand (synthesis of the minus strand of Qβ RNA starts at the penultimate nucleotide of the plus strand; thus, nucleotide 15 from the 5′ end of the minus strand is complementary to nucleotide 16 from the 3′ end of the plus strand), the base analogue directed the incorporation of GMP more efficiently than AMP (about 1:0.7) in the plus strand product,[227] but when present in position 39, the opposite was found (about 1:3.3),[26] in spite of identical neighbor residues at the two sites.[26,227] Thus, the above observations provide indirect, but suggestive, support for the notion that misincorporation rates are not likely to be uniform along the growing RNA chain. Weddell et al.[118] have suggested that secondary structure next to the region encoding a major antigenic site of the FMDV genome could limit the fidelity of the RNA copying machinery, thereby causing an increased mutation rate. Also, the nonrandom location of mutations generated during replication of variant Qβ RNAs (Volume I, Chapter 1) appears to be statistically significant. A critical question is then posed. Have some of the conserved RNA and protein segments known to exist in several groups of viruses (Section III) evolved more slowly because of higher accuracy of the RNA replicases at those sites? Alternatively, is it only the relative disadvantage of variants mutated at those sites acting through selection which keeps them unchanged? It is not possible at present to answer this. However, the few following observations point to selection as the major mechanism to keep certain sequences invariant.

1.   The residues on Qβ RNA and VSV RNA for which mutation frequencies have been calculated[26,208] correspond to strongly conserved nucleotides when related viruses are compared. It is not likely that even more variable nucleotides would show mutation rates much higher than 10^{-3}, although this has yet to be proven.

2.   In highly variable segments of capsid protein VP1 of FMDV some groups of conserved amino acids are found. For example, the sequence Arg–Gly–Asp (positions 145 to 147) is present in many FMDV strains, and it has been suggested that this triplet may constitute a recognition sequence of FMDV for a cellular receptor.[269] The sequence Arg–Gly–Asp plays a key role in the interaction of fibronectin and of other proteins with cells.[270,271] Whatever the function of the Arg–Gly–Asp sequence in FMDV, triplets encoding those amino acids differ in the third base when related strains are compared (see the nucleotide sequences of various A subtypes in Weddell et al.[118]). Either fidelity is much higher when copying the first and second nucleotide than the third one (and than the codons for adjacent variable amino acids) or, more likely, selection acting at the protein level keeps the Arg–Gly–Asp sequence invariant. (Strain A24 Pir includes a synonymous first base exchange CGA, Arg→AGA, Arg).[118]

3.   In instances of extended replication of viruses in unusual environments, novel mutations, not found in the corresponding virus replicated under standard conditions, are fixed. An example is the comparison of viral RNAs from brains of SSPE patients with the standard MV strains[59] (compare Section III.A).

At present it cannot be excluded that RNA synthetic machinery has some repair activity. Ishihama and colleagues have reported that elongation of influenza virus RNA primed by a capped oligonucleotide to which additional 3' guanosine residues were added, proceeded only after removal of the excess GMP residues.[272] It remains to be seen if in the course of normal elongation, an activity to remove misincorporated bases exists. If it does, it will be important to determine its overall contribution to fidelity.

From the observations summarized in this section and various other chapters of this book, the following picture of fidelity of RNA copying emerges: during chain elongation, misincorporations, enzyme slippage, and intra- and intermolecular copy choice events contribute to the generation of variant genomes. By far, point mutations appear to be the most frequent RNA genomic variation at least among viable mutants. Point mutations are mainly caused by misincorporations at the synthetic step, and the misincorporated nucleotides are not efficiently removed because RNA replicases are probably very limited or lacking in the kinds of proofreading and repair activities which act normally on DNA. The microenvironment provided by the enzyme machinery at the catalytic site, as well as the electronic structures of both template and substrate bases, determine the accuracy of copying at each site. This induces a very frequent generation of mutants, each of which will be rated in biological competition with all other variants in the population. This leads us to the proposal of a *population equilibrium model for RNA genomes.*

VII. THE POPULATION EQUILIBRIUM MODEL FOR RNA GENOMES: FORMULATION AND BIOLOGICAL IMPLICATIONS

The experimental results reviewed in Sections III. to VI. indicate that most (perhaps all) natural isolates, as well as clonal populations, of RNA genomes consist of multiple related but nonidentical genomes. In accordance with the concepts and terminology of Eigen and colleagues, these viral genome populations can be termed *quasispecies* (Chapter 12 of this volume). It can be readily calculated that assuming a mutation rate of 10^{-4} per nucleotide (Section VI), and assuming that no selection acted on newly generated variant genomes, the progeny from a cell infected by one viral particle will consist of a distribution of single, double, triple, etc. mutants. The shape of the predicted distribution will be influenced by the number of rounds of replication of each molecule that is encapsidated.[273] However, the main modification of the variant spectrum is caused by differing levels of fitness of individual genomes (i.e., fitness = 0 for lethal mutations; fitness values of up to 1 for increasingly neutral mutations). The relative fitness of the variants will of course depend on the physical and biological environment. At any time a virus population consists of a finite distribution of variants, as exemplified by the experimental analyses reviewed in Section V. The mathematical parameters of mutant distribution are treated by Eigen and Biebricher in Chapter 12 of this volume. A quasispecies distribution is to be expected from high mutation rates, together with the known tolerance of viral genomes for rapid change (even though many variants may function in a deviant way, see below). How can a quasispecies distribution of variant genomes be described? One way is to define an "average" or "consensus" sequence in which each position is represented by the most frequent nucleotide found at the corresponding position when a number of molecules of a sample of the population is sequenced. Except when molecular cloning is involved, most presently available methods of RNA sequence characterization yield such an "average" or "consensus" sequence (compare Section II). Alternatively, the proportion of the most abundant or "master sequence", if known, could constitute a definition. In addition to the "master sequence", the population will include a "mutant spectrum", which results from selective rating of each variant in competition with the others. The nature of the "mutant spectrum" that accompanies any "individual" genome is important for the fate of that "individual" in the evolving population. As shown by Eigen and Biebricher in Chapter 12 of this volume, near the error threshold, an inferior mutant may outgrow another one that has a 10% higher fitness, only because of a better mutant environment. "The target of selection is not the single species, but rather the distribution of the quasispecies as a whole." (See Chapter 12 of this volume, Section VI).

As far as we know from the mechanisms of spontaneous mutagenesis during RNA replication (Section VI), there is an uncertainty regarding which mutants will arise during any replication event. This continuous but unpredictable generation of variants results in a continuum of transient dynamic equilibrium steps. In this view, an RNA virus genome is statistically defined, but individually indeterminate. This concept and its biological implications for RNA viruses have been discussed previously.[9,27,121,197,206-208,214-216,235,274]

It is clear (and we will discuss several examples below) that the biological properties of each of the components from the "mutant spectrum" need not, and frequently will not, coincide with those of the most abundant or "master" molecule. There are instances in which a sudden change in the biological environment of the replicating RNAs provides an extremely high selective advantage for certain previously disadvantaged mutants and a shift in the population equilibrium is readily observed. This occurs with MAb "escape mutants" (Section IV and V.B). This is also very clearly exemplified by variations in the standard VSV genome which allow the virus to replicate efficiently in the presence of a given DI particle RNA. The continuing process of generation of new DI particle genome and of resistant variant infectious VSV genomes can probably proceed indefinitely.[9,204,205,274-276] The temporary dominance is not of a single homogeneous species, but rather of a distribution of variants. This provides RNA genomes with a notable adaptability, not only because many variants are present at a given time, but also because novel distributions are continuously being tested. It is as if an RNA virus encompasses a "range" of phenotypes able to express themselves with a certain probability in a given environment. In this view, SSPE virus, for example, is within the "range" of variant viruses able to be produced by MV during its replication in humans (Section II.A); but of course, the SSPE virus variant distribution and consensus sequences will vary in different individuals and at different times in one host individual. Double or triple variants of coxsackie B_4 selected for resistance to neutralization by two or three MAbs, respectively, were attenuated for mice.[277] Thus, variants present in the "mutant spectrum" had a phenotype which differed from that of the dominant population.

That RNA virus variants can exist which show very different fitness in a given environment is exemplified by the studies of Domingo et al.[26,27] with phage Qβ. The growth of extra-cistronic mutant $A_{-40} \rightarrow G$ was 0.25 that of the "wild-type" virus in competition experiments.[26] In contrast, the corresponding value for those spontaneous variants that contributed to the measurable heterogeneity of Qβ populations was 0.87.[27] Here the molecular basis for the reduced fitness of the extracistronic mutant was a defect in its RNA replication, probably because of reduced binding of the mutated sequence to ribosomal protein S1, a subunit of Qβ replicase (see Volume I, Chapter 1). Variants with such a reduced fitness necessarily will be maintained at a very low level, in spite of their arising at a high rate. In fact, they have never been isolated from "natural" virus populations presumably because of their low fitness and low abundance. Many RNA sequences, because of functional or structural constraints, are expected to show strong conservation: RNA segments encoding the active sites of viral enzymes or nucleic acid binding proteins (such as the cysteine-rich NBPcys domain of retroviruses and other reverse transcribing elements[278]). Also, secondary[279] or higher order RNA structures, such as pseudoknots,[280,281] which occur at least in several plant RNA viruses (see Volume I, Chapter 5, Section IV), are likely to play structural and/or functional roles. There will be severe limitations to the fixation of mutation occurring at those sites. Additional evidence of constraints for variation has been obtained from studies on site-directed mutagenesis of viroid c-DNA (Chapter 4 of this volume, Section III.B). Several point mutations (or even two-point mutations that were expected to preserve the secondary structure at the mutated sites) rendered the viroid noninfectious.

The population equilibrium structure for RNA genomes has, in our view, important theoretical and practical implications that may be summarized as follows:

1. It affords a new description of what an RNA virus really is.
2. Rapid changes in phenotypic properties are explained by selection of variants present or arising in the "mutant spectrum". Among them are antigenic variations with failures of classical or synthetic vaccines, or reversion of viruses to a virulent form. Also, variants that either establish or mediate persistence, or that are able to invade new cells, tissues, or even different host organisms, may be found among the "range" of phenotypes of a virus (several examples were reviewed by Holland et al.[9] and others can be found in several chapters of this volume). An example is the possibility that a virus (or viruses) might be involved in multiple sclerosis[9,282-284] or other chronic human diseases.

3.   It raises difficulties for the correlation of a definite nucleotide sequence with a phenotypic behavior. (As reviewed in Section III.B, a number of different mutations have been associated with attenuation of poliovirus. A similar indetermination was encountered in trying to define mutations that determine alphavirus neurovirulence.[285]) A single infection by any viral clone (including RNA from an infectious cDNA clone) will generate a distribution of genotypes and phenotypes. A claim that a given phenotype is caused by the sequence represented in the initial clone should be substantiated by sequence analysis of those progeny viruses exhibiting that phenotype. For example, it requires sequence examination of the genomes that have colonized a new tissue or organ following infection or transfection.

4.   In the transmission of infectious virus from an infected organism to a new susceptible, often only one or a few infectious particles are involved. This should enhance diversity, and is probably one of the paramenters that influences the extensive diversity among natural isolates of a given virus (Section III).

5.   The biological relevance of RNA heterogeneity becomes apparent if one considers four parameters:[206] (1) the average number of mutations per genome. A value around 1 to 10 is a likely range (Section V); (2) virus population numbers. Individual plaques from cell monolayers contain 10^5 to 10^9 plaque-forming units (p.f.u.). An infected organism contains as much as 10^9 to 10^{12} p.f.u., according to published measurements for animal viruses;[286-288] (3) the genome length of RNA viruses: 3 to 30 kilobases. With those parameters, it can be easily calculated that potentially all possible single and double mutants, as well as decreasing proportions of triple, quadruple, etc. mutants, are present in most natural viral populations;[27,206] and (4) the number of mutations required for a phenotypic change of the virus. Many phenotypic characteristics of viruses are probably polygenic, as documented in Section III.B for the attenuation phenotype of poliovirus.[88] Other phenotypes, however, are brought about by only one or a few genomic alterations and, thus, they are likely to occur in a quasispecies distribution of genomes. This is the case of many variants resistant to neutralization by MAbs (Sections IV and V.B). Several examples were previously discussed[206] and additional ones are included in other chapters of this volume.

Thus, the available experimental evidence suggests that genetic heterogeneity is the key to the adaptability and ubiquity of RNA viruses.

VIII. CONCLUSIONS AND PROSPECTS

The above considerations (together with those in other chapters of these volumes) suggest that RNA genetics differs substantially from classical DNA genetics. In the latter, terms such as "wild-type", "mutant", "revertant", and "gene sequence" can have a clearly defined meaning at an individual level and at a population level. In the former, we must necessarily deal with uncertainty and with probabilities. Replicating RNA populations are indeterminate mixtures of related genomes which undergo biological selection as they compete and rapidly evolve in constant or changing environments.

Much remains to be learned regarding mechanisms of RNA mutations, rates of mutations at different sites and among different viruses and strains, the possible involvement of proofreading in RNA replication, and the relative roles of mutation rates, recombination rates, and selective forces in shaping the evolution of variable and conserved segments of evolving RNA genomes.

As also discussed in Chapter 12 of this volume (Section VII), there has been some confusion in the terminology used to express genetic variability of RNA viruses. It would be desirable to distinguish the following concepts. *Mutation rate*: the frequency of occurrence of a mutation event, i.e., a misincorporation during a single, replicative round of RNA synthesis. *Mutant frequency*: the proportion of a mutated sequence quantitated in an RNA population. *Viable mutation rate or mutant frequency*: the subset of those values defined above and which represent viable genomes in the environment under study. Many quantitations of genetic variability of natural populations of RNA viruses measure yet another subset of the competitive subset of all viable mutants produced. Indeed, *the rate of fixation of mutations* measures the number of mutations which per unit time become dominant among the genomes that replicate within one host organism or that replicate and are transmitted to new susceptible hosts.

Recent developments in molecular genetics will allow new approaches to some of those questions. The elucidation of the three-dimensional structures of more viral proteins and complete virions should permit a better understanding of constraints for variability and of the molecular basis of some novel viral phenotypes. Also, the higher order structure of RNA molecules and clarifications of their functional significance should contribute to define limitations and tolerances of variations at the RNA level. In this respect, expression vectors for prokaryotic and eukaryotic cells may permit complementation of a defective viral gene product by the corresponding functional product expressed in the manipulated cell. This may allow measurements of the mutations occurring in that gene, with selective constraints acting at the RNA level, but not at the protein level. The availability of in vitro systems for viral RNA synthesis should provide new ways to measure fidelities in the test tube and explore factors that influence them. Improved techniques for RNA isolation, cDNA synthesis, and rapid sequencing will help in the precise characterization of the kinds of viral RNA sequences present in tissues of organisms affected by chronic degenerative diseases of a suspected viral etiology.

Certainly the indeterminate and variable nature of RNA genomes will pose many challenges for understanding them as well as for controlling them. Because of the theoretical, as well as the many practical, implications of rapid RNA evolution, the effort is well worth pursuing.

ACKNOWLEDGMENTS

We thank the following scientists for providing us with valuable unpublished information: C. Biebricher, M. Billeter, U. Desselberger, M. Eigen, L. Enjuanes, J. Flanegan, A. Ishihama, G. Jiménez, A. Palmenberg, B. S. Prabhakar, G. Siegl, M. A. M. Stokes, and L. von Vloten-Doting. We also acknowledge the excellent editorial assistance of Carmen Hermoso.

REFERENCES

1. **Granoff, A.,** Induction of Newcastle disease virus mutants with nitrous acid, *Virology,* 13, 402, 1961.
2. **Granoff, A.,** Nature of the Newcastle disease virus population, in *Newcastle Disease Virus, an Evolving Pathogen,* Hanson, R. P., Ed., University of Wisconsin Press, Madison, 1964, 107.
3. **Pringle, C. R.,** Genetic characteristics of conditional lethal mutants of vesicular stomatitis virus induced by 5-fluorouracil, 5-azacytidine, and ethyl methane sulfonate, *J. Virol.,* 5, 559, 1970.
4. **Pringle, C. R. and Duncan, I. B.,** Preliminary physiological characterization of temperature-sensitive mutants of vesicular stomatitis virus, *J. Virol.,* 8, 56, 1971.
5. **Lafay, F.,** Etude des mutants thermosensibles du virus de la stomatite vésiculaire (VSV). Classification de quelques mutants d'après des critères de fonctionnement, *C.R. Acad. Sci.,* Sér. D, 268, 2385, 1969.
6. **Flamand, A. and Pringle, C. R.,** The homologies of spontaneous and induced temperature-sensitive mutants of vesicular stomatitis virus isolated in chick embryo and BHK-21 cells, *J. Gen. Virol.,* 11, 8, 1971.

7. **Pringle, C. R.,** Genetics of rhabdoviruses, in *Comprehensive Virology,* Vol. 9, Fraenkel-Conrat, H. and Wagner, R., Eds., Plenum Press, New York, 1977, 239.
8. **Flamand, A.,** Rhabdovirus genetics, in *Rhabdoviruses,* Vol. 2, Bishop, D. H. L., Ed., CRC Press, Boca Raton, Florida, 1980, 115.
9. **Holland, J. J., Spindler, K., Horodyski, F., Grabau, E., Nichol, S., and Van dePol, S.,** Rapid evolution of RNA genomes, *Science,* 215, 1577, 1982.
10. **Burnet, F. M. and Bull, D. R.,** Changes in influenza virus associated with adaptation to passage in chick embryo, *Aust. J. Exp. Biol. Med. Sci.,* 21, 55, 1943.
11. **Burnet, F. M.,** Variation in animal viruses, in *Principles of Animal Virology,* Academic Press, New York, 1960, 407.
12. **Fields, B. N. and Joklik, W. K.,** Isolation and preliminary genetic and biochemical characterization of temperature-sensitive mutants of reovirus, *Virology,* 37, 335, 1969.
13. **Cooper, P. D.,** Genetics of Picornaviruses, in *Comprehensive Virology,* Vol. 9, Fraenkel-Conrat, H. and Wagner, R. Eds., Plenum Press, New York, 1977, 133.
14. **Kunkel, L. O.,** *Publ. Ann. Assoc. Adv. Sci.,* 12, 22, 1940.
15. **Mundry, K. W. and Gierer, A.,** Die Erzeugung von Mutationen des Tabakmosaikvirus durch chemische Behandlung seiner Nucleinsäure in vitro, *Z. Vererbungsl.,* 89, 614, 1958.
16. **van Vloten-Doting, L.,** Virus genetics, in *The Plant Viruses,* Vol. 1, Francki, R. I. B., Ed., Plenum Press, 1985, 117.
17. **Valentine, R. C., Ward, R., and Strand, M.,** The replication cycle of RNA bacteriophages, *Adv. Virus Res.,* 15, 1, 1969.
18. **Horiuchi, K.,** Genetic studies of RNA phages, in *RNA Phages,* Zinder, N. D., Ed., Cold Spring Harbor Laboratory, Cold Spring Harbor, New York, 1975, 29.
19. **Spiegelman, S., Haruna, I., Holland, I. B., Beaudreau, G., and Mills, D.,** The synthesis of a self-propagating and infectious nucleic acid with a purified enzyme, *Proc. Natl. Acad. Sci. U.S.A.,* 54, 919, 1965.
20. **Mills, D. R., Peterson, R. L., and Spiegelman, S.,** An extracellular darwinian experiment with a self-duplicating nucleic acid molecule, *Proc. Natl. Acad. Sci. U.S.A.,* 58, 217, 1967.
21. **de Wachter, R. and Fiers, W.,** Sequences at the 5'-terminus of bacteriophage Qβ RNA, *Nature (London),* 22, 233, 1969.
22. **Weissmann, C., Billeter, M. A., Goodman, H. M., Hindley, J., and Weber, H.,** Structure and function of phage RNA, *Annu. Rev. Biochem.,* 42, 303, 1973.
23. **Fiers, W., Contreras, R., Duerinck, F., Haegeman, G., Merregaert, J., Min Jou, W., Raeymakers, A., Volckaert, G., Ysebaert, M., Van de Kerckhove, J., Nolf, F., and Van Montagu, M.,** A-protein gene of bacteriophage MS2, *Nature (London),* 256, 273, 1975.
24. **Fiers, W., Contreras, R., Duerinck, F., Haegeman, G., Iserentant, D., Merregaert, J., Min Jou, W., Molemans, F., and Raeymakers, A.,** Complete nucleotide sequence of bacteriophage MS2 RNA: primary and secondary structure of the replicase gene, *Nature (London),* 260, 500, 1976.
25. **Fiers, W.,** Structure and function of RNA bacteriophages, in *Comprehensive Virology,* Vol. 13, Fraenkel-Conrat, H. and Wagner, R., Eds., Plenum Press, New York, 1979, 69.
26. **Domingo, E., Flavell, R. A., and Weissmann, C.,** In vitro site-directed mutagenesis: generation and properties of an infectious extracistronic mutant of bacteriophage Qβ, *Gene,* 1, 3, 1976.
27. **Domingo, E., Sabo, D., Taniguchi, T., and Weissmann, C.,** Nucleotide sequence heterogeneity of an RNA phage population, *Cell,* 13, 735, 1978.
28. **de Wachter, R. and Fiers, W.,** Preparative two-dimentional polyacrylamide gel electrophoresis of ³²P-labeled RNA, *Anal. Biochem.,* 49, 184, 1972.
29. **Kew, O. M., Nottay, B. K., and Obijeski, J. F.,** Applications of oligonucleotide fingerprinting to the identification of viruses, *Methods Virol.,* 8, 41, 1984.
30. **Aaronson, R. P., Young, J. F., and Palese, P.,** Oligonucleotide mapping: evaluation of its sensitivity by computer-simulation, *Nucleic Acids Res.,* 10, 237, 1982.
31. **Ortín, J., Nájera, R., López, C., Dávila, M., and Domingo, E.,** Genetic variability of Hong Kong (H3N2) influenza viruses: spontaneous mutations and their location in the viral genome, *Gene,* 11, 319, 1980.
32. **de la Torre, J. C., Dávila, M., Sobrino, F., Ortín, J., and Domingo, E.,** Establishment of cell lines persistently infected with foot-and-mouth disease virus, *Virology,* 145, 24, 1985.
33. **Sanger, F., Nicklen, S., and Coulson, A. R.,** DNA sequencing with chain-terminating inhibitors, *Proc. Natl. Acad. Sci. U.S.A.,* 74, 5463, 1977.
34. **Maxam, A. M. and Gilbert, W.,** Sequencing end-labeled DNA with base-specific chemical cleavages, *Methods Enzymol.,* 65, 499, 1980.
35. **Messing, J., Crea, R., and Seeburg, P. H.,** A system for shotgun DNA sequencing, *Nucleic Acids Res.,* 9, 309, 1981.

36. **Deininger, P. L.**, Approaches to rapid DNA sequence analysis, *Anal. Biochem.*, 135, 247, 1983.
37. **Fields, S. and Winter, G.**, Nucleotide sequence heterogeneity and sequence rearrangement in influenza virus cDNA, *Gene*, 15, 207, 1981.
38. **Darlix, J. L. and Spahr, P. F.**, High spontaneous mutation rate of Rous sarcoma virus demonstrated by direct sequencing of the RNA genome, *Nucleic Acids Res.*, 11, 5953, 1983.
39. **Winter, E., Yamamoto, F., Almoguera, C., and Perucho, M.**, A method to detect and characterize point mutations in transcribed genes: amplification and overexpression of the mutant c-Ki-ras allele in human tumor cells, *Proc. Natl. Acad. Sci. U.S.A.*, 82, 7575, 1985.
40. **Myers, R. M., Larin, Z., and Maniatis, T.**, Detection of single base substitutions by ribonuclease cleavage at mismatches in RNA:DNA duplexes, *Science*, 230, 1242, 1985.
41. **Ito, Y. and Joklik, W. K.**, Temperature-sensitive mutants of reovirus. II. Anomalous electrophoretic migration of certain hybrid RNA molecules composed of mutant plus strands and wild-type minus strands, *Virology*, 50, 202, 1972.
42. **Smith, F. I., Parvin, J. D., and Palese, P.**, Detection of single base substitutions in influenza virus RNA molecules by denaturing gradient gel electrophoresis of RNA-RNA or DNA-RNA heteroduplexes, *Virology*, 150, 55, 1986.
43. **Reanney, D. C.**, A regulatory role for viral RNA in eukaryotes, *J. Theor. Biol.*, 49, 461, 1975.
44. **Matthews, R. E. F.**, Classification and nomenclature of viruses, *Intervirology*, 17, 1, 1983.
45. **Clewley, J. P., Bishop, D. H. L., Kang, C. Y., Coffin, J., Schnitzlein, W. M., Reichmann, M. E., and Shope, R. E.**, Oligonucleotide fingerprints of RNA species obtained from rhabdoviruses belonging to the vesicular stomatitis virus subgroup, *J. Virol.*, 23, 152, 1977.
46. **Gallione, C. J. and Rose, J. K.**, Nucleotide sequence of a cDNA clone encoding the glycoprotein from the New Jersey serotype of vesicular stomatitis virus, *J. Virol.*, 46, 162, 1983.
47. **Banerjee, A. K., Rhodes, D. P., and Gill, D. S.**, Complete nucleotide sequence of the mRNA coding for the N protein of vesicular stomatitis virus (New Jersey serotype), *Viorology*, 137, 432, 1984.
48. **Rae, R. B. and Elliott, R. M.**, Conservation of potential phosphorylation sites in the NS proteins of the New Jersey and Indiana serotypes of vesicular stomatitis virus, *J. Gen. Virol.*, 67, 1351, 1986.
49. **Gill, D. S. and Banerjee, A. A.**, Vesicular stomatitis virus NS proteins: structural similarity without extensive sequence homology, *J. Virol.*, 55, 60, 1985.
50. **Gopalakrishna, Y. and Lenard, J.**, Sequence alterations in temperature-sensitive M-protein mutants (complementation group III) of vesicular stomatitis virus, *J. Virol.*, 56, 655, 1985.
50a. **Masters, P. S. and Banerjee, A. K.**, Sequences of Chandipura virus N and NS genes: evidence for high mutability of the NS gene within vesiculoviruses, *Virology*, 157, 298, 1987.
51. **ter Meulen, V., Stephenson, J. R., and Kreth, H. W.**, Subacute sclerosing panencephalitis, in *Comprehensive Virology*, Vol. 18, Fraenkel-Conrat, H. and Wagner, R., Eds., Plenum Press, New York, 1983, 105.
52. **ter Meulen, V. and Carter, M. J.**, Measles virus persistency and disease, *Progr. Med. Virol.*, 30, 44, 1984.
53. **Stephenson, J. R. and ter Meulen, V.**, A comparative analysis of measles virus RNA by oligonucleotide fingerprinting, *Arch. Virol.*, 71, 279, 1982.
54. **Hall, W. W. and Choppin, P. W.**, Evidence for lack of synthesis of the M polypeptide of measles virus in brain cells in subacute sclerosing panencephalitis, *Virology*, 99, 443, 1979.
55. **Hall, W. W. and Choppin, P. W.**, Measles virus proteins in the brain tissue of patients with subacute sclerosing panencephalitis, *N. Engl. J. Med.*, 304, 1152, 1981.
56. **Lin, F. H. and Thormar, H.**, Absence of M protein in cell-associated subacute sclerosing panencephalitis virus, *Nature (London)*, 285, 490, 1980.
57. **Sheppard, R. D., Raine, C. S., Bornstein, M. B., and Udem, S. A.**, Measles virus matrix protein synthesized in a subacute sclerosing panencephalitis cell line, *Science*, 228, 1219, 1985.
58. **Baczko, K., Carter, M. J., Billeter, M., and ter Meulen, V.**, Measles virus gene expression in subacute sclerosing panencephalitis, *Virus Res.*, 1, 585, 1984.
59. **Cattaneo, R., Schmid, A., Rebmann, G., Baczko, K., ter Meulen, V., Bellini, W. J., Rozenblatt, S., and Billeter, M. A.**, Accumulated measles virus mutations in a case of subacute sclerosing panencephalitis: interrupted matrix protein reading frame and transcription alteration, *Virology*, 154, 97, 1986.
60. **Rozenblatt, S., Eizenberg, O., Ben-Levy, R., Lavie, V., and Bellini, W. J.**, Sequence homology within the morbilliviruses, *J. Virol.*, 53, 684, 1985.
61. **Barrett, T., Schrimpton, S. B., and Russell, S. E. H.**, Nucleotide sequence of the entire protein coding region of canine distemper virus polymerase-associated (P) protein mRNA, *Virus Res.*, 3, 367, 1985.
62. **Shioda, T., Hidaka, Y., Kanda, T., Shibuta, H., Nomoto, A., Iwasaki, K.**, Sequence of 3,687 nucleotides from the 3' end of Sendai virus genome RNA and the predicted amino acid sequences of viral NP, P and C proteins, *Nucleic Acids Res.*, 11, 7317, 1983.

63. **Sánchez, A., Banerjee, A. K., Furuichi, Y., and Richardson, M. A.,** Conserved structures among the nucleocapsid proteins of the paramyxoviridae: complete nucleotide sequence of human parainfluenza virus type 3 NP mRNA, *Virology,* 152, 171, 1986.

64. **Galinski, M. A., Mink, M. A., Lambert, D. M., Wechsler, S. L., and Pons, M. W.,** Molecular cloning and sequence analysis of the human Parainfluenza 3 virus RNA encoding the nucleocapsid protein, *Virology,* 149, 139, 1986.

65. **Bellini, W. J., Englund, G., Richardson, C. D., Rozenblatt, S., and Lazzarini, R. A.,** Matrix genes of measles virus and canine distemper virus: cloning, nucleotide sequences, and deduced amino acid sequences, *J. Virol.,* 58, 408, 1986.

66. **McGuinnes, L. W. and Morrison, T. G.,** Nucleotide sequence of the gene encoding the Newcastle disease virus fusion protein and comparisons of paramyxovirus fusion protein sequences, *Virus Res.,* 5, 343, 1986.

67. **Nerome, K., Ishida, M., Oya, A., and Bosshard, S.,** Genomic analysis of antigenically related avian paramyxoviruses, *J. Gen. Virol.,* 64, 465, 1983.

68. **Nerome, K., Shibata, M., Kobayashi, S., Yamaguchi, R., Yoshioka, Y., Ishida, M., and Oya, A.,** Immunological and genomic analyses of two serotypes of avian paramyxoviruses isolated from wild ducks in Japan, *J. Virol.,* 50, 649, 1984.

69. **Auperin, D., Compans, R. W., and Bishop, D. H. L.,** Nucleotide sequence conservation at the 3'-termini of the virion RNA sepcies of New World and Old World arenaviruses, *Virology,* 121, 200, 1982.

70. **Romanowski, V. and Bishop, D. H. L.,** Conserved sequences and coding of two strains of lymphocytic choriomeningitis virus (WE and ARM) and Pichinde arenavirus, *Virus Res.,* 2, 35, 1985.

71. **Auperin, D. D., Galinski, M., and Bishop, D. H. L.,** The sequences of the N protein gene and intergenic region of the S RNA of Pichinde arenaviruses, *Virology,* 134, 208, 1984.

72. **Romanowski, V., Matsuura, Y., and Bishop, D. H. L.,** Complete sequence of the S RNA of lymphocytic choriomeningitis virus (WE strain) compared to that of Pichinde arenavirus, *Virus. Res.,* 3, 101, 1985.

73. **Dutko, F. J. and Oldstone, M. B. A.,** Genomic and biologic variation among commonly used lymphocytic choriomeningitis virus strains, *J. Gen. Virol.,* 64, 1689, 1983.

74. **Eshita, Y. and Bishop, D. H. L.,** The complete sequence of the mRNA of snowshoe hare bunyavirus reveals the presence of internal hydrophobic domains in the viral glycoprotein, *Virology,* 137, 227, 1984.

75. **Lees, J. F., Pringle, C. R., and Elliot, R. M.,** Nucleotide sequence of the bunyamwera virus M RNA segment: conservation of structural features in the bunyavirus glycoprotein gene product, *Virology,* 148, 1, 1986.

76. **Toyoda, H., Kohara, M., Kataoka, Y., Suganuma, T., Omata, T., Imura, N., and Nomoto, A.,** Complete nucleotide sequences of all three poliovirus serotype genomes. Implications for genetic relationship, gene function and antigenic determinants, *J. Mol. Biol.,* 174, 561, 1984.

77. **Kew, O. M., Nottay, B. K., Hatch, M. H., Nakano, J. H., and Obijeski, J. F.,** Multiple genetic changes can occur in the oral poliovaccines upon replication in humans, *J. Gen. Virol.,* 56, 337, 1981.

78. **Nottay, B. K., Kew, O. M., Hatch, M. H., Heyward, J. T., and Obijeski, J. F.,** Molecular variation of type 1 vaccine-related and wild type poliovirus during replication in humans, *Virology,* 108, 405, 1981.

79. **Minor, P. D., Schild, G. C., Ferguson, M., Mackay, A., Magrath, D. I., John, A., Yates, P. J., and Spitz, M.,** Genetic and antigenic variation in type 3 polioviruses: chacterization of strains by monoclonal antibodies and T1 oligonucleotide mapping, *J. Gen. Virol.,* 61, 167, 1982.

80. **Stanway, G., Hughes, P. J., Mountford, R. C., Reeve, P., Minor, P. D., Schild, G. C., and Almond, J. W.,** Comparison of the complete nucleotide sequences of the genomes of the neurovirulent poliovirus P3/Leon/37 and its attenuated Sabin vaccine derivative P3/Leon 12a₁b, *Proc. Natl. Acad. Sci. U.S.A.,* 81, 1539, 1984.

81. **Stanway, G., Cann, A. J., Hauptmann, R., Hughes, P., Mountford, R. C., Minor, P. D., Schild, G. C., and Almond, J. W.,** The nucleotide sequence of poliovirus type 3 Leon 12 a₁b: comparison with poliovirus type 1, *Nucleic Acids Res.,* 11, 5629, 1983.

82. **Evans, D. M. A., Dunn, G., Minor, P. D., Schild, G. C., Cann, A. J., Stanway, G., Almond, J. W., Currey, K., and Maizel, J. V.,** Increased neurovirulence associated with a single nucleotide change in a noncoding region of the Sabin type 3 poliovaccine genome, *Nature (London),* 314, 548, 1985.

83. **Jameson, B. A., Bonin, J., Wimmer, E., and Kew, O. M.,** Natural variants of the Sabin type 1 vaccine strain of poliovirus, and correlation with a poliovirus neutralization site, *Virology,* 143, 337, 1985.

84. **Rozhon, E. J., Wilson, A. K., and Jubelt, B.,** Characterization of genetic changes occurring in attenuated poliovirus 2 during persistent infection in mouse central nervous systems, *J. Virol.,* 50, 137, 1984.

85. **Minor, P. D., John, A., Ferguson, M., and Icenogle, J. P.,** Antigenic and molecular evolution of the vaccine strain of type 3 poliovirus during the period of excretion by a primary vaccinee, *J. Gen. Virol.,* 67, 693, 1986.

86. **Magrath, D. I., Evans, D. M. A., Ferguson, M., Schild, G. C., Minor, P. D., Horaud, F., Crainic, R., Stenvik, M., and Hovi, T.,** Antigenic and molecular properties of type 3 poliovirus responsible for an outbreak of poliomyelitis in a vaccinated population, *J. Gen. Virol.,* 67, 899, 1986.

87. **Ferguson, M., Evans, D. M. A., Magrath, D. I., Minor, P. D., Almond, J. W., and Schild, G. C.,** Induction of broadly reactive type specific neutralising antibody to poliovirus type 3 by synthetic peptides, *Virology,* 143, 505, 1985.

87a. **Hughes, P. J., Evans, D. M. A., Minor, P. D., Schild, G. C., Almond, J. W., and Stanway, G.,** The nucleotide sequence of type 3 poliovirus isolated during a recent outbreak of poliomyelitis in Finland, *J. Gen. Virol.,* 67, 2093, 1986.

88. **Omata, T., Kohara, M., Kuge, S., Komatsu, T., Abe, S., Semler, B. L., Kameda, A., Itoh, H., Arita, M., Wimmer, E., and Nomoto, A.,** Genetic analysis of the attenuation phenotype of poliovirus type 1, *J. Virol.,* 58, 348, 1986.

89. **Kohara, M., Abe, S., Kuge, S., Semler, B., Komatsu, T., Arita, M., Itoh, H., and Nomoto, A.,** An infectious cDNA clone of the poliovirus Sabin strain could be used as a stable repository and inoculum for the oral polio live vaccine, *Virology,* 151, 21, 1986.

90. **Racaniello, V. R. and Baltimore, D.,** Cloned poliovirus complementary DNA is infectious in mammalian cells, *Science,* 214, 916, 1981.

91. **Bachrach, H. L.,** Foot-and-mouth disease virus, *Annu. Rev. Microbiol.,* 22, 201, 1968.

92. **Bachrach, H. L.,** Foot-and-mouth disease virus: properties, molecular biology and immunogenicity, in *Beltsville Symp. in Agricultural Research. I. Virology in Agriculture,* Romberger, J. A., Ed., Allanheld, Osmun, Montclair, 1977, 3.

93. **Pereira, H. G.,** Foot-and-mouth disease, in *Virus Diseases of Food Animals,* Vol. 2, Gibbs, E. P. J., Ed., Academic Press, New York, 1981, 333.

94. **Küpper, H., Keller, W., Kurz, C., Forss, S., Schaller, H., Franze, R., Strohmaier, K., Marquardt, O., Zaslavski, V. G., and Hofschneider, P. H.,** Cloning of cDNA of a major antigen of foot and mouth disease virus and expression in E. coli, *Nature (London),* 289, 555, 1981.

95. **Beck, E., Forss, S., Strebel, K., Cattaneo, R., and Feil, G.,** Structure of FMDV translation initiation site and of the structural proteins, *Nucleic Acids Res.,* 11, 7873, 1983.

96. **Carroll, A. R., Rowlands, D. S., and Clarke, B. E.,** The complete nucleotide sequence of the RNA coding for the primary translation product of foot-and-mouth disease virus, *Nucleic Acids Res.,* 12, 2461, 1984.

97. **Forss, S., Strebel, K., Beck, E., and Schaller, H.,** Nucleotide sequence and genomic organization of foot and mouth disease virus, *Nucleic Acids Res.,* 12, 6587, 1984.

98. **Robertson, B. H., Grubman, M. J., Weddell, G. N., Moore, D. M., Welsh, D., Fisher, T., Dowbenko, D. J., Yansura, D. G., Small, B., and Kleid, D. G.,** Nucleotide and amino acid sequence coding for polypeptides of foot-and-mouth disease virus type A12, *J. Virol.,* 54, 651, 1985.

99. **Strebel, K., Beck, E., Strohmaier, K., and Schaller, H.,** Characterization of foot-and-mouth disease virus gene products with antisera against bacterially synthesized fusion proteins, *J. Virol.,* 57, 983, 1986.

99a. **Vakharia, V. N., Devaney, M. A., Moore, D. M., Dunn, J. J., and Grubman, M. J.,** Proteolytic processing of foot-and-mouth disease virus polyproteins expressed in a cell-free system from clone-derived transcripts, *J. Virol.,* 61, 3199, 1987.

100. **Laporte, J., Grosclaude, J., Wantyghem, J., Bernard, S. and Rouze, P.,** Neutralization en culture cellulaire du pouvoir infectieux du virus de la fievre aphteuse par des serums provenant de porcs immunisés à l'aide d'une protéine virale purifiée, *C.R. Acad. Sci.,* 276, 3399, 1973.

101. **Bachrach, H. L., Moore, D. M., McKercher, P. A., and Polatnick, J.,** Immune and antibody responses to an isolated capsid protein of foot-and-mouth disease virus, *J. Immunol.,* 115, 1636, 1975.

102. **Bachrach, H. L., Morgan, D. O., and Moore, D. M.,** Foot-and-mouth disease virus immunogenic capsid protein VP_T: N-terminal sequences and immunogenic peptides obtained by CNBr and tryptic cleavage, *Intervirology,* 12, 65, 1979.

103. **Strohmaier, K., Franze, R., and Adam, K. H.,** Location and characterization of the antigenic portion of the FMDV immunizing protein, *J. Gen. Virol.,* 59, 295, 1982.

104. **Kleid, D. G., Yansura, D., Small, B., Dowbenko, D., Moore, D. M., Grubman, M. J., McKercher, P. D., Morgan, D. O., Robertson, B. H., and Bachrach, H. L.,** Cloned viral protein vaccine for foot-and-mouth disease: responses in cattle and swine, *Science,* 14, 1125, 1981.

105. **Bittle, J. L., Houghton, R. A., Alexander, H., Shinnick, T. M., Sutcliffe, J. G., Lerner, R. A., Rowlands, D. J., and Brown, F.,** Protection against FMDV by immunization with a chemically synthesized peptide predicted from the viral nucleotide sequence, *Nature (London),* 298, 30, 1982.

106. **Pfaff, E., Mussgay, M., Boehm, H. O., Schulz, G. E., and Schaller, H.,** Antibodies against a preselected peptide recognise and neutralize foot-and-mouth disease virus, *EMBO J.,* 1, 869, 1982.

107. **Shire, S. J., Bock, L., Ogez, J., Builder, S., Kleid, D., and Moore, D. M.,** Purification and immunogenicity of fusion VP1 protein of foot-and-mouth disease virus, *Biochemistry,* 23, 6474, 1984.

108. **DiMarchi, R., Brooke, G., Gale, C., Cracknell, V., Doel, T., and Mowat, N.,** Protection of cattle against foot-and-mouth disease by a synthetic peptide, *Science,* 232, 639, 1986.

109. **Cheung, A. K. and Küpper, H.,** Biotechnological approach to a new foot-and-mouth disease virus vaccine, *Biotechnol. Genet. Eng. Rev.,* 1, 223, 1984.

110. **Kurz, C., Forss, S., Kupper, H., Strohmaier, K., and Schaller, H.**, Nucleotide sequence and corresponding amino acid sequence of the gene for the major antigen of foot-and-mouth disease virus, *Nucleic Acids Res.*, 9, 1919, 1981.

111. **Boothroyd, J. C., Harris, T. J. R., Rowlands, D. J., and Lowe, P. A.**, The nucleotide sequence of cDNA coding for the structural proteins of foot-and-mouth disease virus, *Gene*, 17, 153, 1982.

112. **Makoff, A. J., Paynter, C. A., Rowlands, D. J., and Boothroyd, J. C.**, Comparison of the amino acid sequence of the major immunogen from three serotypes of foot-and-mouth disease virus, *Nucleic Acids Res.*, 10, 8285, 1982.

113. **Cheung, A., DeLamarter, J., Weiss, S., and Küpper, H.**, Comparison of the major antigenic determinants of different serotypes of foot-and-mouth disease virus, *J. Virol.*, 48, 451, 1983.

114. **Villanueva, N., Dávila, M., Ortín, J., and Domingo, E.**, Molecular cloning of cDNA from foot-and-mouth disease virus C_1-Santa Pau (CS-8). Sequence of protein VP1 coding segment, *Gene*, 23, 185, 1983.

115. **Beck, E., Feil, G., and Strohmaier, K.**, The molecular basis of the antigenic variation of foot-and-mouth disease virus, *EMBO J.*, 2, 555, 1983.

116. **Rowlands, D. J., Clarke, B. E., Carroll, A. R., Brown, F., Nicholson, B. H., Bittle, J. L., Houghten, R. A., and Lerner, R. A.**, Chemical basis of antigenic variations in foot-and-mouth disease virus, *Nature (London)*, 306, 694, 1983.

117. **Cheung, A., Whitehead, P., Weiss, S., and Küpper, H.**, Nucleotide sequence of the VP1 gene of the foot-and-mouth disease virus strain A Venceslau, *Gene*, 30, 241, 1984.

118. **Weddell, G. N., Yansura, D. G., Dowbenko, D. J., Hoatlin, M. E., Grubman, M. J., Moore, D. M., and Kleid, D. G.**, Sequence variation in the gene for the immunogenic capsid protein VP1 of foot-and-mouth disease virus type A, *Proc. Natl. Acad. Sci. U.S.A.*, 82, 2618, 1985.

119. **Sobrino, F., Palma, E. L., Beck, E., Dávila, M., de la Torre, J. C., Negro, P., Villanueva, N., Ortín, J., and Domingo, E.**, Fixation of mutations in the viral genome during an outbreak of foot-and-mouth disease: heterogeneity and rate variations, *Gene*, 50, 149, 1986.

120. **Beck, E. and Strohmaier, K.**, Subtyping of european FMDV outbreak strains by nucleotide sequence determination, *J. Virol.*, 61, 1621, 1987.

120a. **Martinez, M. A., Carrillo, C., Plana, J., Mascarella, R., Bergada, J., Palma, E. L., Domingo, E., and Sobrino, F.**, Genetic and immunogenic variations among closely related isolates of foot-and-mouth disease virus, *Gene*, in press.

120b. **Piccone, M. E., Kaplan, G., Giavedoni, D., Domingo, E., and Palma, E. L.**, VP1 of foot-and-mouth disease virus of serotype C: long-term conservation of sequences, *J. Virol.*, 1988, in press.

121. **Domingo, E., Dávila, M., and Ortín, J.**, Nucleotide sequence heterogeneity of the RNA from a natural population of foot-and-mouth disease virus, *Gene*, 11, 333, 1980.

122. **King, A. M. Q., Underwood, B. O., McCahon, D., Newman, J. W. I., and Brown, F.**, Biochemical identification of viruses causing the 1981 outbreaks of foot and mouth disease in the U.K., *Nature (London)*, 293, 479, 1981.

123. **Robertson, B. H., Morgan, D. O., Moore, D. M., Grubman, M. J., Card, J., Fisher, T., Weddell, G., Dowbenko, D., and Yansura, D.**, Identification of amino acid and nucleotide sequence of the foot-and-mouth disease virus RNA polymerase, *Virology*, 126, 614, 1983.

124. **Martínez-Salas, E., Ortín, J., and Domingo, E.**, Sequence of the viral replicase gene from foot-and-mouth disease virus C_1-Santa Pau (CS-8), *Gene*, 35, 55, 1985.

124a. **Gebauer, F., de la Torre, J. C., Gomes, I., Mateu, M. G., Barahona, H., Tirabaschi, B., Bergmann, I., Augé de Mello, P., and Domingo, E.**, 1988, submitted.

125. **Takeda, N., Miyamura, K., Ogino, T., Natori, K., Yamazaki, S., Sakwiai, N., Nakazono, N., Ishii, K., and Kono, R.**, Evolution of enterovirus type 70: oligonucleotide mapping analysis of RNA genome, *Virology*, 134, 375, 1984.

126. **Kamahora, T., Itagaki, A., Hattori, N., Tsuchie, H., and Kurimura, T.**, Oligonucleotide fingerprint analysis of coxsackievirus A10 isolated in Japan, *J. Gen. Virol.*, 66, 2627, 1985.

127. **Clewley, J. P., Pullin, J. K., Avery, R. J., and Moore, N. F.**, Oligonucleotide fingerprinting of the RNA species obtained from six Drosophila C virus isolates, *J. Gen. Virol.*, 64, 503, 1983.

128. **Siegl, G., deChastonay, J., and Kronauer, K.**, Propagation and assay of hepatitis A virus in vitro, *J. Virol. Methods*, 9, 53, 1984.

129. **Weitz, M. and Siegl, G.**, Variation among hepatitis A virus strains. I. Genomic variation detected by T_1 oligonucleotide mapping, *Virus Res.*, 4, 53, 1985.

130. **Baroudy, B. M., Ticehurst, J. R., Miele, T. A., Maizel, J. V., Purcell, R. H., and Feinstone, S. M.**, Sequence analysis of hepatitis A virus cDNA coding for capsid proteins and RNA polymerase, *Proc. Natl. Acad. Sci. U.S.A.*, 82, 2143, 1985.

131. **Najarian, R., Caput, D., Gee, W., Potter, S. J., Renard, A., Merrweather, J., Van Nest, G., and Dina, D.**, Primary structure and gene organization of human hepatitis A virus, *Proc. Natl. Acad. Sci. U.S.A.*, 82, 2627, 1985.

132. **Linemeyer, D. L., Menke, J. G., Martín-Gallardo, A., Hughes, J. V., Young, A., and Mitra, S. W.**, Molecular cloning and partial sequencing of hepatitis A viral cDNA, *J. Virol.*, 54, 247, 1985.

132a. **Paul, A. V., Tada, H., von der Helm, K., Wissel, T., Kiehn, R., Wimmer, E., and Deinhardt, F.**, The entire nucleotide sequence of the genome of human hepatitis A virus (isolate MBB), *Virus Res.*, 8, 153, 1987.

133. **Domingo, E. and Sobrino, F.**, unpublished observations.

134. **Venuti, A., Di Russo, C., del Grosso, N., Patti, A. M., Degener, A. M., Midulla, M., Paña, A., and Perez-Bercoff, R.**, Isolation and molecular cloning of a fast-growing strain of human hepatitis A virus from its double-stranded replicative form, *J. Virol.*, 56, 579, 1985.

135. **Stanway, G., Hughes, P. J., Mountford, R. C., Minor, P. D., and Almond, J. W.**, The complete nucleotide sequence of a common cold virus: human rhinovirus 14, *Nucleic Acids Res.*, 12, 7859, 1984.

136. **Callahan, P., Mizutani, S., and Colonno, R. J.**, Molecular cloning and complete sequence determination of human rhinovirus type 14 genome RNA, *Proc. Natl. Acad. Sci. U.S.A.*, 82, 732, 1985.

137. **Skern, T., Sommergruber, W., Blaas, D., Gruendler, P., Fraundorfer, F., Pieler, C., Fogy, I., and Kuechler, E.**, Human rhinovirus 2: complete nucleotide sequence and proteolytic processing signals in the capsid protein region, *Nucleic Acids Res.*, 13, 2111, 1985.

138. **Werner, G., Rosenwirth, B., Bauer, E., Seifert, J. M., Werner, F. J., and Besemer, J.**, Molecular cloning and sequence determination of the genomic regions encoding protease and genome-linked protein of three picornaviruses, *J. Virol.*, 57, 1084, 1986.

139. **Palmenberg, A. C., Kirby, E. M., Janda, M. R., Drake, N. L., Duke, G. M., Potratz, K. F., and Collett, M. S.**, The nucleotide and deduced amino acid sequences of the encephalomyocarditis viral polyprotein coding region, *Nucleic Acids Res.*, 12, 2969, 1984.

140. **Tracy, S., Liu, H. L, and Chapman, N. M.**, Coxsackievirus B3: primary structure of the 5' non-coding and capsid protein, coding regions of the genome, *Virus Res.*, 3, 263, 1985.

140a. **Lindberg, A. M., Stålhanske, P. O. K., and Pettersson, U.**, Genome of coxsackie virus B3, *Virology*, 156, 50, 1987.

140b. **Iizuka, N., Kuge, S., and Nomoto, A.**, Complete nucleotide sequence of the genome of coxsackie virus B1, *Virology*, 156, 64, 1987.

141. **Palmenberg, A. C.**, Sequence alignments and picornaviral relationships: a proposal for a phylogeny-based classification system, in 5th Meet. of the European Group of Molecular Biology of Picornaviruses, Mallorca, Spain, 1987.

142. **Palmenberg, A. C.**, Comparative organization and genome structure in picornaviruses, *J. Cell Biochem.*, 1986, in press.

143. **Staden, R.**, An interactive graphics program for comparing and aligning nucleic acid and amino acid sequences, *Nucleic Acids Res.*, 10, 2951, 1982.

144. **Hogle, J. M., Chow, M., and Filman, D. J.**, Three-dimensional structure of poliovirus at 2.9 Å resolution, *Science*, 229, 1358, 1985.

145. **Rossmann, M. J., Arnold, E., Erickson, J. W., Fraenkenberg, E. A., Griffith, J. P., Hecht, H. J., Johnson, J. E., Kamer, G., Luo, M., Mosser, A. G., Rueckert, R. R., Sherry, B., and Vriend, G.**, Structure of a human cold virus (rhinovirus 14) and functional relationship to other picornaviruses, *Nature (London)*, 317, 145, 1985.

146. **Baltimore, D.**, Picornaviruses are no longer black boxes, *Science*, 229, 1366, 1985.

147. **Luo, M., Vriend, G., Kamer, G., Minor, I., Arnold, E., Rossman, M., Boege, V., Scraba, D., Duke, G., and Palmenberg, A.**, The atomic structure of Mengo Virus at ¹3.0 Å resolution, *Science*, 235, 182, 1987.

148. **Fox, G., Stuart, D., Ravindra Acharya, K., Fry, E., Rowlands, D., and Brown, F.**, Crystallization and preliminary X-ray diffraction analysis of foot-and-mouth disease virus, *J. Mol. Biol.*, 196, 591, 1987.

149. **Trent, D. W., Clewley, J. P., France, J., and Bishop, D. H. L.**, Immunochemical and oligonucleotide fingerprint analyses of Venezuelan equine encephalomyocarditis complex viruses, *J. Gen. Virol.*, 43, 365, 1979.

150. **Trent, D. W. and Grant, J. A.**, A comparison of New World alphaviruses in the western equine encephalomyelitis complex by immunochemical and oligonucleotide fingerprint techniques, *J. Gen. Virol.*, 47, 261, 1980.

151. **Olson, K. and Trent, D. W.**, Genetic and antigenic variations among geographical isolates of Sindbis virus, *J. Gen. Virol.*, 66, 797, 1985.

152. **Garoff, H., Frischauf, A. M., Simons, K., Lehrach, H., and Delius, H.**, Nucleotide sequence of cDNA coding for Semliki Forest membrane glycoproteins, *Nature (London)*, 288, 236, 1980.

153. **Rice, C. M. and Strauss, J. H.**, Nucleotide sequence of the 26S mRNA of Sindbis virus and deduced sequence of the encoded virus structural proteins, *Proc. Natl. Acad. Sci. U.S.A.*, 78, 2062, 1981.

154. **Strauss, E. G., Rice, C. M., and Strauss, J. H.**, Sequence coding for the alphavirus nonstructural proteins is interrupted by an opal termination codon, *Proc. Natl. Acad. Sci. U.S.A.*, 80, 5271, 1983.

155. **Dalgarno, L., Rice, C. M., and Strauss, J. H.,** Ross river-virus 26S RNA: complete nucleotide sequence and deduced sequence of the encoded structural proteins, *Virology,* 129, 170, 1983.
156. **Strauss, E. G., Rice, C. M., and Strauss, J. H.,** Complete nucleotide sequence of the genomic RNA of Sindbis virus, *Virology,* 133, 92, 1984.
157. **Hahn, C. S., Strauss, E. G., and Strauss, J. H.,** Sequence-analysis of three Sindbis virus mutants temperature-sensitive in the capsid protein autoprotease, *Proc. Natl. Acad. Sci. U.S.A.,* 82, 4648, 1985.
158. **Kinney, R. M., Johnson, B. J. B., Brown, V. L., and Trent, D. W.,** Nucleotide sequence of the 26S mRNA of the virulent Trinidad donkey strain of Venezuelan equine encephalitis virus and deduced sequence of the encoded structural proteins, *Virology,* 152, 400, 1986.
159. **Bell, J. R., Kinney, R. M., Trent, D. W., Strauss, E. G., and Strauss, J. H.,** An evolutionary tree relating eight alpha-viruses, based on amino-terminal sequences of their glycoproteins, *Proc. Natl. Acad. Sci. U.S.A.,* 81, 4702, 1984.
160. **Ahlquist, P., Strauss, E. G., Rice, C. M., Strauss, J. H., Haseloff, J., and Zimmern, D.,** Sindbis virus proteins nsP1 and nsP2 contain homology to non-structural proteins from several RNA plant viruses, *J. Virol.,* 53, 536, 1985.
161. **Rice, C. M., Lenches, E. M., Eddy, S. R., Shin, S. J., Sheets, R. L., and Strauss, J. H.,** Nucleotide sequence of yellow fever virus: implications for flavivirus gene expression and evolution, *Science,* 229, 726, 1985.
162. **Hahn, C. S., Dalrymple, J. M., Strauss, J. H., and Rice, C. M.,** Nucleotide sequence of the virulent Asibi strain of yellow fever virus: comparison with the sequence of the 17D strain, 1987, submitted.
163. **Castle, E., Leidner, U., Nowak, T., Wengler, G., and Wengler, G.,** Primary structure of the West Nile flavivirus genome region coding for all nonstructural proteins, *Virology,* 149, 10, 1986.
164. **Wengler, G. and Castle, E.,** Analysis of structural properties which probably are characteristic for the 3'-terminal sequence of the genome RNA of flaviviruses, *J. Gen. Virol.,* 67, 1183, 1986.
165. **Dalgarno, L., Trent, D. W., Strauss, J. H., and Rice, C. M.,** Partial nucleotide sequence of the Murray Valley encephalitis virus genome. Comparison of the encoded polypeptides with yellow fever virus structural and non-structural proteins, *J. Mol. Biol.,* 187, 309, 1986.
166. **Trent, D. W., Grant, J. A., Vorndam, A. V., and Monath, T. P.,** Genetic heterogeneity among Saint Louis encephalitis virus isolates of different geographical origin, *Virology,* 114, 319, 1981.
167. **Trent, D. W., Grant, J. A., Rosen, L., and Monath, T. P.,** Genetic variation among dengue 2 viruses of different geographic origin, *Virology,* 128, 271, 1983.
168. **Deubel, V., Digoutte, J. P., Monath, T. P., and Girard, M.,** Genetic heterogeneity of yellow fever virus strains from Africa and the Americas, *J. Gen. Virol.,* 67, 209, 1986.
169. **Siddell, St., Wege, H., and ter Meulen, V.,** The structure and replication of Coronaviruses, *Curr. Top. Microbiol. Immunol.,* 99, 131, 1982.
170. **Lai, M. M. C. and Stohlman, S. A.,** Comparative analysis of RNA genomes of mouse hepatitis viruses, *J. Virol.,* 38, 661, 1981.
171. **Wege, H., Stephenson, J. R., Koga, M., Wege, H., and ter Meulen, V.,** Genetic variation of neurotropic and non-neurotropic murine coronaviruses, *J. Gen. Virol.,* 54, 67, 1981.
172. **Stohlman, S. A., Brayton, P. R., Fleming, J. O., Weiner, L. P., and Lai, M. M. C.,** Murine coronaviruses: isolation and characterization of two plaque morphology variants of the JHM neurotropic strain, *J. Gen. Virol.,* 63, 265, 1982.
173. **Boursnell, M. E. G., Binns, M. M., Foulds, I. J., and Brown, T. D. K.,** Sequences of the nucleocapsid genes from two strains of avian infectious bronchitis virus, *J. Gen. Virol.,* 66, 573, 1985.
174. **Makino, S., Keck, J. G., Stohlman, S. A., and Lai, M. M. C.,** High-frequency RNA recombination of murine Coronaviruses, *J. Virol.,* 57, 729, 1986.
175. **Franssen, H., Leunissen, J., Goldbach, R., Lomonossoff, G., and Zimmern, D.,** Homologous sequences in non-structural proteins from cowpea mosaic virus and picornaviruses, *EMBO J.,* 3, 855, 1984.
176. **Haseloff, J., Goelet, P., Zimmern, D., Ahlquist, P., Dasgupta, R., and Kaesberg, P.,** Striking similarities in amino acid sequence among nonstructural proteins encoded by RNA viruses that have dissimilar genomic organization, *Proc. Natl. Acad. Sci. U.S.A.,* 81, 4358, 1984.
177. **Argos, P., Kamer, G., Nicklin, M. J., and Wimmer, E.,** Similarity in gene organization and homology between proteins of animal picornaviruses and a plant comovirus suggest common ancestry of these virus families, *Nucleic Acids Res.,* 12, 7251, 1984.
178. **McClure, M. A. and Perrault, J.,** RNA virus genomes hybridize to cellular rRNAs and to each other, *J. Virol.,* 57, 917, 1986.
179. **Yewdell, J. W. and Gerhard, W.,** Antigenic characterization of viruses by monoclonal antibodies, *Annu. Rev. Microbiol.,* 35, 185, 1981.
180. **Almond, J. W., Stanway, G., Cann, A. J., Westrop, G. D., Evans, D. M. A., Ferguson, M., Minor, P. D., Spitz, M., and Schild, G. C.,** New poliovirus vaccines: a molecular approach, *Vaccine,* 2, 177, 1984.

181. **Mufson, M. A., Orvell, C., Rafnar, B., and Norrby, E.,** Two distinct subtypes of human respiratory syncytial virus, *J. Gen. Virol.*, 66, 2111, 1985.

182. **Sheshberadaran, H., Norrby, E., McCullough, K. C., Carpenter, W. C., and Orvell, C.,** The antigenic relationship between measles, canine distemper and rinderpest viruses studied with monoclonal antibodies, *J. Gen. Virol.*, 67, 1381, 1986.

183. **Orvell, C.,** The reactions of monoclonal antibodies with structural proteins of mumps virus, *J. Immunol.*, 132, 2622, 1984.

184. **Iorio, R. M., Borgman, J. B., Glickman, R. L., and Bratt, M. A.,** Genetic variation within a neutralizing domain on the haemagglutinin-neuraminidase glycoprotein of Newcastle disease virus, *J. Gen. Virol.*, 67, 1393, 1986.

185. **Ray, R. and Compans, R. W.,** Monoclonal antibodies reveal extensive antigenic differences between the hemagglutinin-neuraminidase glycoproteins of human and bovine parainfluenza 3 viruses, *Virology*, 148, 232, 1986.

186. **Buchmeier, M. J., Lewicki, H. A., Tomori, O., and Oldstone, M. B. A.,** Monoclonal antibodies to lymphocytic choriomeningitis virus and Pichinde viruses: generation, characterization and cross-reactivity with other arenaviruses, *Virology*, 113, 73, 1981.

187. **Howard, C. R., Lewicki, H., Allison, L., Salter, M., and Buchmeier, M. J.,** Properties and characterization of monoclonal antibodies to Tacaribe virus, *J. Gen. Virol.*, 66, 1383, 1985.

188. **Fleming, J. O., Stohlman, S. A., Harmon, R. C., Lai, M. M. C., Frelinger, J. A., and Weiner, L. P.,** Antigenic relationships of murine coronaviruses: analysis using monoclonal antibodies to JHM (MHV-4) virus, *Virology*, 131, 296, 1983.

189. **Talbot, P. J. and Buchmeier, M. J.,** Antigenic variation among murine coronaviruses: evidence for polymorphism on the peplomer glycoprotein, E2, *Virus Res.*, 2, 317, 1985.

190. **Laude, H., Chapral, J. M., Gelfi, J., Labian, S., and Grosclaude, J.,** Antigenic structure of transmissible gastroenteritis virus. I. Properties of monoclonal antibodies directed against virion proteins, *J. Gen. Virol.*, 67, 119, 1986.

191. **Jimenez, G., Correa, I., Melgosa, M. P., Bullido, M. J., and Enjuanes, L.,** Critical epitopes in transmissible gastroenteritis virus neutralization, *J. Virol.*, 60, 131, 1986.

192. **Delmas, B., Gelfi, J., and Laude, H.,** Antigenic structure of transmissible gastroenteritis virus. II. Domains in the peplomer glycoprotein, *J. Gen. Virol.*, 67, 1405, 1986.

193. **Prabhakar, B. S., Haspel, M. V., McClintock, P. R., and Notkins, A. L.,** High frequency of antigenic variants among naturally occurring human coxsackie B4 virus isolates identified by monoclonal antibodies, *Nature (London)*, 300, 374, 1982.

194. **Prabhakar, B. S., Menegus, M. A., and Notkins, A. L.,** Detection of conserved and nonconserved epitopes on coxsackie virus B4: frequency of antigenic change, *Virology*, 146, 302, 1985.

195. **Webb, S. R., Kearse, K. P., Foulke, C. L., Hartig, P. C., and Prabhakan, B. S.,** Neutralization epitope diversity of coxsackievirus B4 isolates detected by monoclonal antibodies, *J. Med. Virol.*, 12, 1986.

196. **Minor, P. D., Ferguson, M., Evans, D. M. A., Almond, J. W., and Icenogle, J. P.,** Antigenic structure of polioviruses of serotypes 1, 2 and 3, *J. Gen. Virol.*, 67, 1283, 1986.

197. **Mateu, M. G., Rocha, E., Vicente, O., Vayreda, F., Navalpotro, C., Andreu, D., Pedroso, E., Giralt, E., Enjuanes, L., and Domingo, E.,** Reactivity with monoclonal antibodies of viruses from an episode of foot-and-mouth disease, *Virus Res.*, 8, 261, 1987.

198. **Gould, E. A., Buckley, A., Cammack, N., Barret, A. D. T., Clegg, J. C. S., Ishak, R., and Varma, M. G. R.,** Examination of the immunological relationships between flaviviruses using yellow fever virus monoclonal antibodies, *J. Gen. Virol.*, 66, 1369, 1985.

199. **Sato, T. A., Fukuda, A., and Sugiura, A.,** Characterization of major structural proteins of measles virus with monclonal antibodies, *J. Gen. Virol.*, 66, 1397, 1985.

200. **Miyake, T., Shiba, T., and Watanabe, I.,** Grouping of RNA phages by a millipore filtration method, *Jpn. J. Microbiol.*, 11, 203, 1967.

201. **Domingo, E., Sabo, D., and Weissmann, C.,** Nucleotide sequence heterogeneity in the RNA of phage Qβ, *Experientia*, 32, 792, 1976.

202. **Billeter, M. A.,** Sequence and location of large RNAase T1 oligonucleotides in bacteriophage Qβ RNA, *J. Biol. Chem.*, 253, 8381, 1978.

203. **Horodyski, F. M., Nichol, S. T., Spindler, K. R., and Holland, J. J.,** Properties of DI particle resistant mutants of vesicular stomatitis virus isolated from persistant infections and from undiluted passages, *Cell*, 33, 801, 1983.

204. **O'Hara, P. J., Horodyski, F. M., Nichol, S. T., and Holland, J. J.,** Vesicular stomatitis virus mutants resistant to defective-interfering particles accumulate stable 5′-terminal and fewer 3′-terminal mutations in a stepwise manner, *J. Virol.*, 49, 793, 1984.

205. **O'Hara, P. J., Nichol, S. T., Horodyski, F. M., and Holland, J. J.,** Vesicular stomatitis virus defective interfering particles can contain extensive genomic sequence rearrangements and base substitutions, *Cell*, 36, 915, 1984.

206. **Domingo, E., Martínez-Salas, E., Sobrino, F., de la Torre, J. C., Portela, A., Ortín, J., López-Galindez, C., Pérez-Breña, P., Villanueva, N., Nájera, R., VandePol, S., Steinhauer, D., De Polo, N., and Holland, J. J.,** The quasispecies (extremely heterogeneous) nature of viral RNA genome poulations: biological relevance-a review, *Gene*, 40, 1, 1985.

207. **Sobrino, F., Dávila, M., Ortín, J., and Domingo, E.,** Multiple genetic variants arise in the course of replication of foot-and-mouth disease virus in cell culture, *Virology*, 128, 310, 1983.

208. **Steinhauer, D. A. and Holland, J. J.,** Direct method for quantitation of extreme polymerase error frequencies at selected single base sites in viral RNA, *J. Virol.*, 57, 219, 1986.

209. **Fernández-Santos, T. and Domingo, E.,** unpublished experiments.

209a. **de la Torre, J. C., Martinez-Salas, E., Diez, J., Villaverde, A., Gebauer, F., Rocha, E., Dávila, M., and Domingo, E.,** Coevolution of cells and viruses in a persistent infection of foot-and-mouth disease virus in cell culture, 1988, submitted.

210. **Bolwell, C., Parry, N. R., Rowlands, D. J., Brown, F., and Ouldridge, E. J.,** Foot-and-mouth disease virus variants with broad and narrow antigenic spectra, in *Vaccines 86*, Lerner, R. A., Channock, R. M., and Brown, F., Eds., Cold Spring Harbor Laboratory, Cold Spring Harbor, New York, 1986, 51.

211. **Goelet, P., Lomonossoff, G. P., Butler, P. J. G., Akam, M. E., Gait, M. J., and Karn, J.,** Nucleotide sequence of tobacco mosaic virus RNA, *Proc. Natl. Acad. Sci. U.S.A.*, 79, 5818, 1982.

212. **Schubert, M., Harmison, G. G., and Meier, E.,** Primary structure of the vesicular stomatitis virus polymerase (L) gene: evidence for a high frequency of mutations, *J. Virol.*, 51, 505, 1984.

213. **Arias, C. F., López, S., and Espejo, R. T.,** Heterogeneity in base sequence among different DNA clones containing equivalent sequences of rotovirus double-stranded RNA, *J. Virol.*, 57, 1207, 1986.

214. **Reanney, D. C.,** The evolution of RNA viruses, *Annu. Rev. Microbiol.*, 36, 47, 1982.

215. **Reanney, D. C.,** The molecular evolution of viruses, in *The Microbe. I. Viruses*, Mahy, B. W. J. and Pattison, J. R., Eds., Cambridge University Press, Cambridge, Massachusetts, 1984, 175.

216. **Biebricher, C. K.,** Darwinian selection of self-replicating RNA molecules, in *Evolutionary Biology*, Vol. 16, Hechet, M. K., Wallace, B., and Prance, G. T., Eds., Plenum Press, New York, 1983, 1.

217. **Eigen, M.,** Self-organization of matter and the evolution of biological macromolecules, *Naturwissenschaften*, 58, 465, 1971.

218. **Eigen, M. and Schuster, P.,** *The Hypercycle. A Principle of Natural Self-Organization*, Springer-Verlag, Berlin, 1979.

219. **Wiktor, T. J. and Koprowski, H.,** Antigenic variants of rabies virus, *J. Exp. Med.*, 152, 99, 1980.

220. **Portner, A., Webster, R. G., and Bean, W. J.,** Similar frequencies of antigenic variants in Sendai, vesicular stomatitis and influenza A viruses, *Virology*, 104, 235, 1980.

221. **Birrer, M. J., Udem, S., Nathenson, S., and Bloom, B. R.,** Antigenic variants of measles virus, *Nature (London)*, 293, 67, 1981.

222. **Emini, E. A., Jameson, B. A., Lewis, A. J., Larsen, G. R., and Wimmer, E.,** Poliovirus neutralization epitopes: analysis and localization with neutralizing monoclonal antibodies, *J. Virol.*, 43, 997, 1982.

223. **Crainic, R., Couillin, P., Blondel, B., Cabau, N., Bouse, A., and Horodniceanu, F.,** Natural variation of poliovirus epitopes, *Infect. Immun.*, 41, 1217, 1983.

224. **Diamond, D. C., Jameson, B. A., Bonin, J., Kohara, M., Abe, S., Itoh, H., Komatsu, T., Arita, M., Kuge, S., Namoto, A., Osterhaus, A. D. M., Crainic, R., and Wimmer, E.,** Antigenic variation and resistance to neutralization in poliovirus type 1, *Science*, 229, 1090, 1985.

225. **Sheshberadaran, H. and Norrby, E.,** Characterization of epitopes on the measles virus hemagglutinin, *Virology*, 152, 58, 1986.

226. **Lubeck, M. D., Schulman, J. L., and Palese, P.,** Antigenic variants of influenza viruses: marked differences in the frequencies of variants selected with different monoclonal antibodies, *Virology*, 102, 458, 1980.

227. **Flavell, R. A., Sabo, D. L. O., Bandle, E. F., and Weissmann, C.,** Site-directed mutagenesis: generation of an extracistronic mutation in bacteriophage Qβ RNA, *J. Mol. Biol.*, 89, 255, 1974.

228. **Flavell, R. A., Sabo, D. L. O., Bandle, E. F., and Weissmann, C.,** Site-directed mutagenesis: effect of an extracistronic mutation on the in vitro propagation of bacteriophage Qβ RNA, *Proc. Natl. Acad. Sci. U.S.A.*, 72, 367, 1975.

229. **Taniguchi, T. and Weissmann, C.,** Site directed mutations in the initiator region of the bacteriophage Qβ coat cistron and their effect on ribosome binding, *J. Mol. Biol.*, 118, 533, 1978.

230. **Weber, H., Taniguchi, T., Müller, W., Meyer, F., and Weissmann, C.,** Application of site-directed mutagenesis to RNA and DNA genomes, *Cold Spring Harbor Symp. Quant. Biol.*, 43, 669, 1979.

231. **Adams, J. M. and Cory, S.,** Untranslated nucleotide sequence at the 5'-end of R17 bacteriophage RNA, *Nature (London)*, 227, 570, 1970.

232. **Robertson, H. D. and Jeppesen, P. G. N.,** Extent of variation in three related bacteriophage RNA molecules, *J. Mol. Biol.*, 68, 417, 1972.

233. **Min Jou, W. and Fiers, W.,** Studies on the bacteriophage MS2. XXXIII. Comparison of the nucleotide sequences in related bacteriophage RNA's, *J. Mol. Biol.*, 106, 1047, 1976.

234. **Batschelet, E., Domingo, E., and Weissmann, C.,** The proportion of revertant and mutant phage in a growing population, as a function of mutation and growth rate, *Gene,* 1, 27, 1976.

235. **Steinhauer, D. A. and Holland, J. J.,** High mutation rates and population equilibria in RNA virus evolution, in 7th Int. Congr. Virology, Edmonton, Canada, 1987, Abstr. S32.7.

236. **Buonagurio, D. A., Nakada, S., Parvin, J. D., Krystal, M., Palese, P., and Fitch, W. M.,** Evolution of human influenza A viruses over 50 years: rapid uniform rate of change in NS gene, *Science,* 232, 980, 1986.

237. **Parvin, J. D., Moscona, A., Pan, W. T., Leider, J. M., and Palese, P.,** Measurement of the mutation rate of animal viruses: influenza A virus and poliovirus-1, *J. Virol.,* 59, 377, 1986.

238. **Durbin, R. K. and Stollar, V.,** Sequence analysis of the E2 gene of a hyperglycosylated, host restricted mutant of Sindbis virus and estimation of mutation rate from frequency of revertants, *Virology,* 154, 135, 1986.

238a. **Sedivy, J. M., Capone, J. P., Raj Bhandary, U. L., and Sharp, P. A.,** An inducible mammalian amber suppressor: propagation of a poliovirus mutant, *Cell,* 50, 379, 1987.

238b. **Preston, B. D. and Loeb, L. A.,** Retroviral evolution and the fidelity of genomic replication, VII Int. Congr. Virology, Edmonton, Canada, 1987, Abstr. S32.2

239. **Stokes, M. A. M.,** Poliovirus RNA Polymerase: *In Vitro* Enzymatic Activities, Fidelity of Replication and Characterization of a Temperature-Sensitive Mutant, Ph.D. thesis, University of Florida, Gainesville, 1985, 34.

240. **Zarling, D. and Temin, H.,** High spontaneous mutation rate of avian sarcoma virus, *J. Virol.,* 17, 74, 1976.

241. **Coffin, J. M.,** Structure, replication and recombination of retrovirus genomes: some unifying hypotheses, *J. Gen. Virol.,* 42, 1, 1979.

242. **Gojobori, T. and Tokoyama, S.,** Rates of evolution of the retroviral oncogene of Moloney murine sarcoma virus and of its cellular homologues, *Proc. Natl. Acad. Sci. U.S.A.,* 82, 4198, 1985.

243. **Clements, J. E., Pederson, F. S., Narayan, O., and Haseltine, W. A.,** Genomic changes associated with antigenic variation of visna virus during persistent infection, *Proc. Natl. Acad. Sci. U.S.A.,* 77, 4454, 1980.

244. **Salinovich, O., Payne, S. L., Montelaro, R. C., Hussain, K. A., Issel, C. J., and Schnorr, K. L.,** Rapid emergence of novel antigenic and genetic variants of equine infectious anemia virus during persistent infection, *J. Virol.,* 57, 71, 1986.

245. **Hahn, B., Shaw, G., Taylor, M., Redfield, R., Markham, P., Salahuddin, S., Wong-Staal, F., Gallo, R., Parks, E., and Parks, W.,** Genetic variation in HTLV-III/LAV over time in patients with AIDS or at risk for AIDS, *Science,* 232, 1548, 1986.

246. **Coffin, J. M.,** Genetic variation in AIDS viruses, *Cell,* 46, 1, 1986.

247. **Alizon, M., Wain-Hobson, S., Montagnier, L., and Sonigo, P.,** Genetic variability of the AIDS viruses: nucleotide sequence analysis of two isolates from african patients, *Cell,* 46, 63, 1986.

248. **Willey, T., Buckler, C. E., and Martin, M. A.,** Identification of conserved and divergent domains within the envelope gene of the acquired immunodeficiency syndrome retroviruses, *Proc. Natl. Acad. Sci. U.S.A.,* 83, 5038, 1986.

249. **Pedley, S., Hundley, F., Chrystie, I., McCrae, M. A., and Desselberger, U.,** The genomes of rotaviruses isolated from chronically infected immunodeficient children, *J. Gen. Virol.,* 65, 1141, 1984.

250. **Hundley, F., Biryahwaho, B., Gow, M., and Desselberger, U.,** Genome rearrangement of bovine rotavirus after serial passage at high multiplicity of infection, *Virology,* 143, 88, 1985.

251. **Allen, A. M. and Desselberger, U.,** Reassortment of human rotaviruses carrying rearranged genomes with bovine rotavirus, *J. Gen. Virol.,* 66, 2703, 1985.

252. **Biryahwaho, B., Hundley, F., and Desselberger, U.,** Bovine rotavirus with rearranged genome reassorts with human rotavirus, submitted.

253. **Graham, A., Kudesia, G., Allen, A. M., and Desselberger, U.,** Reassortment of human rotavirus possessing genome rearrangements with bovine rotavirus: evidence for host cell selection, *J. Gen. Virol.,* 1986, in press.

254. **Hundley, F. and Desselberger, U.,** manuscript in preparation.

255. **Hundley, F., McIntyre, M., Chrystie, I., Wood, D., and Desselberger, U.,** Heterogeneity of genome rearrangements in rotaviruses isolated from a chronically infected, immunodeficient child — the evolution of a quasispecies, manuscript in preparation.

256. **Wabl, M., Burrows, P. D., von Gabain, A., and Steinberg, C.,** Hypermutation at the immunoglobulin heavy chain locus in a pre-B-cell line, *Proc. Natl. Acad. Sci. U.S.A.,* 82, 479, 1985.

257. **Drake, J. W.,** Comparative rates of spontaneous mutation, *Nature (London),* 221, 1132, 1969.

258. **Loeb, L. A. and Kunkel, T. A.,** Fidelity of DNA synthesis, *Annu. Rev. Biochem.,* 52, 429, 1982.

259. **Singer, B. and Kusmierek, J. T.,** Chemical mutagenesis, *Annu. Rev. Biochem.,* 52, 655, 1982.

260. **Fry, M. and Loeb, L. A.,** Fidelity of DNA synthesis, in *Animal Cell DNA Polymerases,* CRC Press, Boca Raton, Florida, 1986, 157.

261. **Watson, J. D. and Crick, F. H. C.,** A structure of deoxyribose nucleic acid, *Nature (London),* 171, 737, 1953.

262. **Topal, M. D. and Fresco, J. R.,** Complementary base pairing and the origin of substitution mutations, *Nature (London),* 263, 285, 1976.

263. **Wolfenden, R. V.,** Tautomeric equilibria in inosine and adenosine, *J. Mol. Biol.,* 40, 307, 1969.

264. **Pullman, B. and Pullman, A.,** Electronic aspects of purine tautomerism, *Adv. Heterocycl. Chem.,* 13, 77, 1971.

265. **Kwiatrowski, J. S. and Pullman, B.,** Tautomerism and electronic structure of biological pyrimidines, *Adv. Heterocycl. Chem.,* 18, 119, 1975.

266. **Dreyfus, M., Bensaude, O., Dodin, G., and Dubois, J. E.,** Tautomerism in cytosine and 3-methylcytosine. A thermodynamic and kinetic study, *J. Am. Chem. Soc.,* 98 (20), 6338, 1976.

267. **Saenger, W.,** *Principles of Nucleic Acid Structure,* Springer-Verlag, New York, 1984.

268. **Fresco, J. R., Broitman, S., and Lane, A. E.,** Base mispairing and nearest-neighbor effects in transition mutations, in *Mechanistic Studies of DNA Replication and Genetic Recombination,* Vol. 19, ICN-UCLA Symp., Alberts, B. and Cox, C. F., Eds., Academic Press, New York, 1980, 735.

269. **Geysen, H. M., Barteling, S. J., and Meloen, R. H.,** Small peptides induce antibodies with a sequence and structural requirements for binding antigen comparable to antibodies raised against the native protein, *Proc. Natl. Acad. Sci. U.S.A.,* 82, 178, 1985.

270. **Pierschbacher, M. D. and Ruoslahti, E.,** Cell attachment activity of fibronectin can be duplicated by small synthetic fragments of the molecule, *Nature (London),* 309, 30, 1984.

271. **Ruoslahti, E. and Pierschbacher, M. D.,** Arg–Gly–Asp: a versatile cell recognition signal, *Cell,* 44, 517, 1986.

272. **Ishihama, A., Mizumoto, K., Kawakami, K., Kato, A., and Honda, A.,** Proofreading function associated with the RNA-dependent RNA polymerase from influenza virus, *J. Biol. Chem.,* 261, 10417, 1986.

273. **A. Muñoz-Serrano, Díez, J., and Domingo, E.,** manuscript in preparation.

274. **Holland, J. J.,** Continuum of change in RNA virus genomes, in *Concepts in Viral Pathogenesis,* Notkins, A. L. and Oldstone, M. B. A., Eds., Springer-Verlag, New York, 1984, 137.

275. **DePolo, N. and Holland, J. J.,** Very rapid generation/amplification of defective interfering particles by vesicular stomatitis variants isolated from persistent infection, *J. Gen. Virol.,* 67, 1195, 1986.

276. **De Polo et al.,** manuscript in preparation.

277. **Prabhakar, B. S., Srinivasappa, J., and Ray, U.,** Selection of coxsackievirus B₄ variants with monoclonal antibodies results in attenuation, submitted.

278. **Covey, S. N.,** Amino acid sequence homology in gag region of reverse transcribing elements and the coat protein gene of cauliflower mosaic virus, *Nucleic Acids Res.,* 14, 623, 1986.

279. **Newton, S. E., Carroll, A. R., Campbell, O., Clarke, B. E., and Rowlands, D. J.,** The sequence of foot-and-mouth disease virus RNA to the 5' side of the poly(C) tract, *Gene,* 40, 331, 1985.

280. **Pleij, C. W. A., Rietveld, K., and Bosch, L.,** A new principle of RNA folding based on pseudoknotting, *Nucleic Acids Res.,* 13, 1717, 1985.

281. **van Belkum, A., Abrahams, J. P., Pleij, C. W. A., and Bosch, L.,** Five pseudoknots are present at the 204 nucleotides long 3' noncoding region of tobacco mosaic virus RNA, *Nucleic Acids Res.,* 13, 7673, 1985.

282. **Wolinski, J. S. and Johnson, R. T.,** Role of viruses in chronic neurological diseases, in *Comprehensive Virology,* Vol. 16, Fraenkel-Conrat, H. and Wagner, R., Eds., Plenum Press, New York, 1980, 287.

283. **Johnson, R. T.,** *Viral Infections of the Nervous System,* Raven Press, New York, 1982, 263.

284. **Kurtzke, J. F. and Hyllested, K.,** Multiple sclerosis in the Farol Islands. II. Clinical update, transmission, and the nature of MS, *Neurology,* 36, 307, 1986.

285. **Johnson, B. J. B., Kinney, R. M., Kost, C. L., and Trent, D. W.,** Molecular determinants of alphaviruses neurovirulence: nucleotide and deduced protein sequence changes during attenuation of Venezuelan equine encephalitis virus, *J. Gen. Virol.,* 67, 1951, 1986.

286. **Sellers, R. F.,** Quantitative aspects of the spread of foot-and-mouth disease, *Vet. Bull. (London),* 41, 431, 1971.

287. **Halstead, S. B.,** Immunological parameters of Togavirus disease syndromes, in *The Togaviruses. Biology, Structure, Replication,* Schlesinger, R. W., Ed., Academic Press, New York, 1980, 107.

288. **Murphy, R. B. and Webster, R. G.,** Influenza viruses, in *Virology,* Fields, B. N., Ed., Raven Press, New York, 1985, 1179.

Chapter 2

VARIABILITY, MUTANT SELECTION, AND MUTANT STABILITY IN PLANT RNA VIRUSES

L. van Vloten-Doting and J. F. Bol

TABLE OF CONTENTS

I. INTRODUCTION

In the past few years, nucleotide sequence studies have provided us with insight into the structural organization of a number of plant virus RNA genomes as well as with information on the variability in these genomes at the nucleotide level. The data have shed light on the sequence differences between closely related virus strains and on sequence homologies between viruses that were considered to be unrelated previously. Particularly, the finding of significant homologies between proteins encoded by viruses that apparently replicate via different strategies, has confirmed the concept[1] that the present day RNA viruses may have evolved from a few ancestors.[2-4]

As an example, Figure 1 shows a comparison between the genomes of alfalfa mosaic virus (AlMV), tobacco mosaic virus (TMV), and tobacco rattle virus (TRV). These viruses, which will be discussed in some detail in this chapter, all encode at least four proteins of which the information is located in one RNA molecule (TMV), or is distributed over two RNA molecules (TRV), or three RNA molecules (AlMV). The two largest gene products of each virus are believed to be involved in viral RNA replication; homologies between these proteins are indicated by hatched boxes in Figure 1.[3,5,6] A third gene product with molecular weight around 30 kdaltons is believed to play a role in cell-to-cell transport of viral material; there is only limited homology between these proteins.[5,6] The fourth gene product is the viral capsid protein; no homology is detectable between the structural proteins of AlMV, TMV, and TRV. The homologies shown in Figure 1 are also found in other tricornaviruses, such as brome mosaic virus (BMV)[2,3] and cucumber mosaic virus (CMV),[7,8] and in a group of animal viruses, the alpha viruses.[2] An evolutionary relationship between these viruses is further supported by the similarities in genome organization and strategies employed in genome expression. A second group of related RNA viruses contains the animal picornaviruses and the plant comoviruses.[4] Here again, a homology in the amino acid sequence of encoded proteins is paralleled by a similarity in genome organization and expression. Besides the viruses cited above, there are a large amount of plant viruses for which knowledge about the organization and strategy of expression is incomplete, and nothing is known about the nucleotide sequence of the genomes. It may well be that some of these viruses belong to the two clusters described above, but others may form a cluster of their own.

If the approximately 700 plant RNA viruses known today have evolved from a limited number of ancestors, then what is the driving force responsible for these viruses drifting away from each other during evolution? As has been discussed in the previous chapter, the high error rate during RNA-dependent RNA synthesis may be an important factor in this process. The constant generation of base substitutions may enable the virus to select variants encoding proteins with an altered specificity or an altered function. Intramolecular duplications in the viral RNA or the generation of multipartite genomes may permit the virus to extend its genetic content in order to acquire new functional domains or new genes. The present variability that can be observed in plant virus RNA genomes probably reflects a momentary recording of the ongoing process of evolution.

A study of the variability in plant virus RNA genomes not only sheds light on evolutionary aspects, but also permits a functional analysis of coding and noncoding sequences in these genomes. A comparison of the genomic sequences of closely related virus strains may reveal regions that are under a stringent selective pressure in addition to regions where base substitutions are tolerated. Comparison of more distantly related viruses may be helpful to delineate functional domains of gene products (e.g., see Figure 1). An analysis of the variability within a given virus isolate yields quantitative information on the fidelity of RNA replication and allows the isolation of spontaneous mutants with an altered phenotype. Mapping of the mutation on the viral genome identifies the viral gene product involved in

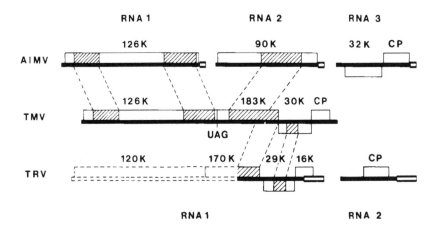

FIGURE 1. Comparison of the genome structure of AlMV, TMV, and TRV. The bars represent the genomic RNAs drawn to scale. Open reading frames are indicated by the molecular weights of the encoded proteins; cp = coat protein. Areas showing amino acid sequence homology are connected by dotted lines.

the mutated function. Particularly useful in this respect are conditional lethal mutants, either obtained spontaneously or induced deliberately. In this chapter we will describe data on the variability in the genomes of closely related virus strains and the variability that can be observed within a given virus strain. In addition, the possibility to select mutants from a given virus population and the stability of such mutants will be discussed.

II. VARIABILITY BETWEEN STRAINS

In this paragraph we will restrict ourselves to virus groups of which the genome or a genome segment of at least two members has been completely sequenced.

A. Tobamoviruses

Tobamoviruses have been classified into ten definite subgroups on the basis of the biochemical and immunological characteristics of their coat proteins.[9] A variable element is the position of the assembly origin, located either in the 30 kdalton cistron (common strain, tomato strain) or in the coat protein cistron (cowpea strain, CGMMV).[10,12] The two origins share a 40% nucleotide sequence homology and are both recognized by coat proteins of the common and CGMMV strain.[13] The genomes of the common strain (vulgare) and the closely related L-strain (TMV-L) have been completely sequenced.[14,15] In addition, a variant has been sequenced, $L_{11}A$, which differs by ten base substitutions from the L-strain.[16] It has been suggested that replacement of a cysteine residue at position 348 of the TMV-L 126-kdalton/183-kdalton protein with tyrosine in strain $L_{11}A$ is responsible for the attenuation of virulence of the variant.

Another variable element is found at the 3′ end of TMV RNAs. The 3′ terminal nucleotide sequences of the RNAs of the vulgare, tomato, and CGMMV strains show a 70 to 80% homology to each other and can be folded into a tRNA-like structure that is aminoacylatable with tyrosine.[17] In the 3′ terminal, 80 nucleotides of RNA of the cowpea strain (Cc) show 60% homology with the corresponding region of the turnip yellow mosaic virus (TYMV) genome, but no similarity to other TMV RNAs. Both the TYMV and Cc genome can be aminoacylated with valine.[17] This may reflect a common lineage of TMV-Cc and TYMV or a recombinational event.

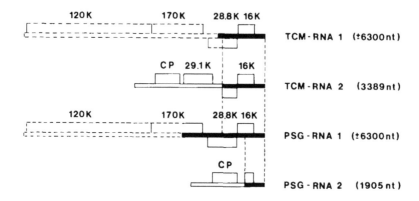

FIGURE 2. Schematic representation of the genetic information encoded in TCM RNAs 1 and 2 and PSG RNAs 1 and 2. Solid bars show the extent of sequence determined on the RNA 1 and the extent of RNA 2 homology with RNA 1. Open bars represent sequences unique to the RNA 2 of each isolate.

B. Tobraviruses

Tobraviruses have been divided into three clusters containing (1) strains of TRV, (2) strains of pea early browning virus (PEBV), and (3) Brazilian isolates represented by the CAM strain.[18] Within each cluster there is from 40 to over 90% homology between the RNA 1 molecules of different strains, whereas only 1 to 12% homology exists between RNA 1 molecules of different clusters. Figure 2 shows the parts of the genomes of two cluster 1 viruses (strains PSG and TCM) that have been sequenced.[5,19] In addition, the 3′ terminal sequence of SYM RNA 1 (cluster 1) and the complete sequence of CAM RNA 2 (cluster 3) are known.[6,20] These sequence data have revealed several variable elements in tobravirus genomes. It is surprising to see that within each strain, the 3′ terminal sequences of RNA 2 molecules are 100% identical to the 3′ termini of the corresponding RNA 1 molecules over a variable length. In strains CAM, PSG, and TCM this length is 459, 497, and 1099 nucleotides, respectively. The genome segments of cluster 1 viruses can be freely exchanged to form pseudorecombinants. In an inoculum containing PSG RNA 1 and TCM RNA 2, the homologous 3′ termini of these two RNA molecules would differ in 63 positions. It has been suggested that a copy choice mechanism permits sequence exchange between RNAs 1 and 2, restoring 100% identity between the two 3′ termini.[19] RNA 2 molecules with 3′ termini identical to RNA 1 may be replicated by the RNA 1 encoded replicase in preference over molecules with nonidentical termini. In view of the low fidelity of RNA replicases there must be considerable selection pressure to maintain 100% identity over a region of up to 1100 nucleotides.

A second variable element is the number of genes. TCM RNA 2 contains a reading frame for a 29.1-kdalton protein that has no homologous counterpart in the genome of strain PSG. It is possible that this gene is equivalent to the RNA 1-encoded 28.8-kdalton protein and is derived from an RNA 1 molecule of a tobravirus from a yet unknown cluster during an earlier symbiotic period of TCM RNA 2.[19] In this view, the 29.1-kdalton cistron may be redundant, just as the 16-kdalton cistron in TCM RNA 2 is redundant. The exchange of information between TRV RNAs 1 and 2 may reflect a mechanism that enables the virus to extend its genetic content.

A third variable element is found in the leader sequence of CAM RNA 2. This 5′ terminal sequence of 573 nucleotides preceding the coat protein cistron, contains two different sets of direct repeats, one of 119 nucleotides and the other of 76, which are absent in RNA 2 of PSG and TCM. These repeats may have been generated by polymerase molecules that have prematurely terminated transcription of a minus strand RNA 2 template, reinitiating again on the same template with their nascent chain still attached.

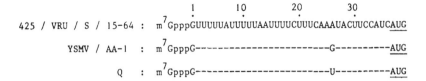

FIGURE 3.   Comparison of the 5' untranslated sequences of RNA 4 of seven strains of AlMV.

C. Como- and Nepoviruses

The complete nucleotide sequence of B-RNA (5889 nucleotides) and M-RNA (3841 nucleotides) of cowpea mosaic virus (CPMV, the type member of the comoviruses) has been deduced.[21,22] Hybridization studies failed to detect any homology of these RNAs with the genome of other comoviruses (CPSMV, RCMV, SMV).[23] Sequencing of M-RNA of red clover mottle virus (RCMV), however, showed that the polyprotein encoded by this genome segment has an overall amino acid sequence homology of 64% with the corresponding CPMV protein.[24] The polyprotein encoded by M-RNA of tomato black ring virus (TBRV, a nepovirus) contains small homologies with the corresponding CPMV protein and with the TMV-Cc 30-kdalton protein.[25]

D. Tricornaviruses

The genomes of the type members of the bromo-, cucumo, and ilarvirus groups, BMV, CMV, and AlMV, respectively, have been completely sequenced.[2,3,7,8,26] The 3' termini of the RNAs of bromo- and cucumoviruses show extensive homology in primary and three-dimensional structure and can be aminoacylated with tyrosine.[17,27] No such tRNA-like structure is present at the 3' end of ilarvirus RNAs. Instead, these termini contain specific binding sites for viral coat protein, and each genome segment of an ilarvirus has to be complexed with a few coat protein molecules to initiate infection.[28] There is no obvious homology between the amino acid sequences of the coat proteins of AlMV and tobacco streak virus (TSV).[29] However, the two proteins bind to the same sequences of AlMV and TSV RNAs and are able to activate the genome of both viruses.[28] This illustrates that extreme variation in primary sequence does not necessarily preclude maintenance of a highly specific function in a viral protein.

At the 5' termini of tricornavirus RNAs 1 and 2 there is a 90% homology over a length ranging from 11 to 55 nucleotides, whereas homology with the 5' end of RNA 3 is much lower. Figure 3 shows a comparison of the leader sequences of RNA 4, the subgenomic coat protein messenger, of seven AlMV strains.[30] The sequences are identical except for position 26 where three out of the four possible bases are found. Probably, this sequence conservation reflects the selective pressure imposed on this 5' noncoding region. On the other hand, the leader sequence of AlMV RNA 3 contains a number of variable elements. The RNA 3 sequences of AlMV strains from Madison (M), Leiden (L), and Strasbourg (S) have been deduced.[2,31] The M and L strain are believed to originate from the same isolate (AlMV 425), but the sequences differ as much from each other as they differ from the S-strain (about 2.5%). Figure 4 shows a schematic representation of the three 5' untranslated sequences. A stretch of 27 to 30 nucleotides occurs at least three times in all strains. In addition, sequences of 56 and 75 nucleotides are repeated in the leaders of strains S and L, respectively. Moreover, the 5' terminal 38 nucleotides of RNA 3-S show no homology with the RNA 3-L or 3-M sequence, but are identical to the 5' end of RNA 1-S. Possibly, the *copy choice mechanism discussed for tobraviruses is responsible for the recombination of RNA 1 and RNA 3 sequences,* as well as for the generation or loss of repeats in the RNA 3 leader sequences. Another variable element is the A-rich sequence that is extended from 10 nucleotides in strains S and M to 40 nucleotides in strain L.

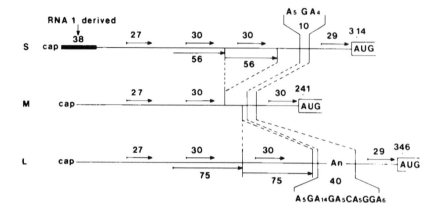

FIGURE 4.   Schematic comparison of the 5' untranslated sequences of RNA 3 of three strains of AlMV. Repeated sequences are indicated by arrows; the figures give the number of nucleotides. (From Langereis, K., Mugnier, M.-A., Cornelissen, B. J. C., Pinck, L., and Bol, J. F., *Virology*, 154, 409, 1986. With permission.)

Experimental proof that genetic recombination between RNA components of a mutipartite plant virus is indeed possible has been presented recently. A deletion in the 3' end of BMV RNA 3 was found to be repaired during the development of infection by recombination with the homologous region of either of the two remaining wild-type BMV RNA components.[63]

III. VARIABILITY WITHIN STRAINS

Compared to DNA replication, the fidelity of RNA replication is much lower.[32] Studies with RNA phages and animal viruses[33] suggest an error frequency around 10^{-4}. Available data indicate that the error rate in plant virus RNA replication is of the same order of magnitude.[34] As early as 1940, Kunkel[35] observed that one out of every 200 lesions induced by TMV in tobacco was a mutant. In a more recent study, variants that occur naturally in single lesion-passaged preparations of the common strain (U1) of TMV were analyzed by cDNA hybridization and RNA fingerprinting.[36] Two classes of variants with symptoms differing from TMV-U1 were obtained at high frequency. The ''a'' variants contained RNA with very distinct sequences when compared to U1-RNA and were considered to represent a distinct strain contaminating the U1-inoculum. The ''b variants were found to be closely related to the U1 strain and probably had arisen by mutation from the parental strain. Sequencing the cDNA clones from TMV vulgare by Goelet et al.[14] showed that the 5' end of the RNA was represented by two substantially different variants. Later studies by Meshi et al.[37] showed that one of the two variants in fact represented the tomato strain of TMV. These data indicate that despite single-lesion passage, isolates of a given virus may be contaminated with highly divergent variants or strains (see also paragraph IV).

In addition to the possible presence of contaminating strains, obtaining a genetically homogenous virus population is blocked by the intrinsic polymorphism of individual strains, due to the naturally high mutation rate observed with RNA viruses. In our sequence studies on the genomes of AlMV[3] and TRV,[5,19] we observed polymorphism between corresponding cDNA clones at a frequency of approximately one per 1000 nucleotides. Similar results have been reported by others.[26] In an attempt to analyze the mutation in an AlMV coat protein mutant, *Tbts 7*, we have cloned cDNA of RNA 4 of this mutant. Sequence studies showed that of three cDNA clones of the coat protein cistron that were analyzed all differed from each other by at least one nucleotide substitution.[38] This illustrates that putative mutations mapped by sequencing cDNA should be confirmed by sequencing the RNA population or by transcribing the cDNA into biologically active RNA.

A third source for sequence heterogeneity in a given virus isolate results from the variability in the length of poly(A) tracts in plant virus RNAs. Poly(A) stretches of variable length have been found in the 5' leader sequence of AlMV RNA 3-L,[31] the intercistronic region of RNA 3 of bromoviruses,[27] and the 3' noncoding regions of TMV[15] and barley stripe mosaic virus[39] RNAs. This may be due to a "stuttering" of the viral replicase on poly(U) stretches in a minus strand template.

In summary, the available data support the view that under specific growth conditions, a plant virus isolate contains one major genotype in addition to a minor pool of variants that are kept at a low level because of selective pressure. Three factors that contribute to the genomic variation are (1) the generation of base substitutions, (2) slipping of the polymerase on homopolymer tracts in the template, and (3) recombinational events. In the next paragraphs we will consider the consequences when the state of equilibrium is disturbed by introducing mutations deliberately or changing the growth conditions.

IV. BIOLOGICAL CLONING OF STRAINS AND MUTANTS

An essential step in virus genetics is the biological cloning of strains or mutants. If a plant virus itself is to be cloned, rather than its cDNA, the only available method is single lesion passage. This method is based on the assumption that at a high dilution of the inoculum, there is a high probability that an infection center is caused by one RNA (in the case of a virus with a monopartite genome) or by one genome equivalent (in the case of a virus with a multipartite genome). The infection dilution curves found for the different types of viruses are in agreement with this assumption.[40-42] Furthermore, it has been shown that for a large number of viruses biologically pure lines have been obtained by repeated single-lesion transfer.[43]

Recently,[36] the value of this method has been challenged by the demonstration that a preparation of TMV, passed through a single lesion once, contained variants which differed so much from the dominant strain, that it seems unlikely that they were derived from this strain by mutations during the few rounds of replication after the single-lesion passage. It is more likely that these variants were already present in the inoculum and were not eliminated by the single step of single-lesion passage employed in this study (see also Section III). However, these variants could be eliminated by the usual procedure of *repeated* single-lesion transfer.[36] One of the unexplained results of this study[36] is that, although the dominant and variant strains were equally infectious and hybridization data showed that the starting preparation contained less than 0.01% of the variants, more than 25% of the lesions induced by this preparation showed variant symptoms. At present it is unknown whether the discrepancy observed between expression of dominant and variant strains is specific for the TMV strains involved, or represents a more general phenomenon.

V. SELECTION OF MUTANTS

A. Selection of Symptom Mutants

Symptom mutants were the first plant virus mutants to be recognized.[43] In systemically infected plants atypical symptoms are frequently observed. Repeated single-lesion transfer at high dilution is normally sufficient to obtain a virus isolate which induces only symptoms differing from those induced by the parental strains. Occasionally, a systemic infection is observed in plants which normally react hypersensitively to the virus under investigation. From such systemically infected leaves a mutant can easily be isolated. Usually symptom mutants are, at least phenotypically, very stable.

B. Selection of Mutants by Host Shift

It has been known for a long time that when a plant virus isolate cultivated in, and fully

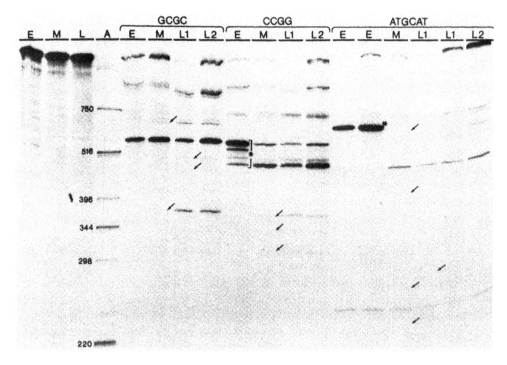

FIGURE 5.   Autoradiograph of partial cleavage products of serially passaged STNV RNA after shift from tobacco to mung bean. STNV RNAs isolated from early (3rd + 4th) passage (lanes E), middle (7th) passage (lanes M), and late (12th) passage (L1) virus stocks were 5' end labeled, purified, then hybridized to DNA oligomers GCGC, GCGG, or ATGCAT, and partially cleaved with calf thymus RHase H. The reaction products were separated in size by polyacrylamide gel electrophoresis. Lanes L2, late passage (12th, 14th, 15th) STNV RNA isolated from an independent series of passages with the same starting inoculum. The lengths of the DNA fragments used as size markers are indicated in nucleotides. Arrows indicate fragments found in middle and late passage STNV RNA, but not found in early passage STNV RNA. (*) Designates fragments common to all STNV RNAs isolated (E, M, L1, L2) which vary in band intensity (among the different isolates) on the autoradiograph. (From Donis-Keller, H., Browning, K. S., Clarck, J. M., Jr., *Virology*, 110, 43, 1981. With permission.)

adapted to, one particular host is repeatedly passaged through another host, a change in virus traits may take place. This phenomenon has been called host adaptation or host passage effect.[44] In molecular terms, it means that the nucleotide sequence of the mutant selected differs on at least one position from that of the parental strain. It is difficult to prove whether this mutant sequence was already present or was generated during host passage (see Section III). In the plant there will be a competition between the mutants (newly generated or preexisting) and the parental strain. If the parental strain is fully adapted to the host, it would be expected to outcompete most mutants. However, when the virus is transferred to another host, it might be expected that some of the mutants would have a better chance of competing successfully with the parental strain. The molecular basis for this selection is unknown, but speed of systemic movement may be an important factor. When mutations are generated during virus replication, positive mutations may accumulate during host passage and a gradual change in character of the isolate is observed. Such a gradual change in character has been seen in several cases[45] and has been documented on the molecular level for the satellite virus of tobacco necrosis virus (STNV).[46] It is noteworthy that independent STNV single-lesion isolates, made after host shift from tobacco to mungbean, showed a number of identical base sequence changes (Figure 5). The appearance of reproducible gradual changes after host shift probably reflects the selection of STNV sequences optimally adapted to replication in the new host. That the host may be instrumental in selecting particular sequences is also indicated by the observation that a mutant of AlMV strain 425, selected via bean,[47] showed similarities in amino acid sequence of the capsid protein and in structural properties with an AlMV strain (VRU), also isolated via bean.[38,47,48]

C. Selection by Gradual Temperature Shift

Selection of thermoresistant plant virus mutants by gradual temperature shift has only been employed for AlMV.[49] Virus infected tobacco leaf discs were incubated close to the nonpermissive temperature, 28°C for the wild type. After a number of transfers at this temperature, the temperature was raised to 30°C, then to 33°C, and finally to 34°C. The isolate obtained in this way could mutiply reasonably well at 34°C and hardly at 22°C. The isolate was found to be heterogeneous. The fact that by such a temperature regime mutants can be selected that are able to multiply at 34°C, while such mutants cannot be selected when virus-infected tobacco leaf discs are placed directly at 34°C, argues for the accumulation of advantageous mutations during virus replication under this selection pressure and against the selection of preexisting variants.

D. Selection by Differential Development of Symptoms

Dawson and Jones[50,51] inoculated plants with mutagen treated TMV or CCMV, respectively. The infected plants were first incubated at the nonpermissive temperature. All lesions developed at this temperature (which presumably contained wild-type virus) were marked before the plants were transferred to the lower temperature. Any lesions subsequently developed after shifting to the lower permissive temperature might therefore be due to thermosensitive (*ts*) mutants. After passage of such potential *ts* isolates through repeated single-lesion transfer, less than 10% of the TMV isolates and less than 1% of the CCMV isolates were found to be *ts*.

VI. MOLECULAR BASIS AND STABILITY OF MUTATIONS

Several mutagenic treatments have been used to induce mutations in plant virus RNAs. The first, and most widely employed mutagen is nitrous acid.[52-54] Analysis of the capsid protein of nitrous acid-induced mutants of TMV[45] and CCMV[45,55] showed that mutants often contained several mutations. These mutations are only partly consistent with the known mutagenic action of nitrous acid, indicating that, besides the chemically induced mutations, spontaneous mutations are also present. Since the coat protein cistron, which represents only 10% of the genome, sometimes contained five mutations leading to amino acid substitutions,[54] the total number of base changes in such mutants may be over 50.

The alkylating agent *N*-methyl-*N'*-*N*-nitrosoguanidine has also been used as a mutagen for plant viruses.[51,56] However, the number of mutants obtained was very low and not significantly above that found for nontreated preparations. No data about the number of base changes present in these mutants are known.

A large number of mutants has been isolated after UV irradiation of AlMV nucleoproteins[45,57-59] In these experiments, one of the three genome parts was irradiated and recombined with the complementing untreated genome parts. A high percentage of the single-lesion isolates obtained from such preparations showed mutant characteristics (changes in symptoms, thermosensitivity, and/or component composition). By supplementation tests it was shown[45,57-59] that the mutations, responsible for the altered phenotypes, always mapped on the UV irradiated component indicating that the UV irradiation was involved in mutant induction. For one particular symptom mutant, which mapped on RNA 2, it was shown by S1 mapping that the mutant RNA 2 differed on at least four positions from the parental sequence (Figure 6). How many of these base changes were due to UV irradiation is unknown.

Mutations may be induced by mutagenic treatments as mentioned above, or by manipu-

mutant   wild type

FIGURE 6.   Comparison of the nuclease S1 sensitivity of hybrids consisting of AlMV cDNA 2 and wild-type AlMV RNA 2 or mutant M *syst
1* RNA 2. Hybrids were incubated for 1 hr at 37°C with 1000 units S1
nuclease. Digestion products were precipitated with ethanol and run on an
agarose gel (Neeleman, L., unpublished results). The degradation pattern
of the mutant RNA indicates the presence of at least four mutations.

lation of cDNA clones (see Chapter 3 of this volume) or they may arise spontaneously by
replication or recombination errors. If a mutant is to be propagated as a virus stock rather
than as a cDNA clone, then whatever the origin of the first mutation may be, the mutated
sequence will always have to be amplified and, in most cases, separated from other sequences
by repeated single-lesion transfer. Recently, it has become evident that during RNA replication a large number of additional mutations are introduced.[32-34] Whether or not such
additional mutations become fixed in the virus population depends on whether or not they
are advantageous to the virus under the selection pressure employed during single lesion
transfer and/or virus mass culture. It is quite feasible that under the selection pressure where
the parental strain is completely stable (which means that it represents a stable dominant
nucleotide sequence), additional mutations will accumulate in a mutant sequence. This may
be explained in the following way: the dominant sequence found for a particular virus strain
in a host to which it is fully adapted, is one of the optimal sequences for this virus in this
host. Nearly every mutation in this sequence will represent a handicap under these conditions
(Figure 7). Some changes will be lethal, others will decrease the viability. The reduction
in viability depends on the position and type of base change. The range of the reduction in
viability is indicated in Figure 7 by the hatched area. Upon replication of a mutant sequence,
a new mutation on the same site, but more likely elsewhere in the genome, may appear.
Such an additional mutation, which would handicap the parental sequence, may benefit the

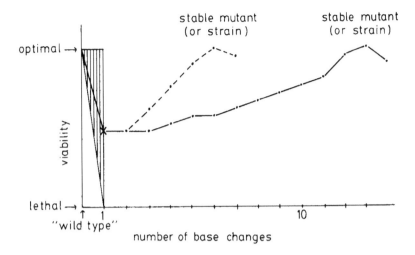

FIGURE 7. Schematic representation of the hypothetical relationship between the number of nucleotide changes and viability of viral sequences. In one particular host, every stable strain "wild type" is thought to represent an optimal viable sequence. Changes in this sequence will effect the viability. The change in viability depends on the position and type of nucleotide change and may vary from 0 to 100% (hatched area). Lethal mutations will be lost from the population of sequences. When a mutant sequence (e.g., X) is propagated, additional mutations enhancing the viability will accumulate until a new optimal sequence has come into being. When starting from one mutant sequence, several single-lesion isolates are made, each isolate may accumulate different additional mutations, —— and ----.

mutant sequence. When the additional mutation enhances the viability, the new sequence will outcompete the original mutant sequence and all other sequences carrying less advantageous mutations. During the following rounds of replication other beneficial mutations will accumulate until a new optimal sequence has emerged. When starting from one particular mutant sequence (e.g., X in Figure 7) several single-lesion isolates are made, each isolate may carry, besides the original mutation, a varying number of mutations, but this variation will be limited by the selection pressure. For several viruses[45,51,54,59] indications for the accumulation of mutations during cultivation has been obtained. When the processes of mutation and selection take place in the laboratory, the isolates are called mutants, but it is quite feasible that the same processes in nature have led to the appearance of the different virus strains. Theoretically, an immense variation in sequences is possible. However, only a limited number will be successful in nature.

The assumptions described above could explain a number of observations made during our studies on *ts* mutants of AlMV. Our type strain AlMV 425 is fully adapted to tobacco, which is evident from the finding that a number of cDNA clones all showed (nearly) the same nucleotide sequence (see Section III). Spontaneously arising symptom mutants were isolated by single-lesion transfer on tobacco.[45,47] A large percentage of these mutants showed a thermosensitive defect in virus replication in tobacco leaf discs. Two mutants with temperature-sensitive defects in virus replication in tobacco leaf discs on RNA 1 (*Bts4*) and RNA 2 (*Mts3*), respectively, were analyzed in more detail. In cowpea protoplasts, the two mutants each expressed *ts* mutations in both RNA 1 and RNA 2, while the *ts* behavior of *Mts3* in bean plants was due to mutations in both RNA 2 and RNA 3.[68] Apparently these mutants carried mutations in all genome parts. Given the stability of the parental strain under these conditions, it seems unlikely that each of the original atypical symptoms carried mutations in all genome parts. It is more likely that, after the initial "labilizing" effect (in these two cases a spontaneous mutation), other mutations accumulated during the 2- to 3-week period required for biological cloning.

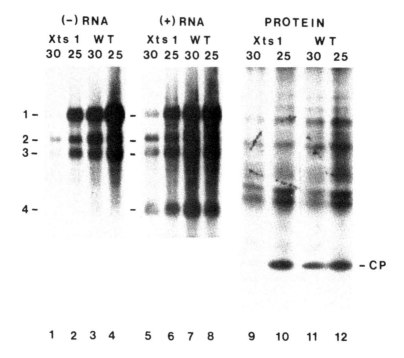

FIGURE 8.   RNA and protein synthesis induced by *Xts1* and wild-type AlMV at 25 and 30°C in cowpea protoplasts. Protoplasts were inoculated with *Xts1* (lanes 1, 2, 5, 6, 9, and 10) or wild-type AlMV (WT) (lanes 3, 4, 7, 8, 11, and 12) and incubated for 24 hr at 25 (lanes 2, 4, 6, 8, 10, and 12) or 30°C (lanes 1, 3, 5, 7, 9, and 11). Viral minus strand RNAs (lanes 1 to 4) and plus strand RNAs (lanes 5 to 8) were analyzed on Northern blots that were hybridized to strand-specific probes. ^{35}S-labeled proteins (lanes 9 to 12) were analyzed by polyacrylamide gel electrophoresis. The positions of AlMV-RNAs 1, 2, 3, and 4 and coat protein (CP) are indicated. (From Huisman, M. J., Sarachu, A. N., Alblas, F., Broxterman, H. J. G., van Vloten-Doting, L., and Bol, J. F., *Virology*, 154, 401, 1986. With permission.)

From both *Bts4* and *Mts3* pseudorecombinants with wild type were made. All pseudo-recombinants, containing mutant RNA 1 or mutant RNA 2, showed the same phenotype: a defect in the synthesis of viral minus RNA *and* in the translation of RNA 4 into coat protein at the restrictive temperature. Plus strand RNA synthesis was only slightly affected, probably because of the reduced availability of minus strand templates. It is remarkable that out of our collection of 48 *ts* mutants (partly spontaneous and partly UV-induced mutants) 11 mutants showed the phenotype described above. Figure 8 shows the results obtained with such a mutant. (This mutant is called *Xts* 1, indicating that the *ts* defect has not yet been mapped). In all these 11 mutants we found that the defect in synthesis of minus strand RNA was coupled with a defect in translatability of RNA 4. That these two defects can be uncoupled is illustrated by the observation that upon prolonged cultivation and reisolation of *Bts4*, we obtained a batch of virus that only showed a *ts* defect in minus strand RNA synthesis. Apparently the mutation(s) affecting the coat protein synthesis had been lost or, more likely, a new mutation was introduced that suppressed the *ts* character of coat protein synthesis. For several other mutants it was found that upon cultivation and reisolation the phenotype had changed. The change in mutant character was most noticeable for *ts* mutants; e.g., upon recultivation of 34 AlMV *ts* mutants, it was found that only 21 mutants showed the same thermosensitivity as before, six isolates had completely lost the *ts* phenotype, while seven isolates showed an intermediate phenotype because they were significantly less *ts* than before, but more *ts* than wild type.[62] The fact that the loss of *ts* phenotype was often only partial, indicates that it was not due to contaminating wild type or an exact reversion of the original mutation, but to the acquisition of additional mutation(s) suppressing the *ts* phenotype. The rather rapid loss of a *ts* phenotype is probably unavoidable. It is well known that proteins nonfunctional at the nonpermissive temperature are often significantly less active at the permissive temperature than their wild-type counterparts. In such mutants there is a continuous selection pressure for phenotypic reversion of the *ts* character. In other words, *ts* mutants will seldomly represent "optimal sequences" (Figure 7), which means that they are apt to change upon cultivation.

VII. OUTLOOK

Although virus mutants are extremely important in unraveling the different steps in virus replication,[45] the extreme variability of RNA genomes will remain a complicating factor in these studies. Insight into the effect of selection pressures acting on the viral sequences may help to avoid some of the pitfalls (like host shift and prolonged cultivation of mutants) in this type of work.

One of the reasons for the detailed analysis of virus mutants is the hope of obtaining insight into the correlation between the structure and the function of virus-coded molecules, both on the RNA and on the protein level. Those studies are also complicated by the high variability of RNA viruses leading to differences between individual batches of mutant preparations and to the occurrence of numerous differences between mutants and their corresponding parental strain (Figures 5, 6, 7).[45,52-55,59] Determination of the nucleotide sequence change(s) responsible for a mutated function will be dependent on the possibility of obtaining infectious in vitro transcripts of cloned cDNA (see Chapter 3 of this volume). When this system is operational for a given virus one can construct chimeric cDNA molecules containing mainly the wild-type sequence and only a small part of the mutant cDNA. Analysis of the *ts* character of in vitro transcripts of these constructs will indicate the mutation(s) responsible for the *ts* character. This type of work will be very labor intensive, but it is worthwhile doing since it will eventually give insight into the molecular events leading to virus replication and make a start of understanding the molecular basis of plant virus disease symptoms.

REFERENCES

1. **van Vloten-Doting, L.,** Advantages of multipartite genomes of single-stranded RNA plant viruses in nature, for research, and genetic engineering, *Plant Mol. Biol. Rep.,* 1, 55, 1983.
2. **Haseloff, J., Goelet, P., Zimmern, D., Ahlquist, P., and Dasgupta, R., and Kaesberg, P.,** Striking similarities in amino acid sequence among nonstructural proteins encoded by RNA viruses that have dissimilar genomic organization, *Proc. Natl. Acad. Sci. U.S.A.,* 81, 4358, 1984.
3. **Cornelissen, B. J. C. and Bol, J. F.,** Homology between the proteins encoded by tobacco mosaic virus and two tricorna viruses, *Plant Mol. Biol.,* 3, 379, 1984.
4. **Franssen, H., Leunissen, J., Goldbach, R.,Lomonosoff, G., and Zimmern, D.,** Homologous sequences in non-structural proteins from cowpea mosaic virus and picorna viruses, *EMBO J.,* 3, 855, 1984.
5. **Cornelissen, B. J. C., Linthorst, H. J. M., Brederode, F. Th., and Bol, J. F.,** Analysis of the genome structure of tobacco rattle virus strain PSG, *Nucleic Acids Res.,* 14, 2157, 1986.
6. **Boccara, M., Hamilton, W. D. O., and Baulcombe, D. C.,** The organisation and interviral homologies of genes at the 3' end of tobacco rattle virus RNA 1, *EMBO, J.,* 5, 223, 1986.
7. **Rezaian, M. A., Williams, R. H. V., and Symons, R. H.,** Nucleotide seqeunce of cucumber mosaic virus RNA 1, Presence of a sequence complementary to part of the viral satellite RNA and homologies with other viral RNAs, *Eur. J. Biochem.,* 150, 331, 1985.

8. **Rezaian, M. A., Williams, R. H. J., Gordon, K. H. J., Gould, A. R., and Symons, R. H.,** Nucleotide sequence of cucumber mosaic virus RNA 2 reveals a translation product significantly homologous to corresponding proteins of other viruses, *Eur. J. Biochem.,* 143, 277, 1984.

9. **Gibbs, A. J.,** Tobamo virus group, in *CMI/AAB Descriptions of Plant Viruses,* No. 184, 1977.

10. **Zimmern, D.,** The nucleotide sequence at the origin for assembly on tobacco mosaic virus RNA, *Cell,* 11, 463, 1977.

11. **Fukuda, M., Okada, Y., Otsuki, Y., and Takebe, I.,** The site of initiation of rod assembly on the RNA of a tomato and a cowpea strain of tobacco mosaic virus, *Virology,* 101, 493, 1980.

12. **Fukudas, M., Meshi, T., Okada, Y., Otsuki, Y., and Takebe, I.,** Correlation between particle multiplicity and location of virion RNA of the assembly initiation site for viruses of the tobacco mosaic virus group, *Proc. Natl. Acad. Sci. U.S.A.,* 78, 4231, 1981.

13. **Kurisu, M., Ohno, T., Okada, Y., and Nozu, Y.,** Biochemical characterization of cucumber green mottle virus ribonucleic acid, *Virology,* 70, 214, 1976.

14. **Goelet, P., Lomonosoff, G. P., Butler, P. J. G., Akain, M. E., Gait, M. J., and Karn, J.,** Nucleotide sequence of tobacco mosiac virus RNA, *Proc. Natl. Acad. Sci. U.S.A.,* 79, 5818, 1982.

15. **Ohno, T., Aoyagi, M., Yamanashi, Y., Saito, H., Ikowa, S., Meshi, T., and Okada, Y.,** Nucleotide sequence of the tobacco mosaic virus (tomato strain) genome and comparison with the common strain genome, *J. Biochem.,* 96, 1915, 1984.

16. **Nishiguchi, M., Kiduchi, S., Kiho, Y., Ohno, T., Meshi, T., and Okada, Y.,** Molecular basis of plant viral virulence; the complete nucleotide sequence of an attenuated strain of tobacco mosaic virus, *Nucleic Acids Res.,* 13, 5585, 1985.

17. **Rietveld, K., Linschooten, K., Pleij, C. W. A., and Bosch, L.,** The three-dimensional folding of the tRNA-like structure of tobacco mosaic virus RNA. A new building principle applied twice, *EMBO J.,* 3, 2613, 1984.

18. **Robinson, D. J. and Harrison, B. D.,** Unequal variation in the two genome parts of tobraviruses and evidence for the existence of three separate viruses, *J. Gen. Virol.,* 66, 171, 1985.

19. **Angenent, G. C., Linthorst, H. J. M., Van Belkum, A. F., Cornelissen, B. J. C., and Bol, J. F.,** RNA 2 of tobacco rattle virus strain TCM encodes an unexpected gene, *Nucleic Acids Res.,* 14, 4673, 1986.

20. **Bergh, S. T., Koziel, M. G., Huang, S.-C., Thomas, R. A., Gilley, D. P., and Siegel, A.,** The nucleotide sequence of tobacco rattle virus RNA 2 (CAM strain), *Nucleic Acids Res.,* 13, 8507, 1985.

21. **Lomonosoff, G. P. and Shanks, M.,** The nucleotide sequence of cowpea mosaic virus B-RNA, *EMBO J.,* 2, 2153, 1983.

22. **Van Wezenbeek, P., Verver, J., Harmsen, J., Vos, P., and Van Kammen, A.,** Primary structure and genome organization of the middle component RNA of cowpea mosaic virus, *EMBO J.,* 2, 941, 1983.

23. **Goldbach, R.,** Comoviruses: molecular biology and replication, in *The Plant Viruses: Viruses with Bipartite RNA Genomes and Isometric Particles,* Harrison, B. D. and Murant, A. F., Eds., Plenum Press, New York, in press.

24. **Shanks, M., Stanley, J., and Lomonosoff, G. P.,** Primary structure of middle component RNA of red clover mottle virus, paper presented at EMBO Workshop on Molecular Plant Virology, Wageningen, Netherlands, July 6 to 10, 1986.

25. **Fritsch, C., Meijer, M., Hemmer, O., and Mayo, M. A.,** Nucleotide sequence of TBRV RNA 2; speculations about the organization of the coding sequence and search for homologies with satellite RNA, paper presented at EMBO Workshop on Molecular Plant Virology, Wageningen, Netherlands, July 6 to 10, 1986.

26. **Gould, A. R. and Symons, R. H.,** Cucumber mosaic virus RNA 3; determination of the nucleotide sequence provides the amino acid sequences of protein 3A and viral coat protein, *Eur. J. Biochem.,* 126, 217, 1982.

27. **Ahlquist, P., Dasgupta, R., and Kaesberg, P.,** Near identity of 3' RNA secondary structure in bromoviruses and cucumber mosaic virus, *Cell,* 23, 183, 1981.

28. **Jaspars, E. M. J.,** Interaction of alfalfa mosaic virus nucleic acid and protein, in *Molecular Plant Virology,* Vol. 1, Davies, J. W., Ed., CRC Press, Boca Raton, Florida, 1985, 155.

29. **Cornelissen, B. J. C., Janssen, H., Zuidema, D., and Bol, J. F.,** Complete nucleotide sequence of tobacco streak virus RNA 3, *Nucleic Acids Res.,* 12, 2427, 1984.

30. **Swinkels, P. P. H. and Bol, J. F.,** Limited sequence variation in the leader sequence of RNA 4 from several strains of alfalfa mosaic virus, *Virology,* 106, 145, 1980.

31. **Langereis, K., Mugnier, M.-A., Cornelissen, B. J. C., Pinck, L., and Bol, J. F.,** Variable repeats and poly (A)-stretches in the leader sequence of alfalfa mosaic virus RNA 3, *Virology,* 154, 409, 1986.

32. **Holland, J., Spindler, K., Horodyski, F., Grabau, E., Nichol, S., and Van de Pol, S.,** Rapid evolution of RNA genomes, *Science,* 215, 1577, 1982.

33. **Domingo, E., Martinez-Salas, E., Sobrino, F., De la Torre, J. C., Portela, A., Ortin, J., Lopez-Galindez, C., Perez-Brena, P., Villanueva, N., Najera, R., Van de Pol, S., Steinhauer, D., DePolo, N., and Holland, J.,** The quasispecies (extremely heterogeneous) nature of viral RNA genome populations: biological relevance — a review, *Gene,* 40, 1, 1985.

34. **van Vloten-Doting, L., Bol, J. F., and Cornelissen, B. J. C.,** Plantvirus-based vectors for gene transfer will be of limited use because of the high error frequency during viral RNA synthesis, *Plant Mol. Biol.,* 4, 323, 1985.

35. **Kunkel, L. O.,** Genetics of viruses pathogenic to plants, *Publ. Am. Assoc. Adv. Sci.,* 12, 22, 1940.

36. **Garcia-Arenal, F., Palukaitis, P., and Zaitlin, M.,** Strains and mutants of tobacco mosaic virus are both found in virus derived from single-lesion-passaged inoculum, *Virology,* 132, 131, 1984.

37. **Meshi, T., Ishikawa, M., Takamatsu, N., Ohno, T., and Okada, Y.,** The 5′-terminal sequence of TMV RNA: question on the polymorphism found in vulgare strain, *FEBS Lett.,* 162, 282, 1983.

38. **Cornelissen, B. J. C.,** unpublished results.

39. **Agranovsky, A. A., Dolja, V. V., and Atabekov, J. G.,** Structure of the 3′ extremity of barley stripe mosaic virus RNA: evidence for internal poly (A) and a 3′-terminal tRNA-like structure, *Virology,* 119, 51, 1982.

40. **Furomoto, W. and Mickey, R.,** A mathematical model for the infectivity dilution curve of tobacco mosaic virus: theoretical considerations, *Virology,* 32, 216, 1967.

41. **Furomoto, W. and Mickey, R.,** A mathematical method for the infectivity dilution curve of tobacco mosaic virus: experimental tests, *Virology,* 33, 224, 1967b.

42. **Van Vloten-Doting, L., Kruseman, J., and Jaspars, E. M. J.,** The biological function and mutual dependence of bottom component and top component *a* of alfalfa mosaic virus, *Virology,* 34, 728, 1968.

43. **Matthews, R. E. F.,** *Plant Virology* 2nd ed., Academic Press, New York, 1981.

44. **Yarwood, C. E.,** Host passage effects with plant viruses, *Adv. Virus Res.,* 25, 169, 1979.

45. **van Vloten-Doting, L.,** Virus genetics, in *The Plantviruses,* Vol. 1, Francki, R. I. B., Ed., Plenum Press, New York, 1985, 117.

46. **Donis-Keller, H., Browning, K. S., and Clarck, J. M., Jr.,** Sequence heterogeneity in satellite tobacco necrosis virus RNA, *Virology,* 110, 43, 1981.

47. **Roosien, J. and van Vloten-Doting, L.,** A mutant of alfalfa mosaic virus with an unusual structure, *Virology,* 126, 155, 1983.

48. **Hull, R.,** Alfalfa mosaic virus, *Adv. Virus Res.,* 15, 365, 1969.

49. **Franck, A. and Hirth, L.,** Temperature-resistant strains of alfalfa mosaic virus, *Virology,* 70, 283, 1976.

50. **Dawson, W. O. and Jones, G. E.,** A procedure for specifically selecting temperature-sensitive mutants of tobacco mosaic virus, *Mol. Gen. Genet.,* 145, 307, 1976.

51. **Dawson, W. O.,** Isolation and mapping of replication-deficient, temperature-sensitive mutants of cowpea chlorotic mottle virus, *Virology,* 90, 112, 1978.

52. **Robinson, D. J.,** Inactivation and mutagenesis of tobacco rattle virus by nitrous acid, *J. Gen. Virol.,* 18, 215, 1973.

53. **Atabekov, L.,** Defective and satellite plant viruses, in *Comprehensive Virology,* Vol. 11, Fraenkel-Conrat H , and Wagner, R. R., Eds., Plenum Press, New York, 1977, 143.

54. **Bancroft, J. B., McDonald, J. G., and Rees, M. W.,** A mutant of cowpea chlorotic mottle virus with a perturbed assembly mechanism, *Virology,* 75, 293, 1976.

55. **Rees, M. W. and Short, M. N.,** The primary structure of cowpea chlorotic mottle virus coat protein, *Virology,* 119, 500, 1982.

56. **Hartman, D., Mohier, E., Leroy, C., and Hirth, L.,** Genetic analysis of alfalfa mosaic virus, *Virology,* 74, 470, 1976.

57. **Van Vloten-Doting, L., Hasrat, J. A., Oosterwijk, E., Van't Sant, P., Schoen, M. A., and Roosien, J.,** Description and complementation analysis of 13 temperature-sensitive mutants of alfalfa mosaic virus, *J. Gen. Virol.,* 46, 415, 1980.

58. **Roosien, J. and Van Vloten-Doting, L.,** Complementation and interference of ultraviolet-induced Mts mutants of alfalfa mosaic virus, *J. Gen. Virol.,* 63, 189, 1982.

59. **Roosien, J.,** Mutants of alfalfa mosaic virus, Ph.D. thesis, University of Leiden, Netherlands, 1983.

60. **Sarachu, A. N., Nassuth, A., Roosien, J., Van Vloten-Doting, L., and Bol, J. F.,** Replication of temperature-sensitive mutants of alfalfa mosaic virus in protoplasts, *Virology,* 125, 64, 1983.

61. **Sarachu, A. N., Huisman, M. J., Van Vloten-Doting, L. and Bol, J. F.,** Alfalfa mosaic virus temperature-sensitive mutants. I. Mutants defective in viral RNA and protein synthesis, *Virology,* 141, 14, 1984.

62. **Huisman, M. J., Sarachu, A. N., Alblas, F., Broxterman, H. J. G., van Vloten-Doting, L., and Bol, J. F.,** Alfalfa mosaic virus temperature-sensitive mutants. III. Mutants with a putative defect in cell-to-cell transport, *Virology,* 154, 401, 1986.

63. **Bujarski, J. J. and Kaesberg, P.,** Genetic recombinant between RNA components of a multipartite plant virus, *Nature (London),* 321, 528, 1986.

Chapter 3

MOLECULAR GENETIC APPROACHES TO REPLICATION AND GENE EXPRESSION IN BROME MOSAIC AND OTHER RNA VIRUSES

Paul Ahlquist and Roy French

TABLE OF CONTENTS

I. INTRODUCTION

The natural genetic variability of RNA viruses, the major theme of this volume, coupled with their relatively small genome sizes, would seem to make them particularly favorable objects for productive genetic study. Indeed, phenotypically interesting mutants of RNA viruses were isolated and studied well before the nature of viral RNA as genetic material was demonstrated.[1-3] However, despite some notable successes, the difficulties in direct manipulation of large viral RNA molecules present substantial technical barriers to some of the most basic steps in genetic analysis, such as mapping mutations and separating the phenotypic effects of multiple mutations. For some time, these barriers largely prevented the effective use of RNA virus mutants for many kinds of potential genetic studies.

For a growing number of RNA viruses, most of the difficulties inherent in direct RNA manipulation have now been circumvented by the development of infectious viral cDNA clones. Such clones provide a DNA form in which a wild-type virus genome and its mutants can be isolated, stabilized, and then manipulated freely by an expanding recombinant DNA technology which already allows virtually any desired DNA sequence to be constructed. The consequent ability to facilitate genetic study of any assayable RNA virus characteristic endows such clones with far-reaching potential for RNA virology. In addition to studying clones of mutants isolated at the RNA level, an even broader selection of mutants can be generated at the cDNA level. Sequence rearrangements can be made at will, and a powerful spectrum of in vitro DNA mutagenesis procedures can be applied to study the function of viral RNA regulatory sequences and encoded gene products.

Although infections have been successfully expressed from cDNA clones of only a limited number of RNA viruses to date, these represent a fairly diverse group of bacterial, insect, mammalian, and plant viruses. They include bacteriophage Qβ,[4] the related picornaviruses poliovirus,[5] coxsackie B3 virus,[6] and rhinovirus,[7] brome mosaic virus (BMV),[8] black beetle virus,[9] tobacco mosaic virus (TMV),[10,11] and Sindbis virus (Rice, C., Levis, R., Strauss, J., and Huang, H., personal communication). All of these are positive or messenger-sense ssRNA viruses, which package a directly infectious RNA. The existence in these viruses of a nucleic acid stage which is fully and independently competent for initiating infection has clearly contributed to the early success of expression experiments in these systems. This characteristic and the variety of successful instances to date suggest that it will be possible to express infections of most, if not all, positive strand RNA viruses from cloned cDNA. Projects are also underway in some laboratories to construct cDNA clones which can be expressed to initiate or contribute to infections of dsRNA or negative sense ssRNA viruses. An interesting preliminary to such work is the recent demonstration that negative sense in vitro transcripts of cloned cDNA to bacteriophage Qβ, a positive strand RNA phage, are noninfectious to normal *Escherichia coli*, but infectious to cells in which the viral replicase gene is expressed from a suitably engineered plasmid (Shaklee, P., Miglietta, J., Palmenberg, A., and Kaesberg, P., personal communication).

Despite their relatively recent introduction, the use of infectious cDNA clones has already contributed significant advances in a diverse range of studies of virus functions at both the RNA and protein level. These include mapping poliovirus neurovirulence determinants,[12,13] identification of sequences directing viral RNA amplification in both animal and plant virus systems[14,15,76] (French and Ahlquist, unpublished results), novel demonstration of viral RNA recombination,[16] and development of RNA-based gene expression systems.[15,17,18] This early experience with a variety of viruses and experimental applications makes it clear that direct alteration of RNA viruses through cloned cDNA can be profitably extended to many aspects of virus research and is likely to increase in use as a major experimental tool of RNA virology. This article will discuss general perspectives on the expression of RNA virus infections from cloned cDNA, the use of such expressible clones as a genetic tool for molecular studies of RNA viruses, and some of the broad conclusions arising to date from these studies. The discussion will focus on in vivo analyses of positive strand RNA viruses in natural host cells and, in particular, on the example of brome mosaic virus, which is under study in the authors' laboratory. Related questions concerning expression and manipulation of cDNA clones of viroid and viroid-like satellite RNAs will not be covered in this chapter, and for these topics the reader is referred to the chapters by Owens and Hammond, Kaper and Collmer, and Bruening et al. in these volumes.

II. CONSTRUCTION AND EXPRESSION OF DIRECTLY AND INDIRECTLY INFECTIOUS VIRAL cDNA CLONES

Cloned viral cDNAs have been used to generate infections by two approaches: directly by transfection of the cloned cDNA itself, and indirectly by inoculation of transcripts synthesized in vitro from the cloned cDNA. For simplicity, in later sections we will refer to cDNA clones capable of either kind of expression simply as infectious cDNA clones. However, it is appropriate at this juncture to briefly consider some aspects of the two approaches.

A. Directly Infectious Viral cDNA Clones

For several RNA viruses, plasmids containing complete viral cDNA copies have been shown to be directly infectious upon transfection into appropriate host cells. RNA bacteriophage Qβ,[4] poliovirus,[5] and coxsackie B3 virus[6] are examples of such viruses. As discussed elsewhere in these volumes, directly infectious cDNA clones have also been constructed for viroids[19] and plant virus satellite RNAs such as the satellites of the tobacco necrosis and ringspot viruses.[20,21] Presumably, virus expression from such clones involves transcription of the cDNA insert within the host cell and, consistent with this view, the addition of a eukaryotic episomal replication origin and fusion of an active eukaryotic promoter to the cDNA sequences substantially increases the infectivity of poliovirus cDNA clones.[22] However, for most directly infectious cDNA clones, important aspects of expression remain unresolved, including whether effective cDNA transcription occurs on free or chromosomally integrated DNA, the structure of the initial transcript, the relation of possible RNA processing to its eventual successful expression, and the efficiency of presumed RNA transport from nucleus to cytoplasm.

For many other eukaryotic viruses, clones containing complete viral cDNAs have been constructed which are not directly infectious, even though the cDNA inserts can be shown to be infectious by other expression routes (see below). Viruses in this class include brome mosaic virus,[8,23] rhinovirus,[7] black beetle virus,[9] tobacco mosaic virus,[10,11] and Sindbis virus (Rice, C., Levis, R., Strauss, J., and Huang, H., personal communication) and one of its defective interfering RNAs.[14] Whether this lack of direct cDNA infectivity is due to inherent aspects of the biology of these viruses, or whether it could be altered by changes in plasmid structure, is not known. Such questions are particularly intriguing in the case of the reported rhinovirus clone, since directly infectious cDNA clones have been generated for other closely related picornaviruses.

B. Synthesis of Infectious In Vitro Transcripts from Cloned Viral cDNA

Despite the failure to demonstrate direct infectivitiy for cloned cDNA from the viruses mentioned in the preceeding paragraph, infections with all of them have been successfully initiated by messenger-sense in vitro transcripts from their cloned cDNA. In addition to its high general success rate, this approach presents several distinct advantages over direct DNA transfection. First, the specific infectivity of in vitro transcripts is generally higher than that obtained with directly transfected DNA, approaching 5 to 10% of the infectivity of natural viral RNA for several viruses.[8,24,25] For poliovirus, where both options have been successfully developed, the molar infectivity of suitably designed in vitro transcripts[25] is over two orders of magnitude higher than that of directly transfected cDNA clones, even when direct expression from plasmid clones has been enhanced by addition of mammalian DNA virus replication and transcription signals.[22] Additionally, use of infectious in vitro transcripts avoids the poorly understood in vivo DNA transcription events inherent in expression of directly infectious cDNA clones. This presumably nuclear DNA transcription stage has no counterpart in the natural life cycle of these ssRNA viruses and may induce temporal delays and possibly other aberrations into infections. Infectious transcripts thus provide a better mode for the initiation of infection by natural virus RNA, allowing questions to be examined concerning early as well as late stages of infection. The use of infectious transcripts also has intrinsic advantages in experiments concerning genetic variation arising during viral RNA replication, because the potential for rearrangement of viral sequences by recombination at the DNA level is removed, as well as the possibility of added genetic instability due to nuclear processing of primary transcripts from the transfected cDNA.

Studies with poliovirus, BMV, and TMV show that the infectivity of in vitro transcripts can be influenced substantially by their terminal structures. In all three cases, addition of extra nonviral bases to the ends of the transcripts reduces infectivity. Polio transcripts with 60 5′ nonviral bases and 632 bases of extra 3′ sequence are 60-fold less infectious than transcripts with only two extra 5′ G residues and no more than seven extra 3′ bases.[25] Compared to transcripts initiated with the same 5′ sequence as virion RNA, TMV transcripts with six extra bases are 300 times less infectious,[10] while BMV transcripts with seven extra 5′ bases are 20 to 100 times less infectious.[75] Moreover, on whole plants, uncapped BMV in vitro transcripts have no detectable infectivity, while transcripts bearing the natural capped 5′ end have about 10% of the infectivity of natural virion RNA.[8]

A variety of useful in vitro transcription systems are now available to meet the structural demands for efficient expression of viral cDNA clones. For many applications, systems based on RNA polymerases encoded by bacteriophages T7, T3, and SP6 provide particularly effective approaches because of the high specificity of each polymerase for its respective promoters and the ease and reliability with which extremely highly active polymerase preparations can be produced.[25-28] For instances where the initiation site requirements of these enzymes are unsuitable, or for occasionally encountered sequences not transcribed well by these bacteriophage polymerases,[29] other systems are also available.[23,30] This variety should provide sufficient flexibility to allow most transcription requirements to be met, both for the expression of cDNA clones of additional viruses and for improved expression of existing clones. However, for most of the successfully expressed clones cited above, the biological activity of transcripts is already high enough to allow routine and reliable production of infection, reducing the actual in vitro transcription to a simple step in the inoculation procedure.

III. MOLECULAR GENETIC APPROACHES TO RNA VIRUS GENE FUNCTIONS

For positive strand RNA viruses, the functions of most nonstructural genes have not yet been established, and this lack of knowledge constitutes a considerable challenge to further progress in RNA virology. To date, the major insight into such genes and their functions has come from direct biochemical analyses which have been somewhat hindered by the usually low concentration of the relevant viral proteins in infected cells and by frequent difficulties in solubilizing these proteins in active form. Recently, genes for a number of these nonstructural proteins have been sequenced, and their comparison with other known nucleotide and amino acid sequences has become increasingly informative as sequence data accumulate for larger numbers of viral and cellular genes. Such comparisons have already led to the definition of several classes of structurally similar genes and proteins in surprisingly broad groups of viruses, forming an important basis for molecular taxonomy and allowing useful pooling of information on the functions of similar genes[31-34] (see also the chapter by Zimmern in Volume II). Sequence comparisons have also suggested functions for some viral genes by identification of cellular homologues.[35,36]

RNA virus gene functions have also been investigated by genetic approaches. In the alphavirus genome, e.g., temperature-sensitive mutants have been used to define seven viral complementation groups, including three in structural genes and four in nonstructural genes.[37] Mutants with altered phenotypes in many important properties have been isolated and studied in a number of other positive strand RNA viruses including the picornaviruses[38] and many plant viruses[39-41] (see also Chapter 2 in this volume). However, as noted in the introduction, current technology imposes significant limitations on generation and study of such mutants at the level of the natural RNA genome. These obstacles include restriction of mutagenesis to certain classes of point substitutions, an inability or poor ability to limit mutagenesis to a defined target region within the RNA genome, occasional difficulties in obtaining pure or stable mutant stocks, and difficulties in mapping mutations as well as detecting and separating multiple mutations. Infectious virus cDNA clones provide ways to surmount these limitations, and examples of studies which have already utilized this potential are discussed in the sections below.

A. Recombination Studies with Natural Virus Mutants and Strains

One genetic approach to virus function is the construction of recombinants between wild-type virus and phenotypic mutants or strains. The most extensive work of this kind to date has been with poliovirus. Poliovirus recombinants can be isolated from mixed infections without the use of cDNA clones (see chapter by King in Volume II) and this approach has been applied with significant results to studies of neurovirulence and other viral properties.[42-44] However, such recombination sites cannot be predicted or well controlled, and the variety of recombinant types is limited by the need to retain a selectable marker from each parent. In practice, recombinants produced by this approach have generally been single crossovers near the middle of the genome. To produce a much greater variety of predesigned recombinants, poliovirus cDNA clones have been used. At the cDNA level, viable recombinants have been produced between different poliovirus strains of the same serotype, between different poliovirus serotypes, and between poliovirus and coxsackie B3 virus.

To map the determinants of poliovirus neurovirulence, Nomoto and colleagues have studied recombinants between the neurovirulent type 1 Mahoney strain and its attenuated vaccine derivative, the Sabin 1 strain.[12,13] The availability of infectious clones for both strains, which differ by 55 nucleotide substitutions spread throughout the 7.4-kilobase viral RNA, allows much more detailed mapping studies than would be possible by in vivo recombination. Examination of seven in vitro cDNA-generated recombinants showed that determinants of neurovirulence are located at multiple sites within the 5' 5.6 kilobases of polio RNA.[12] Moreover, it was found that a number of phenotypic characteristics of the Sabin 1 strain, which have been used in cell culture tests to qualify batches of live vaccine for clinical use, do not always segregate with attenuation of neurovirulence. In a similar mapping study, Pincus and Wimmer[45] substituted selected cDNA fragments from guanidine-resistant and -dependent mutants of poliovirus type 1 into an infectious wild-type poliovirus cDNA clone. The major effect of guanidine on wild-type poliovirus appears to be inhibition of viral RNA synthesis, and this study demonstrated that the guanidine resistance or dependence characteristics of the mutants resulted from nucleotide changes in the viral polypeptide 2C coding region.

A viable recombinant has also been constructed at the cDNA level between poliovirus types 1 and 3, which are only 70% homologous in RNA sequence.[46] This recombinant carried the 0.74-kilobase 5' untranslated sequence and first 11 polyprotein codons from type 3, with the remainder of the sequence from type 1. The 5' noncoding region of poliovirus significantly influences neurovirulence,[12,47] and this approach is of considerable practical interest since the reversion to neurovirulence of the type 1 live vaccine strain is much lower than that for types 2 and 3. It is therefore hoped that safer vaccine strains might be produced by combining the antigenic properties of poliovirus types 2 and 3 with the attenuation stability of the type 1 vaccine strain. A second recombinant inserting a portion of the VP1 capsid protein gene from type 3 into a type 1 context was nonviable, however, showing that at this level of divergence some care is needed to obtain functional compatibility of different regions in the recombinants.[46]

An additional example of successful recombination between divergent picornavirus genomes is the construction of a hybrid between poliovirus type 1 and coxsackie B3 virus, another enterovirus of significance in clinical cardiology.[48] In this case, a 0.4-kilobase segment of the coxsackie B3 5' noncoding region was inserted in a poliovirus context, replacing a segment with which it had 70% homology. The resulting hybrid virus was viable, but showed a temperature-sensitive replication phenotype. Study of this and other mutants described below may provide some insight into the function of the long 5' noncoding sequence of picornaviruses.

Many situations exist in RNA virology where recombinants analogous to these picornavirus hybrids would be valuable, and as additional RNA viruses are expressed from cDNA clones, the opportunity to carry out such studies becomes more accessible. One such opportunity exists now with the L strain of TMV and its pathogenically attenuated derivative, $L_{11}A$. Complete sequences and infectious clones have been derived for both strains, providing the basis for mapping pathogenicity determinants in this plant virus in the same way that has been initiated with poliovirus.[11,49]

B. Mutational Studies at the Cloned cDNA Level

The investigations described above all utilized virus strains isolated at the RNA level by phenotypic characteristics, and subsequently sequenced and cloned for manipulation. An additional approach made possible by infectious viral cDNA clones is that of "reverse genetics", in which mutations created at the DNA level are first isolated according to the presence of a genotypic change, and then screened for phenotype. This approach, which has been widely applied to cloned segments of DNA genomes, offers several advantages: first, a considerable spectrum of mutant types, including rearrangements, deletions, and insertions, as well as substitutions, can be readily generated at preselected sites or over preselected regions. Secondly, mutations are generally mapped prior to phenotypic testing, and a variety of approaches can be used to assure freedom from unwanted mutations outside of the intended target site. Finally, mutants can be made and isolated without reference to phenotype, so that both lethal and phenotypically wild-type mutant genotypes can be recognized and characterized, which constitutes a major benefit in many studies.

Initial results with mutational analyses of infectious cDNA clones for several RNA viruses suggest that this approach will yield biochemically informative mutants at diverse loci in many RNA viruses. For poliovirus, the first eukaryotic RNA virus expressed from cloned cDNA, a number of insertion mutants in the viral polyprotein coding sequences have been made which show altered plaque morphology, temperature response, and/or differential host responses. One of these, a small plaque variant having the insertion of a single amino acid in the 2A nonstructural protein, has been shown to give nonspecific premature suppression of both host and viral mRNA translation.[36] The same mutant also fails to induce cleavage of the cap-binding protein, p220, which precedes specific shutdown of host mRNA translation in normal polio infections. The phenotypic defects in other cDNA-generated poliovirus mutants are currently under investigation.[36,50]

Mutants have also been constructed through cloned cDNA for several plant RNA viruses. A number of linker insertions have been made and tested in an infectious cDNA clone of the 1.2-kilobase satellite tobacco necrosis virus (STNV) RNA in both coding and noncoding regions.[20] Frameshift insertions in the coding region, and certain insertions in the 3' noncoding region, block normal infection of cowpea plants coinfected with tobacco necrosis virus, yielding little or no STNV coat protein, and only double-stranded STNV RNA with little or no excess positive strand RNA. However, certain in-frame six-codon insertions in the STNV coat gene, and other insertions in the 3' noncoding region give an essentially wild-type infection. Similarly, two viable single amino acid insertions have also been made at the cDNA level in the readthrough domain of the tobacco mosaic virus (TMV) L strain p180 nonstructural protein.[51] These variants give small, delayed local lesions compared to wild-type virus on test plants. In the same study, alterations were made to the leaky, amber termination codon which separates the TMV p130 protein coding sequence from the p180 readthrough domain. The behavior of these mutants showed that both the p130 and p180 proteins are necessary for normal virus replication, and a mutant with an ochre rather than an amber codon at the readthrough position was also found to be viable. A variety of insertion and substitution mutations have also been made in the nonstructural genes of brome mosaic virus (Richards, D., Kroner, P., and Ahlquist, P., unpublished results). In phenotypic tests to date, these mutants have given a wide variety of responses ranging from lethality to apparently wild-type infection. Some of the mutants appear to have conditional or partial inhibition of infection which may be particularly illuminating in further biochemical investigation.

IV. STUDIES OF REGULATORY SEQUENCES IN BROME MOSAIC VIRUS

The genomic RNAs of positive strand RNA viruses not only function as mRNAs to direct synthesis of viral proteins, but also contain *cis*-acting recognition and regulatory sequences which allow them to interact with these proteins in virus-specific processes such as RNA replication and encapsidation. Such recognition sequences are thus one of the major characteristics that differentiate functional viral RNAs from cellular mRNAs, and also are amenable to study by genetic manipulation. For poliovirus, cDNA-generated mutants have been produced in both the 5' and 3' noncoding regions. Sarnow et al.[50] have described an eight-base insertion mutant in the 3' noncoding region which has a strongly temperature-sensitive phenotype in cultured cells. Presumably, this character arises from inhibition of some *cis*-active function of the 3' noncoding segment involved in, e.g., RNA synthesis or encapsidation. Curiously, two- and ten-base insertions at the same site both result in a wild-type phenotype, possibly because the observed mutant defect results from formation of an unsuitable RNA secondary structure rather than from interruption of the primary sequence. Mutations have also been generated in the 5' noncoding region of poliovirus. In addition to the temperature-sensitive poliovirus/coxsackie B3 recombinant virus mentioned above,[48] Nomoto has described viable deletion mutants in the 5' noncoding region, as well as four-base insertion mutants with a range of phenotypes.[52]

Current studies of regulatory sequences in poliovirus are limited, though, by the interdependence of all aspects of infection on a single viral RNA component. This makes it difficult to assess whether regulatory sequences might exist within coding sequences, as well as noncoding sequences, since possible *cis*- and *trans*-acting effects of coding region mutations cannot easily be separated. Moreover, the use of plaque and other infection assays which allow detection of viral replication only when multiple infection cycles are completed makes it difficult to distinguish defects in infection subprocesses such as viral RNA replication and encapsidation. The availability of high-infectivity in vitro transcripts from poliovirus cDNA clones may circumvent this difficulty by making single cycle infection experiments practical.[25]

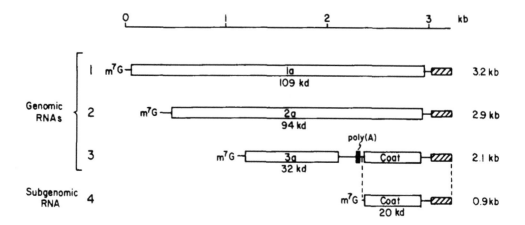

FIGURE 1. Schematic diagram of BMV genomic RNAs 1 to 3 and subgenomic RNA4. All are capped, single-stranded mRNAs, with open bars showing the coding regions for nonstructural proteins 1a, 2a, and 3a, as well as the single coat protein. Molecular weight (kdaltons) of the encoded protein is shown below each gene. Crosshatched boxes map the approximately 200-base 3' terminal regions of high homology,[57] and the 16- to 22-base heterogeneous intercistronic poly(A) sequence of RNA3[54] is shown by the small filled box. (From Ahlquist, P., French, R., and Bujarski, J., *Adv. Virus Res.*, 32, 215, 1987. With permission.)

Another notable example of molecular genetic characterization of *cis*-acting sequences are the studies by Schlesinger and colleagues on defective interfering (DI) RNAs of Sindbis virus.[14] This system possesses the advantage that the dispensable nature of the DI RNA ensures that all mutational defects can be assigned to *cis*-acting characters.

One more virus in which *cis*-acting regulatory sequences have been examined is brome mosaic virus, whose genome is divided among RNAs 1, 2, and 3, of 3.2, 2.9, and 2.1 kilobases, respectively.[53-55] Each of these genomic RNAs serves directly as mRNA for a different nonstructural protein, while the coat protein encoded by the 3' portion of RNA3 is translated from a subgenomic RNA, called RNA4 (Figure 1). The 3' ends of these RNAs are aminoacylatable[56] and highly homologous over their last 200 bases.[57] A template-dependent RNA polymerase which selectively copies BMV (+)ssRNA has been isolated from BMV-infected barley,[58,59] and in vitro, the last 134 bases of BMV RNA direct (−)-strand RNA synthesis by this extract as efficiently as the full-length RNA.[60,61] The effects of mutations on this process and other interactions of the viral RNA 3' end have been investigated in vitro[62,63] (see also Hall and Dreher, Volume I, Chapter 5) and in vivo.[64] Complete cDNA clones of RNAs 1 to 3 have been constructed whose in vitro transcripts are highly infectious to both whole barley plants and protoplasts, making BMV the first multicomponent RNA virus expressed from cloned cDNA.[8,23,65,75]

A. Sequences Required for Stable Amplification of RNA3 In Vivo

With the development of infectious BMV cDNA clones, we have investigated the requirements for *cis*-acting sequences in viral RNA amplification in vivo. Sequences required in *cis* for efficient amplification of viral genomic RNA might be involved in one or more of three possible functions: initiation of (−)-strand RNA synthesis, initiation of (+)-strand genomic RNA synthesis, and stabilization of product RNA. The initiation and regulation of subgenomic RNA synthesis is discussed in a separate section below.

Isolated protoplasts present several advantages for in vivo studies of BMV RNA replication, including a high primary infection frequency. With appropriate inoculation techniques,[66] greater than 90% of treated cells can be infected using authentic BMV virion RNA, and greater than 50% using transcripts from BMV cDNA clones. This makes it practical to follow the course of a single, quasisynchronous infection cycle, allowing the viral genetic requirements for productive RNA replication to be examined independently of other constraints such as those for encapsidation and successful systemic spread in whole plant infections. In addition, replication of BMV RNAs 1 and 2 in protoplasts is independent of the very presence of RNA3,[15,67] while synthesis of RNAs 3 and 4 in BMV-infected cells is independent of the two known gene products of RNA3 (see below). All factors required for synthesis of BMV RNAs are therefore encoded either by RNA1, RNA2, or the host cell, and mutations causing altered replication characteristics in RNA3 must reside in *cis*-acting sequences directing RNA synthesis or stability of the progeny RNA.

Because of this freedom from *trans*-mediated effects, we have utilized RNA3 as an initial test object in in vivo studies of *cis*-acting sequences in BMV RNA replication. Deletions, rearrangements, and other changes have been made in RNA3, and the behavior of these mutants examined in barley protoplasts coinoculated with wild-type RNAs 1 and 2. These studies show that efficient accumulation of RNA3 in vivo requires segments from each of its three noncoding regions, but that outside of these regions, surprising amounts of sequence can be deleted with little or no effect on accumulation.[76] These dispensable regions correspond essentially to the two coding regions of RNA3, plus small amounts of flanking sequence. Thus, as mentioned above, neither of the known proteins encoded by RNA3 is required for its replication or for synthesis of subgenomic RNA.

Although coding sequences in RNA3 can be removed, deletions which extend across or too far into any of the three noncoding regions seriously inhibit RNA3 accumulation. Attempts to define these required noncoding regions by fine-structure deletion mapping do not show sharp boundaries between deletions which preserve normal function and those which are completely nonfunctional. Rather, in each case, reduction of RNA3 accumulation occurs gradually as deletion endpoints are extended through a region of perhaps 20 to 50 bases bordering the required segments. This may reflect a functional requirement for a particular higher order RNA structure in each region, rather than simply a given primary sequence recognition site.

Requirements for subsets from the 3' and 5' ends of the RNA were anticipated because of the expectation than the viral RNA polymerase must interact with these sequences or their complements in initiation of (−) and (+) strand RNA synthesis. Moreover, as noted above, a 3' subset of BMV RNA is necessary and sufficient to direct (−) strand synthesis in vitro by a BMV polymerase extract.[60,61] The 5' and 3' ends may also contribute to stabilization of the viral RNA by in vivo directing the covalent addition of a capping group and an amino acid, respectively, and by other noncovalent interactions.[56] It is noteworthy, however, that wild-type levels of RNA3 accumulation in vivo require more than the 134 bases which are able to direct (−) strand synthesis and aminoacylation in vitro. In vivo, the additional sequence may contribute a previously unknown function, or may facilitate a known function of the 3' end. One possibility in the latter class is that the additional sequences contribute a structural context in which sequence elements within the last 134 bases are accessible and able to function without steric hindrance from the remainder of the viral RNA.

Thus, while detailed interactions of the 5' and 3' ends contributing to RNA3 amplification require considerable additional investigation, some aspects of their function are known, and other likely interactions have been suggested. In contrast, there are no prior data which explain the requirement for intercistronic sequences in RNA3 accumulation in vivo. The observed stability of wild-type subgenomic RNA4 and a variety of RNA4 derivatives (see below), which do not contain the required intercistronic region, suggests that this sequence may not be required for RNA3 stability. The other evident possibility is that the intercistronic region is required for RNA3 replication. Although there is no evidence for a particular role within this context, the ability of 3' proximal sequences of BMV (+) strand RNA to independently direct (−) strand production in an in vitro system suggests that the intercistronic region might function in (+) strand synthesis.

Interestingly, the required intercistronic region contains a sequence homologous to the consensus element GGUUCAAyCCCU (y = pyrimidine), which is conserved not only at the 5' ends of both BMV RNAs 1 and 2, but also at analogous positions in all three RNAs of cucumber mosaic virus (CMV), a related tripartite RNA virus.[68] The possibility that the in vivo requirement for BMV RNA3 intercistronic sequences might relate specifically to this conserved element is currently under investigation.

B. Initiation Sites for Subgenomic RNA

An important feature of gene expression in many RNA viruses is the production of subgenomic RNAs which allow translation of internal cistrons not accessible to ribosomes in intact genomic RNA. Such subgenomic RNAs allow expression of viral genes to be differentially regulated in time and amount. In BMV, coat protein is translated from subgenomic RNA4 (Figure 1). RNA4 competes more effectively for ribosomes than genomic RNAs 1 to 3, leading to preferential expression of coat protein when large amounts of viral RNA are present and ribosomes are limiting.[69] The dependence of RNA4 synthesis on prior production of viral polymerase subunits may also lead to a useful initial delay in coat protein production after infection begins.

In vitro results show that BMV subgenomic RNA is made by partial transcription, rather than cleavage,[70] and this also appears to be true of the related alphaviruses.[71] In the template-dependent in vitro BMV polymerase system, artificially truncated (−) strand RNA3 templates extending only around 20 bases 3' to the RNA4 initiation site direct subgenomic RNA synthesis.[70] However, further in vitro studies show that additional upstream sequences 3' to this "core" promoter are required for efficient subgenomic RNA synthesis when the natural full-length (−) strand RNA3 template is used (Marsh et al., unpublished results; see also Volume I, Chapter 5). Similar to considerations in the section above, these additional upstream sequences may function to present the core promoter sequence in a structural context where it is directly accessible to the polymerase. Alternatively, the additional 3' sequences may provide a separate polymerase entry or attachment site necessary when the core promoter is located in the interior of the RNA template, but dispensable when the promoter is adjacent to a 3' end, as in the artificially truncated templates.

Recently, we have studied sequences directing and regulating BMV subgenomic RNA production in vivo. Deletions and foreign gene insertions in RNA3 (see below) show that sequences 17 or more bases downstream of the RNA4 transcription start site can be altered without significant effect on subgenomic RNA production.[15] Investigating the contribution to subgenomic RNA production of sequences upstream of the RNA4 initiation site was initially hampered by the involvement of nearby intercistronic sequences in efficient accumulation of both (+) and (−) strand genomic RNA3, as mentioned above. Consequently, in natural RNA3, the direct effects of sequence alterations in this region on subgenomic RNA production could not readily be distinguished from indirect effects due to reductions in synthesis of the (−) strand RNA3 template for subgenomic RNA. To circumvent this constraint, attempts were made to construct sequence duplications which would introduce the subgenomic RNA initiation site to a new context in RNA3, so that the sequences controlling production of the new subgenomic RNA could be examined in greater independence from sequences controlling genomic RNA accumulation. This has now been accomplished, and mapping studies on the functional subgenomic promoter so isolated are now in progress (French, R. and Ahlquist, P., unpublished results).

Present results (French and Ahlquist, manuscript in preparation) show that a small segment of BMV RNA3 will direct production of a new subgenomic RNA when transferred to any of at least several different, widely separated sites in RNA3. This segment need not contain the conserved intercistronic GGUUCAAyCCCU element,[68] showing that this element does not need to be directly linked to the subgenomic initiation site for subgenomic RNA to be made. To date, the segment has functioned at all sites in RNA3 which have been tested. The segment can be duplicated in RNA3, and in this way RNA3 derivatives have been constructed which synthesize multiple subgenomic RNAs, in contrast to the single subgenomic RNA of wild-type virus. Both in single and multiple insertions, the efficiency of subgenomic RNA synthesis from novel sites is often comparable to the production of subgenomic RNA4 from wild-type virus. However, in multiple promoter constructs, the efficiency of subgenomic RNA production generally varies between different promoter insertion sites. Conversely, the efficiency of a given initiation site varies in different constructs depending on the presence or absence of other subgenomic promoters 5' and 3' to the site in question. These results suggest that the efficiency of subgenomic RNA synthesis from a given site is more influenced by distal interactions or polar effects among different promoters than the local sequence context within several hundred bases of the promoter.

C. Expression of Foreign Genes Inserted in RNA Viruses

The present picture of BMV RNA function, emerging from only a short period of study by direct genetic manipulation, is clearly incomplete and doubtless naive in certain perspectives. Nevertheless, it is interesting that genetic mapping to date shows that a number of different functions are separately compartmentalized, allowing the viral RNA to be described in at least some respects as a mosaic of functional modules. One significant division of function in BMV RNA3, as described above, is that the sequences required for viral RNA amplification and gene expression, such as those for subgenomic RNA synthesis, are outside of the coding regions. This has considerable import for possible evolutionary paths of the virus, since it suggests that the now established process of RNA recombination[16] could insert or substitute new coding regions into the viral RNA, and rather readily generate functional viral components able to express the newly acquired genes. An alternative view would be that the apparently modular structure of the viral RNA might reflect past evolutionary events which involved gene recruitment from cellular or other viral sources into the genome of a precursor virus bearing functional units from other sources (see Chapter 10 in Volume II).

To test whether new genes could be readily introduced into viral RNA and successfully expressed, we have made and examined a number of hybrids between BMV RNA3 and foreign sequences. These experiments serve to test models for natural virus gene expression, potentially useful virus expression vectors, and possible modes of viral evolution. The hybrids examined most thoroughly to date involve insertion of the bacterial chloramphenicol acetyltransferase or CAT gene just 3' to the initiation codon of the highly expressed viral coat gene.[15,72]

These studies largely confirm the expectations discussed above: simple substitutions and insertions, with fusion points simply selected from convenient preexisting restriction sites, are sufficient to make replicatable hybrid virus RNAs carrying new sequences. Specifically, insertion of the CAT gene around 17 to 21 bases 3' to the subgenomic RNA initiation site in either orientation and with or without deletion of the coat gene sequences, generates a series of hybrid RNA3s which are replicated in BMV-infected cells and direct synthesis of CAT-containing subgenomic RNAs. Expression of a functional CAT protein from these constructs simply requires that the gene be inserted in the same orientation as the natural viral genes, making it available for translation from the hybrid subgenomic RNA4. An example of such an RNA3/CAT hybrid, its replication, and expression are shown in Figure 2. In this construct the CAT coding sequences are inserted in phase with the coat gene initiation codon, directing synthesis of a fusion protein with the N terminal residues of BMV coat protein joined to the CAT enzyme. Frameshifted CAT gene insertions in the remaining two reading frames downstream of the BMV coat protein initiation codon were also tested, and these also gave significant CAT expression.[15] The level of CAT expression varied, however, over a roughly sevenfold range, depending on the overlap and phase relationship

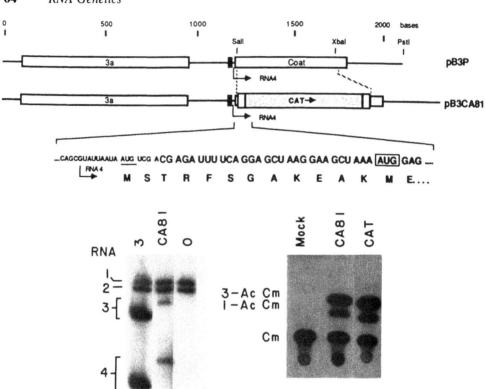

FIGURE 2.   Example of chloramphenicol acetyltransferase (CAT) gene replication and expression by a BMV RNA3 derivative. The diagram at top shows the BMV cDNA regions of plasmids pB3P, a complete clone of wild-type BMV RNA3 cDNA, and pB3CA81, in which a 0.77-kilobase Taq I fragment of pBR325 containing the CAT coding sequences (shaded) has been substituted for BMV coat protein sequences. The BMV cDNA region in pB3CA81 is identical to that of pB3CA42,[15] but is transcribed in vitro from a more efficient promoter (Janda, M., French, R., and Ahlquist, P., unpublished results). The lower left panel shows RNA blot hybridization analysis of BMV progeny RNA in gel-electrophoresed protoplast RNA samples. Protoplasts were inoculated with transcripts of BMV RNA 1 and 2 cDNA clones and transcripts of a wild-type RNA3 clone (3), pB3CA81 (CA81), or no RNA3 transcript (0), and were incubated 20 hr before RNA extraction. The probe used was a radiolabeled in vitro transcript complementary to the conserved 200 bases at the 3' end of all BMV virion RNAs. The lower right panel shows CAT enzyme assays[15] of extract from mock-inoculated protoplasts (mock), extract from protoplasts inoculated with transcripts of wild-type RNA1 and 2 cDNA clones and pB3CA81 (CA81), and purified CAT enzyme from bacterial cells.

between the CAT gene reading frame and the upstream reading frame initiated by the coat gene AUG.

Other coding sequences have been similarly inserted in BMV RNA and, at the RNA level at least, give similar expression results (Ahlquist, P., Gatenby, A., Sacher, R., unpublished results). Therefore, there appears to be nothing unusual in the CAT sequences to promote their expression in BMV. Moreover, the CAT gene has now also been inserted in, replicated, and expressed from both tobacco mosaic virus RNA[18] and a defective interfering RNA of Sindbis virus.[17]

V. LABORATORY MANIPULATIONS AND NATURAL EVOLUTION OF RNA VIRUSES

As introduced in the preceeding section, it is appropriate in the context of this volume on RNA variability to consider the relationship between laboratory manipulation and natural evolution of RNA viruses. All of the major types of changes proposed as significant pathways of RNA virus evolution can and have now been successfully duplicated in the laboratory, including point substitutions, sequence rearrangements, and hybrid viruses. This growing diversity introduced into RNA viruses in the laboratory offers a new perspective on the diversity and evolution of viruses in nature.

A significant basic lesson from work to date is that the replication machinery of some RNA viruses will tolerate dramatic, arbitrary changes in the viral genome. Previously, this conclusion was not obvious, since easily disrupted, long-range tertiary interactions in viral RNA might be required for basic functions such as replication. However, as described above, a surprisingly wide variety of extensive deletions, rearrangements, duplications, insertions, and block substitutions can be created without interfering with the ability of at least some viral RNAs to be replicated and to express genes. These results show that many of the products of natural recombination events in RNA virus infections will have the potential to be maintained in infection long enough to compete for preferential amplification under selection pressure, and for secondary genetic events to improve the fitness of such variants further.

Another significant point for virus evolution is the observed lack of extensive requirements for expression of new genes inserted in RNA viruses, as shown by the BMV/CAT model hybrids. Consequently, virus recombinants which simply have a complete new gene inserted in the proper orientation may frequently be able to express any selectable advantage which that gene confers. Moreover, as with BMV/CAT hybrids, reading frame effects on translation efficiency would allow at least one basis for evolutionary selection for appropriate expression levels of newly acquired genes. Gene recruitment by evolving viruses, through recombination with other viral or cellular RNAs, is thus a reasonable possibility.

Lastly, observations by Nomoto et al.[52] suggest that some sequences of wild-type poliovirus may be deletable with no apparent loss of virus fitness. Similarly, preliminary results suggest that some sequences of wild-type BMV RNA are also dispensable without obvious reduction in fitness for whole plant infections (Sacher, R. and Ahlquist, P., unpublished results). It is possible that the sequences in these examples have selectable functions for the virus in their natural, nonlaboratory environments. However, even in this case, it is noteworthy that the dispensable sequences are maintained stably over multiple propagation cycles in the laboratory under apparently nonselective conditions. The potential for fixation of selectively neutral sequences may thus apply to RNA as well as DNA genomes,[73,74] and it may not be proper to conclude from sequence conservation alone that a given viral RNA sequence necessarily encodes a strongly selected function.

VI CONCLUSION

The introduction of an artificial DNA stage into the life cycle of many RNA viruses offers new possibilities for genetic study of most aspects of the infection process, complementing the bridging biochemical and structural approaches which have been emphasized in past investigations. As discussed above, this approach has already proven productive in the study of virus functions at both the RNA and protein level. Direct control over virus genomes has also moved RNA molecular genetics beyond descriptive study into the deliberate manipulation of RNA viruses for potential practical goals, as in the use of RNA virus elements for directed gene expression and the controlled exchange of sequences between poliovirus strains to improve vaccines. The future power of RNA molecular genetics will doubtless increase as additional understanding is gained of the major parameters determining viability of deliberately altered viruses, as new experimental designs are developed to ask questions about the unique aspects of RNA genetic systems, and as expression of infection from cloned cDNA is extended to ($-$)ssRNA and dsRNA viruses by more sophisticated expression strategies.

ACKNOWLEDGMENTS

Research in the authors' laboratory was supported by grants to P.G.A. from the National Institutes of Health (No. GM35072), the National Science Foundation (No. DMB-8451884), the Shaw Fund of the Milwaukee Foundation, and Agrigenetics Research Associates.

REFERENCES

1. **Bawden, F. C.**, *Plant Viruses and Virus Diseases,* Chronica Botanica, Waltham, Massachusetts, 1950, 106.
2. **Gierer, A. and Schramm, G.**, Infectivity of ribonucleic acid from tobacco mosaic virus, *Nature (London),* 177, 702, 1956.
3. **Fraenkel-Conrat, H., Singer, B., and Williams, R. C.**, Infectivity of viral nucleic acid, *Biochim. Biophys. Acta,* 25, 87, 1957.
4. **Taniguchi, T., Palmieri, M., and Weissmann, C.**, Qβ DNA-containing hybrid plasmids giving rise to Qβ phage formation in the bacterial host, *Nature (London),* 274, 223, 1978.
5. **Racaniello, V. R. and Baltimore, D.**, Cloned poliovirus complementary DNA is infectious in mammalian cells, *Science,* 214, 916, 1981.
6. **Kandolf, R. and Hofschneider, P. H.**, Molecular cloning of the genome of a cardiotropic Coxsackie B3 virus: full-length reverse-transcribed recombinant cDNA generates infectious virus in mammalian cells, *Proc. Natl. Acad. Sci. U.S.A.,* 82, 4818, 1985.
7. **Mizutani, S. and Colonno, R. J.**, *In vitro* synthesis of an infectious RNA from cDNA clones of human rhinovirus type 14, *J. Virol.,* 56, 628, 1985.
8. **Ahlquist, P., French, R., Janda, M., and Loesch-Fries, L. S.**, Multicomponent RNA plant virus infection derived from cloned viral cDNA, *Proc. Natl. Acad. Sci. U.S.A.,* 81, 7066, 1984.
9. **Dasmahapatra, B., Dasgupta, R., Saunders, K., Selling, B., Gallagher, T., and Kaesberg, P.**, Infectious RNA derived by transcription from cloned cDNA copies of the genomic RNA of an insect virus, *Proc. Natl. Acad. Sci. U.S.A.,* 83, 63, 1986.
10. **Dawson, W. O., Beck, D. L., Knorr, D. A., and Grantham, G. L.**, cDNA cloning of the complete genome of tobacco mosaic virus and production of infectious transcripts, *Proc. Natl. Acad. Sci. U.S.A.,* 83, 1832, 1986.
11. **Meshi, T., Ishikawa, M., Motoyoshi, F., Semba, K., and Okada, Y.**, *In vitro* transcription of infectious RNAs from full-length cDNAs of tobacco mosaic virus, *Proc. Natl. Acad. Sci. U.S.A.,* 83, 5043, 1986.
12. **Omata, T., Kohara, M., Kuge, S., Komatsu, T., Abe, S., Semler, B. L., Kameda, A., Itoh, H., Arita, M., Wimmer, E., and Nomoto, A.**, Genetic analysis of the attenuation phenotype of poliovirus type 1, *J. Virol.,* 58, 348, 1986.
13. **Kohara, M., Omata, T., Kameda, A., Semler, B. L., Itoh, H., Wimmer, E., and Nomoto, A.**, *In vitro* phenotypic markers of a poliovirus recombinant constructed from infectious cDNA clones of the neurovirulent mahoney strain and the attenuated Sabin 1 strain, *J. Virol.,* 53, 786, 1985.
14. **Levis, R., Weiss, B. G., Tsiang, M., Huang, H., and Schlesinger, S.**, Deletion mapping of Sindbis virus DI RNAs derived from cDNAs defines the sequences essential for replication and packaging, *Cell,* 44, 137, 1986.
15. **French, R., Janda, M., and Ahlquist, P.**, Bacterial gene inserted in an engineered RNA virus: efficient expression in monocotyledonous plant cells, *Science,* 231, 1294, 1986.
16. **Bujarski, J. J. and Kaesberg, P.**, Genetic recombination between RNA components of a multipartite plant virus, *Nature (London),* 321, 528, 1986.
17. **Schlesinger, S., Levis, R., Weiss, B. G., Tsaing, M. and Huang, H.**, Replication and packaging sequences in defective interfering RNAs of Sindbis Virus, in *Positive Strand RNA Viruses,* Brinton, M. and Rueckert, R., Eds., Alan R. Liss, New York, 1987, 241.
18. **Takamatsu, N., Ishikawa, M., Meshi, T., and Okada, Y.**, Expression of bacterial chloramphenicol acetyltransferase gene in tobacco plants mediated by TMV RNA, *EMBO J.,* 6, 307, 1987.
19. **Cress, D. E., Kiefer, M. C., and Owens, R. A.**, Construction of infectious potato spindle tuber viroid cDNA clones, *Nucleic Acids Res.,* 11, 6821, 1983.
20. **van Emmelo, J., Ameloot, P., and Fiers, W.**, Expression in plants of the cloned satellite tobacco necrosis virus genome and of derived insertion mutants, *Virology,* 157, 480, 1987.
21. **Gerlach, W. L., Buzayan, J. M., Schneider, I. R., and Bruening, G.**, Satellite tobacco ringspot virus RNA: biological activity of DNA clones and their *in vitro* transcripts, *Virology,* 151, 172, 1986.

22. **Semler, B. L., Dorner, A. J., and Wimmer, E.,** Production of infectious poliovirus from cloned cDNA is dramatically increased by SV40 transcription and replication signals, *Nucleic Acids. Res.,* 12, 5123, 1984.

23. **Ahlquist, P. and Janda, M.,** cDNA cloning and in vitro transcription of the complete brome mosaic virus genome, *Mol. Cell. Biol.,* 4, 2876, 1984.

24. **Kaplan, G., Lubinski, J., Dasgupta, A., and Racaniello, V. R.,** *In vitro* synthesis of infectious poliovirus RNA, *Proc. Natl. Acad. Sci. U.S.A.,* 82, 8424, 1985.

25. **van der Werf, S., Bradley, J., Wimmer, E., Studier, F. W., and Dunn, J. J.,** Synthesis of infectious poliovirus RNA by purified T7 RNA polymerase, *Proc. Natl. Acad. Sci. U.S.A.,* 83, 2330, 1986.

26. **Melton, D. A., Krieg, P. A., Rebagliati, M. R., Maniatis, T., Zinn, K., and Green, M. R.,** Efficient *in vitro* synthesis of biologically active RNA and RNA hybridization probes from plasmids containing a bacteriophage SP6 promoter, *Nucleic Acids Res.,* 12, 7035, 1984.

27. **Davanloo, P., Rosenberg, A. H., Dunn, J. J., and Studier, F. W.,** Cloning and expression of the gene for bacteriophage T7 RNA polymerase, *Proc. Natl. Acad. Sci. U.S.A.,* 81, 2035, 1984.

28. **Klement, J. F., Ling, M.-L., and McAllister, W. T.,** Sequencing of DNA using T3 RNA polymerase and chain terminating ribonucleotide analogs, *Gene Anal. Tech.,* 3, 59, 1986.

29. **Simon, A. E. and Howell, S.,** Synthesis *in vitro* of infectious RNA copies of a virulent satellite of turnip crinkle virus, *Virology,* 156, 146, 1987.

30. **Contreras, R., Cheroutre, H., Degrave, W., and Fiers, W.,** Simple efficient *in vitro* synthesis of capped RNA useful for direct expression of eukaryotic genes, *Nucleic Acids Res.,* 10, 6353, 1982.

31. **Haseloff, J., Goelet, P., Zimmern, D., Ahlquist, P., Dasgupta, R., and Kaesberg, P.,** Striking similarities in amino acid sequence among nonstructural proteins encoded by RNA viruses that have dissimilar genomic organization, *Proc. Natl. Acad. Sci. U.S.A.,* 81, 4358, 1984.

32. **Cornelissen, B. J. C. and Bol, J. F.,** Homology between the proteins encoded by tobacco mosaic virus and two tricornaviruses, *Plant Mol. Biol.,* 3, 379, 1984.

33. **Franssen, H., Leunissen, J., and Goldbach, R.,** Homologous sequences in non-structural proteins from cowpea mosaic virus and picornaviruses, *EMBO J.,* 3, 855, 1984.

34. **Ahlquist, P., Strauss, E. G., Rice, C. M., Strauss, J. H., Haseloff, J., and Zimmern, D.,** Sinbis virus proteins nsP1 and nsP2 contain homology to nonstructural proteins from several RNA plant viruses, *J. Virol.,* 53, 536, 1985.

35. **Gorbalenya, A. E., Blinov, V. M., and Koonin, E. V.,** Prediction of nucleotide binding properties of virus-specific proteins from their primary structure, *Mol. Genet.,* 11, 30, 1985.

36. **Bernstein, H. D., Sonenberg, N., and Baltimore, D.,** Poliovirus mutant that does not selectively inhibit host cell protein synthesis, *Mol. Cell. Biol.,* 5, 2913, 1985.

37. **Strauss, E. G. and Strauss, J. H.,** Mutants of alphaviruses: genetics and physiology, in *The Togaviruses,* Schlesinger, R. W., Ed., Academic Press, New York, 1980, 393.

38. **Cooper, P. D.,** Genetics of picornaviruses, in *Comprehensive Virology,* Vol. 9, Fraenkel-Conrat, H. and Wagner, R. R., Eds., Plenum Press, New York, 1977, 133.

39. **Sarachu, A. N., Nassuth, A., Roosien, J., van Vloten-Doting, L., and Bol, J.,** Replication of temperature sensitive mutants of alfalfa mosaic virus in protoplasts, *Virology,* 125, 64, 1983.

40. **Leonard, D. A. and Zaitlin, M.,** A temperature-sensitive strain of tobacco mosaic virus defective in cell-to-cell movement generates an altered viral-coded protein, *Virology,* 117, 416, 1982.

41. **Dawson, W. O.,** Effect of temperature-sensitive, replication-defective mutations on RNA synthesis of cowpea chlorotic mottle virus, *Virology,* 115, 130, 1981.

42. **Agol, V. I., Grachev, V. P., Drozdov, S. G., Kolesnikova, M. S., Kozlov, V. G., Ralph, N. M., Romanova, L. I., Tolskaya, E. A., Tyufanov, A. V., and Viktorova, E. G.,** Construction and properties of intertypic poliovirus recombinants: first approximation mapping of the major determinants of neurovirulence, *Virology,* 136, 41, 1984.

43. **Agol, V. I., Drozdov, S. G., Grachev, V. P., Kolesnikova, M. S., Kozlov, V. G., Ralph, N. M., Romanova, L. I., Tolskaya, E. A., Tyufanov, A. V., and Viktorova, E. G.,** Recombinants between attenuated and virulent strains of poliovirus type 1: derivation and characterization of recombinants with centrally located crossover points, *Virology,* 143, 467, 1985.

44. **Agol, V. I., Drozdov, S. G., Frolova, M. P., Grachev, V. P., Kolesnikova, M. S., Kozlov, V. G., Ralph, N. M., Romanova, L. I., Tolskaya, E. A., and Viktorova, E. G.,** Neurovirulence of the intertypic poliovirus recombinant v3/a1-25: characterization of strains isolated from the spinal cord of diseased monkeys and evaluation of the contribution of the 3' half of the genome, *J. Gen. Virol.,* 66, 309, 1985.

45. **Pincus, S. and Wimmer, E.,** Production of guanidine-resistant and -dependent poliovirus mutants from cloned cDNA: mutations in polypeptide 2C are directly responsible for altered guanidine sensitivity, *J. Virol.,* 60, 793, 1986.

46. **Stanway, G., Hughes, P. J., Westrop, G. D., Evans, D. M. A., Dunn, G., Minor, P. D., Schild, G., and Almond, J. W.,** Construction of poliovirus intertypic recombinants by use of cDNA, *J. Virol.,* 57, 1187, 1986.

47. **Evans, D. M. A., Dunn, G., Minor, P. D., Schild, G. C., Cann, A. J., Stanway, G., Almond, J. W., Kurrey, K., and Maizel, J. V., Jr.,** Increased neurovirulence associated with a single nucleotide change in a noncoding region of the Sabin type 3 poliovaccine genome, *Nature (London),* 314, 548, 1985.

48. **Semler, B. L., Johnson, V. H., and Tracy, S.,** A chimeric plasmid from cDNA clones of poliovirus and coxsackievirus produces a recombinant virus that is temperature-sensitive, *Proc. Natl. Acad. Sci. U.S.A.,* 83, 1777, 1986.

49. **Nishiguchi, M., Kikuchi, S., Kiho, Y., Ohno, T., Meshi, T., and Okada, Y.,** Molecular basis of plant viral virulence; the complete nucleotide sequence of an attenuated strain of tobacco mosaic virus, *Nucleic Acids Res.,* 13, 5585, 1985.

50. **Sarnow, P., Bernstein, H. D., and Baltimore, D.,** A poliovirus temperature-sensitive RNA synthesis mutant located in a noncoding region of the genome, *Proc. Natl. Acad. Sci. U.S.A.,* 83, 571, 1986.

51. **Ishikawa, M., Meshi, T., Motoyoshi, F., Takamatsu, N., and Okada, Y.,** *In vitro* mutagenesis of the putative replicase genes of tobacco mosaic virus, *Nucleic Acids Res.,* 14, 8291, 1986.

52. **Kuge, S. and Nomoto, A.,** Construction of viable deletion and insertion mutants of Sabin strain of type I poliovirus: function of the 5′ noncoding sequence in viral replication, *J. Virol.,* 61, 1478, 1987.

53. **Lane, L. C.,** Bromoviruses, in *Handbook of Plant Virus Infections and Comparative Diagnosis,* Kurstak, E., Ed., Elsevier/North Holland, Amsterdam, 1981, 334.

54. **Ahlquist, P., Luckow, V., and Kaesberg, P.,** Complete nucleotide sequence of brome mosaic virus RNA3, *J. Mol. Biol.,* 153, 23, 1981.

55. **Ahlquist, P., Dasgupta, R., and Kaesberg, P.,** Nucleotide sequence of the brome mosaic virus genome and its implications for viral replication, *J. Mol. Biol.,* 172, 369, 1984.

56. **Haenni, A.-L., Joshi, S., and Chapeville, F.,** tRNA-like structures in the genomes of RNA viruses, in *Progress in Nucleic Acid Research and Molecular Biology,* Vol. 27, Academic Press, New York, 1982, 85.

57. **Ahlquist, P., Dasgupta, R., and Kaesberg, P.,** Near identity of 3′ RNA secondary structure in bromoviruses and cucumber mosaic virus, *Cell,* 23, 183, 1981.

58. **Hardy, S. F., German, T. L., Loesch-Fries, L. S., and Hall, T. C.,** Highly active template-specific RNA-dependent RNA polymerase from barley leaves infected with brome mosaic virus, *Proc. Natl. Acad. Sci. U.S.A.,* 76, 4956, 1979.

59. **Miller, W. A. and Hall, T. C.,** Use of micrococcal nuclease in the purification of highly template-dependent RNA-dependent RNA polymerase from brome mosaic virus-infected barley, *Virology,* 125, 236, 1983.

60. **Ahlquist, P., Bujarski, J. J., Kaesberg, P., and Hall, T. C.,** Localization of the replicase recognition site within brome mosaic virus RNA by hybrid-arrested RNA synthesis, *Plant Mol. Biol.,* 3, 37, 1984.

61. **Miller, W. A., Bujarski, J. J., Dreher, T. W., and Hall, T. C.,** Minus-strand initiation by brome mosaic virus replicase within the 3′ tRNA-like structure of native and modified RNA templates, *J. Mol. Biol.,* 187, 537, 1986.

62. **Dreher, T. W., Bujarski, J. J., and Hall, T. C.,** Mutant viral RNAs synthesized *in vitro* show altered aminoacylation and replicase template activities, *Nature (London),* 311, 171, 1984.

63. **Bujarski, J. J., Dreher, T. W., and Hall, T. C.,** Deletions in the 3′-terminal tRNA-like structure of brome mosaic virus RNA differentially affect aminoacylation and replication, *in vitro, Proc. Natl. Acad. Sci. U.S.A.,* 82, 5636, 1985.

64. **Bujarski, J. J., Ahlquist, P., Hall, T. C., Dreher, T. W., and Kaesberg, P.,** Modulation of replication, aminoacylation, and adenylation *in vitro* and infectivity *in vivo* of BMV RNAs containing deletions within the multifunctional 3′ end, *EMBO J.,* 5, 1769, 1986.

65. **Ahlquist, P.,** In vitro transcription of infectious viral RNA from cloned cDNA, in *Methods in Enzymology,* Vol. 118, Weissbach, A. and Weissbach, H., Eds., Academic Press, Orlando, Florida, 1986, 704.

66. **Samac, D. A., Nelson, S. E., and Loesch-Fries, L. S.,** Virus protein synthesis in alfalfa mosaic virus infected alfalfa protoplasts, *Virology,* 131, 455, 1983.

67. **Kiberstis, P. A., Loesch-Fries, L. S., and Hall, T. C.,** Viral protein synthesis in barley protoplasts inoculated with native and fractionated brome mosaic virus RNA, *Virology,* 112, 804, 1981.

68. **Rezaian, M. A., Williams, R. H. V., and Symons, R. H.,** Nucleotide sequence of cucumber mosaic virus RNA1, *Eur. J. Biochem.,* 150, 331, 1985.

69. **Shih, D. S. and Kaesberg, P.,** Translation of brome mosaic viral ribonucleic acid in a cell-free system derived from wheat embryo, *Proc. Natl. Acad. Sci. U.S.A.,* 70, 1799, 1973.

70. **Miller, W. A., Dreher, T. W., and Hall, T. C.,** Synthesis of brome mosaic virus subgenomic RNA *in vitro* by internal initiation of (−)-sense genomic RNA, *Nature (London),* 313, 68, 1985.

71. **Strauss, E. G. and Strauss, J. H.,** Replication strategies of the single stranded RNA viruses of eukaryotes, *Curr. Top. Microbiol. Immunol.,* 105, 1, 1983.

72. **Ahlquist, P., French, R., and Bujarski, J.,** Molecular studies of brome mosaic virus using infectious transcripts from cloned cDNA, *Adv. Virus Res.,* 32, 215, 1987.

73. **Loomis, W. F. and Gilpin, M. E.,** Multigene families and vestigal sequences, *Proc. Natl. Acad. Sci. U.S.A.,* 83, 2143, 1986.
74. **Kimura, M.,** The neutral theory of molecular evolution, *New Sci.,* 11, 41, 1985.
75. **Janda, M., French, R., and Ahlquist, P.,** High efficiency T7 polymerase synthesis of infectious RNA from cloned brome mosaic virus cDNA and effects of 5′ extensions on transcript infectivity, *Virology,* 158, 259, 1987.
76. **French, R. and Ahlquist, P.,** Intercistronic as well as terminal sequences are required for efficient amplification of brome mosaic virus RNA3, *J. Virol.,* 61, 1457, 1987.

Chapter 4

SEQUENCE VARIABILITY IN PLANT VIROID RNAs

P. Keese, J. E., Visvader, and R. H. Symons

TABLE OF CONTENTS

I. INTRODUCTION

Viroids are the smallest known pathogens in nature, and to date have been found only in flowering plants. Only 12 have been described so far (Table 1), and they consist of single-stranded, circular RNA molecules which vary in length from 246 to 375 nucleotides and are not encapsidated. The names that have been given to them are generally descriptive of the symptoms they cause. In the field, they continue to cause diseases of varying severity and economic importance. A historical perspective is given in the book by Diener[17] while several recent reviews[18-22] provide a detailed coverage of the determination of the sequence, secondary structure, and properties of viroids.

This chapter presents the most recent data on the sequence and structure of viroids and of approaches and results in the analysis of structure in relation to function. All evidence indicates that viroids do not code for any polypeptides in vivo, so that their biological effects must be exerted via sequence and structural signals, and their replication is completely dependent on host factors.[18-22]

Four definitions are central to this chapter:

Sequence homology — The percent sequence homology between two viroids was determined by the computer method of Wilbur and Lipman[23] using the parameters of $k = 4$ (matches of length k or greater), window size $w = 100$, and gap penalty $g = 4$.

Viroid species — A population of one or more independently replicating sequence variants which show more than 90% sequence homology. All members have less than 80 to 90% sequence homology with the members of any other viroid species.

Sequence variant — An individual viroid molecule of defined sequence. A viroid species contains one or more sequence variants, each of which varies in sequence by one or more nucleotides from other sequence variants, but all show more than 90% sequence homology.

Isolate of viroid — Viroid containing one or more viroid species isolated from a single infected plant. Each viroid species may consist of one or more sequence variants in each isolate.

II. APPROACHES TO INVESTIGATION OF RELATION OF STRUCTURE TO FUNCTION

A. Limited Approaches Available

There is a scarcity of suitable biological parameters for an analysis of structure/function relationships in viroids, and this severely limits the approach that can be used, especially since viroids apparently do not code for any polypeptides. About the only two parameters that can be used are infectivity (viable or not) and the variation in symptom expression on

Table 1
LIST OF VIROIDS

Viroid	Abbreviation	Number of nucleotides	Ref.
Avocado sunblotch viroid	ASBV	247	1, 2
Chrysanthemum stunt viroid	CSV	356, 354	3, 4
Citron variable viroid	CVaV	N.D.[a]	5
Citrus exocortis viroid	CEV	370—375	4, 6—8
Coconut cadang-cadang viroid[b]	CCCV	246, 247	9
Coconut tinangaja viroid	CTiV	254	10, 11
Columnea viroid	CV	N.D.	12
Cucumber pale fruit viroid[c]	CPFV	303	13
Hop stunt viroid	HSV	297	14
Potato spindle tuber viroid	PSTV	359	15
Tomato apical stunt viroid[d]	TASV	360	16
Tomato planto macho viroid	TPMV	360	16

[a] N.D.: not determined.

[b] CCCV infections produce four major RNA components, all derived from the infectious monomeric small form of 246 or 247 nucleotides. These include monomers and dimers of both the D-O form and any of a set of larger forms (CCCV 287, 296, 301, and 346) which contain a repeat involving 41, 50, or 55 nucleotides, or a double repeat totalling 100 nucleotides.

[c] CPFV is a sequence variant of HSV since the two viroids share 95% sequence homology.

[d] In early reports, TASV was referred to as tomato bunchy top viroid.[17]

a given host plant, e.g., mild, intermediate, or severe symptoms. These phenotypic characters are quite crude in relation to the detailed sequence and structural analysis that is required. Variation of host range between two sequence variants is also a possibility, but no examples are known and experiments could be impractical unless infection is being assessed between a limited number of plant species.

B. Approaches Being Used
1. Sequence Comparisons of Naturally Occurring Sequence Variants of the Same Viroid

These are of importance in defining the conserved and variable features of the viroid molecule, and they may indicate regions that have a role in viroid replication and symptom expression. This approach, which has been successful for CEV[8,24,25] and for PSTV,[26] has relied on nature to provide a range of viable mutants. However, a significant problem in relating a particular sequence variant to a particular symptom is the recent recognition that naturally occurring viroid infections often contain more than one sequence variant in a single host plant[8] and even a mixture of viroid species, at least in citrus.[27] Hence, it is imperative that infection is carried out with a single sequence variant. The finding that full-length cDNA clones of PSTV containing a tandem dimeric insert in a bacterial vector are infectious when inoculated on to tomato seedlings[28] makes such experiments feasible. Infectious cDNA clones have also been prepared for HSV,[29,30] CEV,[24] and TASV.[31]

As a rule, it appears that plasmid clones containing only monomeric viroid inserts are likely to be noninfectious.[8,29,32] The exceptions so far have been for monomeric BamHI cDNA inserts of PSTV and CEV cloned into the BamHI site of the polylinker of bacterial vectors, followed by inoculation with the whole plasmid.[24,31,32] The results of Visvader et al.[24] for CEV showed that this was due to a serendipitous combination of the choice of BamHI cloning site and the nucleotide sequence adjacent to the BamHI site of the bacterial vector. They explained the unexpected infectivity in terms of the processing of longer-than-unit-length precursors in vivo at a site in the CEV molecule close to the BamHI site (Section VII.A.2).

The usual noninfectivity of monomeric plasmid clones can, however, be overcome by using the excised inserts for inoculation, as shown for HSV,[30] PSTV,[32] and CEV.[24] Presumably, the monomeric inserts, which contained "sticky ends" because of the restriction enzyme used, were circularized after inoculation and prior to transcription by host RNA polymerases to give a longer-than-unit-length transcript which can enter the rolling circle replication mechanism predicted for viroids (Section VII.A).[33-35]

Further flexibility in the use of cloned viroid sequences is shown by the unexpected infectivity of two subgenomic AvaI fragments of PSTV of 167 and 192 base-pairs derived from the whole genome of 359 nucleotides.[32] The fragments were infectious when inoculated together, but noninfectious when inoculated separately. These data provide further evidence for the intracellular ligation of viroid cDNA fragments.

2. Site-Directed Mutagenesis

The ability to prepare infectious cDNA clones of viroids makes this approach feasible. However, except in the special case of an infectious CEV clone with a single base change,[24] other single base changes in CEV[25] and PSTV[31] produced noninfectious clones. A similar lack of success was obtained with short deletions and insertions in HSV[36] and a small deletion in CEV.[25]

The general lack of success so far may preclude the extensive use of this approach in the future. It appears that regions of the viroid molecule have evolved in such a way that rigorous conservation of sequence and, hence, structure is maintained by the nonviability of any spontaneous mutants that arise. This is considered further in Section VI.

3. Preparation of Chimeric Viroids

The approach here is to construct chimeric viroids by joining complete sections from two different sequence variants of the one viroid or from two different viroids. In this way, the effect of whole regions on viability and symptom expression can be investigated. The method has been successful so far for chimeric constructs within the same viroid species,[25] but not for those between two different viroids.[25,31] These results are considered further in Sections IV.A.2 and VI.

III. VARIATION OF SEQUENCE AND STRUCTURE BETWEEN VIROIDS

A. Sequence and Structure

The predicted secondary structures of eight of the viroids which have been sequenced are given in Figure 1. Each of the circular molecules contains intramolecular base-paired regions separated by small internal loops to give a linear, rod-like structure. The sequence of PSTV was determined in 1978 and its secondary structure predicted on the basis of maximizing the number of base-pairs together with data indicating sites of single-stranded regions susceptible to single strand-specific RNAses and to bisulfite modification.[15] It was then straightforward to derive a secondary structure for related viroids, as they were sequenced on the basis of regions of sequence homology. One viroid, ASBV, is unrelated to the rest, and it was necessary to use the base-pairing matrix procedure of Tinoco et al.[37] as a first step in the development of a secondary structure.[2]

Further details of the proposed secondary structures are given in Table 2. The percent nucleotides base-paired range from 60 to 73%, while G:U base-pairs are present in significant amounts in all viroids. It is feasible that calculation of the thermodynamically most stable structure using a set of stability constants, reaction enthalpies and entropies for base-pair formation[38-40] could be used to add refinements to the predicted structures. This work would be aided by computer programs which make use of a given set of parameters to calculate the most thermodynamically stable structure.[40-43]

None of these considerations take account of any tertiary structure in the viroid molecule.

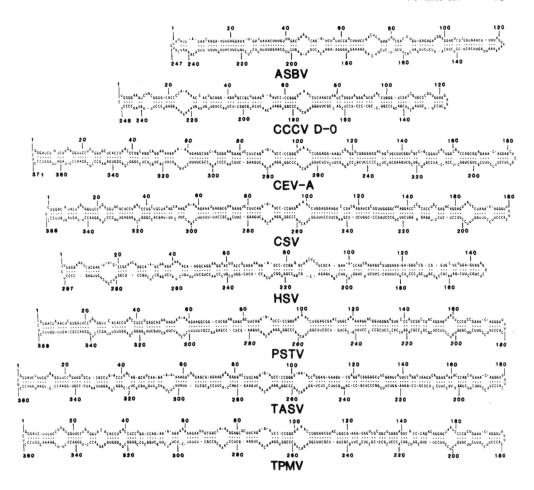

FIGURE 1. Sequences and proposed secondary structures of ASBV,[2] CCCV D-O (the basic 246-nucleotide species),[9] CEV-A,[6] CSV,[3] HSV,[14] PSTV,[15] TASV,[16] and TPMV.[16] CPMV[13] (not given) is a sequence variant of HSV (Table 1). (From Keese, P. and Symons, R. H., in *Viroids and Viroid-Like Pathogens*, Semancik, J. S., Ed., CRC Press, Boca Raton, Fla., 1987, 1.)

Table 2
PROPERTIES OF PROPOSED SECONDARY STRUCTURES OF VIROIDS

Viroid	No. of Nucleotides	Nucleotides				% Nucleotides base-paired	Base-pairs as % of total			Ref.
		A	U	G	C		A:U	G:C	G:U	
ASBV	247	68	85	51	43	67	51	34	14	2
CSV	356	75	93	89	99	70	35	52	13	3
CEV-A	371	72	75	112	112	69	28	56	16	6
CCCV D-O	246/7	53	47	73	73/4	65	24	69	8	9
CPFV	303	64	70	81	88	69	31	65	4	13
HSV	297	61	69	79	88	67	29	64	7	14
PSTV	359	73	77	101	108	70	29	58	13	15
TASV	360	70	90	101	99	73	32	57	11	16
TPMV	360	72	81	99	108	68	31	60	9	16

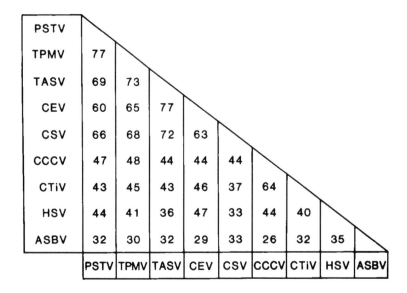

	PSTV	TPMV	TASV	CEV	CSV	CCCV	CTiV	HSV	ASBV
TPMV	77								
TASV	69	73							
CEV	60	65	77						
CSV	66	68	72	63					
CCCV	47	48	44	44	44				
CTiV	43	45	43	46	37	64			
HSV	44	41	36	47	33	44	40		
ASBV	32	30	32	29	33	26	32	35	

PSTV vs. 25 random sequences of same size and base composition
Sequence homology = 29.5 ± 4.2%

FIGURE 2.   Sequence homology between pairs of viroids. Sequences of viroids in Figures 1 and 10 were compared using the computer program of Wilbur and Lipman[23] and the parameters of $k = 4$, $w = 100$, and $g = 4$.

That such additional intramolecular interactions occur is clearly indicated by the specific cross-linking induced in PSTV by irradiation with UV light.[44,45] An extensive characterization of all tertiary interactions within each viroid molecule will be essential in sorting out the fine details of how viroid structure determines function.

B. Sequence Homology Between Viroids

The overall sequence homology between pairs of viroids as calculated using the method of Wilbur and Lipman[23] is given in Figure 2. The viroids fall into two groups on the basis of these homologies, as well as of sequence comparison between the predicted secondary structures.[21,22,46] One group contains only ASBV which shares between 26 and 35% sequence homology with any other viroid (Figure 2). Pair-wise comparisons between the other eight viroids show sequence homologies ranging from 33 to 77%; this group is called the PSTV-like viroids. In such comparisons it is important to note that PSTV shows a 29.5 ± 4.2% sequence homology with 25 random sequences of the same size and same base composition (Figure 2).

C. Structural Domains and Functional Implications

From an examination of sequence homology with more than 40 sequence variants of the PSTV-like viroids, Keese and Symons[46] developed a general model of viroid structure. This model is not applicable to ASBV. Five structural domains are distinguished (Figure 3), and these can be related to functional aspects of viroid replication and expression. In addition, they allow an examination of evolutionary relationships between viroids.

A schematic diagram of the positions of these domains in each viroid[46] is given in Figure 4, while the number of nucleotides in each domain is in Table 3. The PSTV-like viroids share at least 33% sequence homology between any two members (Figure 2) and they all contain a conserved, U-bulged helix in the C domain, an oligo-(A) sequence in the P domain

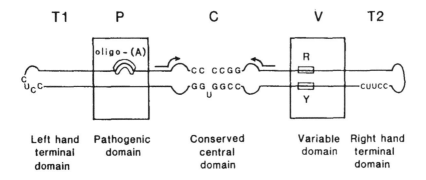

FIGURE 3. Model of viroid domains. The five domains, T1, P, C, V, and T2, were determined from sequence homologies between the viroids.[46] The arrows depict an inverted repeat sequence which can form the stem loop depicted in Figure 5A. R, Y: a short oligopurine, oligopyrimidine helix. The conserved bulged helix (Figure 5B) is given in the C domain. (Modified from Keese, P. and Symons, R. H., *Proc. Natl. Acad. Sci. U.S.A.*, 82, 4582, 1985. With permission.)

that occurs in the same relative position, about 35 residues 5′ of the central conserved sequence, GGA$_G^U$CCCCGGGG$_C^A$AAC, a CCUC sequence in the T1 domain, and a CCUUC sequence in the T2 domain in the positions indicated (Figure 3). The general features of these domains can be summarized as follows with more specific details for individual viroids being given later.

1. C Domain

The central domain contains about 95 nucleotides and is the most highly conserved region between viroids (Table 4).[3,21,22] In pairwise comparisons between viroids, the sequence homology varies between 35 and 99% (Table 4), values which are equal to or appreciably higher than the overall sequence homology (Table 4, Figure 2). The C domain of CCCV D-O, which constitutes 40% of the molecule, shows 75% sequence homology with PSTV, a value greater than the 72% sequence homology for this domain between PSTV and two other closely related viroids, CEV and TASV. On the basis of this and other comparisons (Table 4), CCCV D-O is included in the PSTV-like group. The boundaries of the C domain were determined from pairwise sequence comparisons of viroids containing highly homologous central domains; e.g., between PSTV, TPMV, and CCCV or between CEV, TASV, and CSV.[46]

Several intriguing features of the C domain point to an important role in viroid replication. The highly conserved, bulged helix (Figure 5B), the 16-nucleotide conserved sequence which includes the uppermost strand of the bulged helix, GGA$_G^U$CCCCGGGG$_C^A$AAC, and the nine nucleotide inverted repeat sequence on either side of this sequence (long arrows in Figure 3) have been predicted to have two separate, mutually exclusive, as yet poorly defined functions.[46] The nine-nucleotide inverted repeat sequence can form a stem loop (Figure 5A) which is observed during the thermal denaturation of PSTV, CEV, CSV, and CCCV.[38,39] It is feasible that this stem loop can also arise in vivo and hence provide an alternative structural signal to that provided by the bulged helix and neighboring nucleotides. Hence, structural switching in the C domain could control the change between two phases of the replication cycle.[21,22]

In the predicted native state, UV irradiation of PSTV causes a specific stable cross-link between G98 and U260 (Figure 5B), which is strongly indicative of an activated tertiary structure.[44,45] It may be of significance that nucleotides G96 to G98 correspond to the site in the closely related CEV that genetic evidence indicates is the site of processing of longer-than-unit-length precursors, a key step in the rolling circle model of viroid replication.[24]

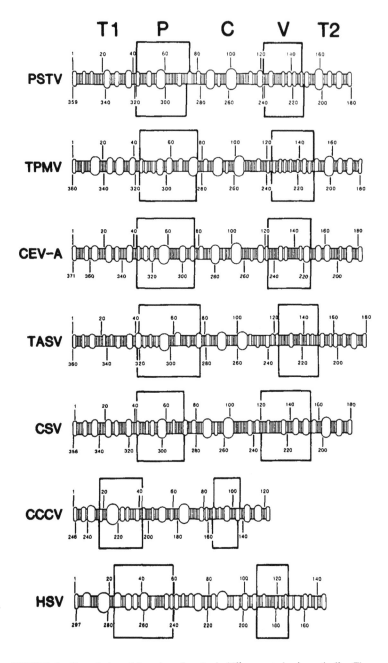

FIGURE 4. Boundaries of domains of each viroid[46] presented schematically. The domains of CTiV are given in Figure 11, and the sizes of the domains for each viroid in Table 3.

2. P Domain

The boundaries of the P domain (Figure 4) were based on sequence homologies between the P domain of HSV and other viroids such as PSTV and by certain pairwise comparisons, e.g., CEV-A and TASV, in which there is significant change from relatively low sequence homology in the P domain to higher sequence homology in the adjacent T1 and C domains. In seven viroids, it ranges in size from 63 to 71 nucleotides (Table 3) and contains, except in CCCV, a long adenine-dominated oligopurine sequence of 15 to 17 nucleotides in one

Table 3
SIZE OF VIROID DOMAINS[a]

	Number of nucleotides in each domain					
	T1	P	C	V	T2	Total
PSTV	83	63	95	54	64	359
TPMV	88	63	95	52	62	360
TASV	82	70	97	49	62	360
CEV-A	84	69	97	57	64	371
CSV	83	60	95	66	52	356
CCCV (246)	33	49	95	28	41	246
CTiV	32	47	95	39	41	254
HSV	45	71	97	37	47	297

[a] Data obtained from Figure 4 and Keese and Symons.[46] The boundaries of the T1 and P domains have been changed slightly from the originally published.[46] Data for CTiV taken from Figure 10.

Table 4
SEQUENCE HOMOLOGY BETWEEN DOMAINS OF DIFFERENT VIROIDS[a]

Viroids used for pair-wise comparison		% Sequence homology					
		Domains					Overall
1	2	T1	P	C	V	T2	
PSTV	CEV-A	71	64	72	49	48	60
	TPMV	70	76	98	38	94	77
	TASV	72	60	72	41	90	69
	CSV	78	47	79	45	83	66
	CCCV (246)	26	27	75	37	35	47
	CTiV	28	36	63	28	36	43
	HSV	34	64	44	35	29	44
CEV-A	TPMV	74	64	72	46	48	65
	TASV	93	72	99	45	51	77
	CSV	71	48	84	39	38	63
	CCCV (246)	31	32	61	35	46	44
	CTiV	38	38	67	40	38	46
	HSV	42	67	45	32	34	47
TPMV	TASV	74	57	74	38	98	73
	CSV	69	52	74	59	88	68
TASV	CSV	72	56	82	52	89	72
CCCV (246)	CTiV	68	50	66	75	59	64
	HSV	64	45	43	34	43	44
CTiV	HSV	60	39	47	32	41	40

[a] Sequence homologies calculated from sequences and domains in Figures 1, 3, and 10, using the computer program of Wilbur and Limpan[23] and the parameters of $k = 4$, $w = 100$, and $g = 4$.

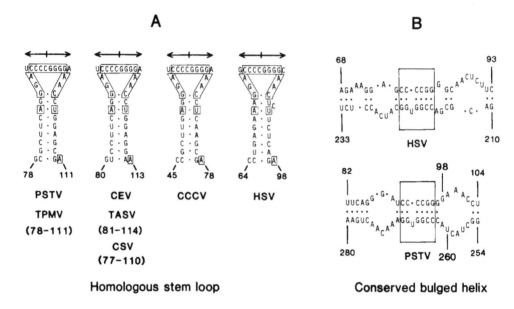

Homologous stem loop   Conserved bulged helix

FIGURE 5.   Alternative structures that may form in the C domain of PSTV-like viroids. (A) Stem loop as described by Riesner et al.[38] with a nine-base-pair stem and a 10-nucleotide, self-complementary sequence in the loop (arrowed). The same structure can form for all PSTV-like viroids except HSV where there is a nonbase-paired C nucleotide in the stem.[46] Conserved nucleotides are boxed. (B) Conserved bulged helix postulated for the native structure of purified viroids.[46] The upper strand corresponds to part of the stem loop predicted in A. The nucleotides of PSTV involved in the UV-induced covalent cross-linking (G98 to U260)[44,45] are indicated. (From [A] Keese, P. and Symons, R. H., *Proc. Natl. Acad. Sci. U.S.A.*, 82, 4582, 1985; and [B] Keese, P. and Symons, R. H., in *Viroids and Viroid-Like Pathogens*, Semancik, J. S., Ed., CRC Press, Boca Raton, Fla., 1987, 1. With permissions.)

strand, and an oligo-(U_{4-7}) sequence in the opposite strand. CCCV retains only an oligo-(A_5) sequence in the P domain. It is interesting to speculate if this significant difference is related to the distinctive host range of CCCV which appears to be restricted to members of the palm family,[47] whereas the other PSTV-like viroids have a much wider host range, infecting members of several families of dicotyledons.[48]

The P domain is the only one of the five domains which have been clearly shown to have an effect on symptom expression. Studies of naturally occurring sequence variants of PSTV and CEV which vary in the P domain have shown a correlation with severity of symptom expression in tomato; these aspects are considered in detail in Sections IV.A and B.

3. V Domain

This is the smallest domain of 28 to 66 nucleotides (Table 3), and also the most variable, showing less than 60% sequence homology between any pair of viroids except between CCCV and CTiV (Table 4). The only significant sequence relationship between viroids in this domain appears to be the presence of an oligo-purine:oligopyrimidine helix, usually with a minimum of three G:C base-pairs (Figure 3). The boundaries were defined by a change from low sequence homology to higher sequence homology in the adjacent C and T2 domains. This domain contains a smaller region, P_R, in CEV in which there is high sequence variation in 17 sequence variants of CEV, and which may have a role in determining the level of CEV in infected plants (see Section IV.A.3).

4. T Domains

These domains are intriguing in the roles they may play in viroid replication and in the evolutionary development of the PSTV-like viroids. Sequence homologies in the domains of these viroids include a strictly conserved CCUC sequence in the end loop of the T1 domain and a CCUUC sequence in the T2 domain (Figure 3). The sequence homology in the T1 domain becomes longer when the whole end loop and neighboring nucleotides are compared between two or more viroids. For example, the sequence is CCCUCGGG in CEV, CSV, TASV, and TPMV, and CCCUCUGGG in HSV and CCCV, while the terminal sequence of PSTV, UCCUCGGA, shows strong homology. Purified tomato DNA-dependent RNA polymerase binds to the ends of PSTV as shown by electron microscopy,[49] but there is no evidence that the conserved sequence of the T1 domain is involved in this binding or is of any relevance to the in vivo replication of PSTV.

Comparison of the sequences of the T1 and T2 domains of several viroids revealed homologies which were the starting point for the development of the domain model of viroids.[46] For example, TASV shares 77% overall sequence homology with CEV-A, but the T2 domains are only 51% homologous (Table 4). In contrast, TASV shares less overall sequence homology with PSTV (69%), but the T2 domains show 90% homology. TASV can therefore be considered to be a recombinant between the T2 domain of a PSTV-ancestral viroid (or a TPMV-ancestral viroid, since TASV and TPMV share 98% T2 domain sequence homology) and all but the T2 domain of a CEV-ancestral viroid, with the crossing over occurring in the variable V domain. Of relevance here is the high homology between TASV and CEV-A in the T1 and C domains (Table 4).

CCCV shows two potential sets of terminal domain arrangements. In one, the left half of the T1 domain of CCCV is almost identical to HSV, but the C domain is more homologous to PSTV (75%) than to HSV (43%). CCCV could therefore be a recombinant between a viroid with partial PSTV lineage and the T1 domain of a HSV ancestral viroid. However, the low homology of the T1 domain of CCCV and PSTV suggest complex origins for CCCV. The other well-characterized rearrangements of the T2 domain of CCCV are considered in Section IV.C.

IV. SEQUENCE VARIABILITY IN DOMAINS OF SEQUENCE VARIANTS

Most of the data considered in Section III.C to support the model of viroid domains were obtained from only one sequence variant of each viroid. The sequence analysis of sequence variants of a single viroid would be expected to provide information on the conserved and variable sequences in relation to these domains and to provide an indication of the possible functional significance of each domain. Most data have been obtained with CEV and PSTV, while CCCV has provided an intriguing example of sequence variation by rearrangements in the T2 domain.

A. Sequence Variants of CEV
1. Two Classes of Sequence and Severity of Symptom Expression
Most work has been carried out with five Australian isolates extracted from citrus and propagated on chrysanthemum plants and on tomato seedlings.[6-8,25] When nucleic acid extracts of infected tomato plants were used for inoculation, symptoms were only of two types: mild with symptoms barely visible, and severe where there was extensive epinasty (leaf curling) and stunting. Intermediate forms of severity as observed for PSTV were not found. The results were:

Severe symptoms   Isolates CEV-A, CEV-DE25, CEV-J
Mild symptoms   Isolates CEV-DE26, CEV-DE30

Earlier sequence analysis of purified isolates propagated on chrysanthemum showed that isolate CEV-A contained at least two sequence variants, CEV-A and CEV-AM, which

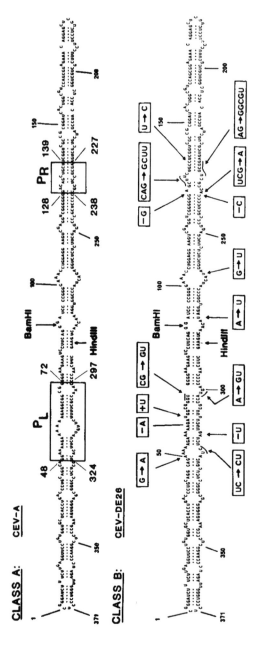

FIGURE 6. The primary and proposed secondary structures of the Class A and Class B reference sequences of CEV, CEV-A, and CEV-DE26.[23] The nucleotide changes in CEV-A necessary to give the sequence of CEV-DE26 are boxed in CEV-DE26. The P_L and P_R domains of CEV[8] are indicated for CEV-A, together with the nucleotide positions on the outside of the boundaries of the two domains. It is important to note that these P_L and P_R domains of CEV are smaller than and enclosed within the general P and V domains, respectively, as defined by Keese and Symons[46] for the PSTV-like group of viroids (Figure 7). The BamHI and HindIII sites used for the preparation of various cDNA clones are indicated. (From Visvader, J. E. and Symons, R. H., *EMBO J.*, 5, 2051, 1986. With permission.)

differed by only a few nucleotides,[7] whereas CEV-DE25 and CEV-DE26 appeared to contain only one variant.[7] Sequence analysis was done directly on the purified viroid as well as after cloning of cDNA fragments. The sequences of CEV-A, CEV-AM, and CEV-DE25 were very similar to a Californian isolate, CEV-C, sequenced by Gross et al.,[4] except for a few base changes, whereas CEV-DE26 contained 27 nucleotide differences relative to CEV-A, which included exchanges, insertions, and deletions (Figure 6). The three Australian isolates all induced similar symptoms on chrysanthemum of reduced growth rate, plus yellow spots on the leaves.

The first attempts at direct RNA sequencing of viroid purified from tomato plants inoculated with an extract of isolates CEV-DE30 and CEV-J indicated that they each contained a mixture of sequence variants. Hence, a method was developed[8] for the rapid preparation of full-length cDNA clones of sequence variants in the single strand vector M13mp9 and their sequencing. From a total of 19 cDNA clones examined, 11 new sequence variants were found, ranging in size from 370 to 375 nucleotides. This was a surprising result and suggested that the sequencing of a larger number of cDNA clones would produce even more sequence variants.

In spite of the fact that no infectivity studies were done because the full-length cDNA clones were not infectious on tomato as a result of the cloning site used and the method of preparation of cDNA,[8] it was still possible to correlate the Class A sequence of the type member CEV-A with severe symptoms on tomato seedlings and the Class B sequence of the type member CEV-DE26 with mild symptoms. This was because only Class B sequences were found in cDNA clones of isolate CEV-DE30, while cDNA clones of isolate CEV-J contained a mixture of Class A and Class B sequences.[8]

Direct proof of this correlation was provided by the preparation of infectious monomeric BamHI cDNA clones in the BamHI site of the vector pSP64 of Class A sequence variants CEV-A and CEV-AM, and of one of the Class B sequence variants of CEV-DE30. As predicted, the Class A clones gave severe symptoms when inoculated onto tomato seedlings, whereas the Class B clone gave only mild symptoms.[25]

2. P Domain Determines Severity of Symptom Expression

The sequence variations found in the 15 Australian sequence variants, plus the two Californian variants, were mostly located in the P and V domains common to all PSTV-like viroids (Figures 6 and 7). In the case of CEV, two smaller regions or domains, P_L and P_R, were defined within the P and V domains, respectively, that contained most of the sequence variations observed so far (Figures 6 and 7). The P_R domain, of only 20 to 24 nucleotides (6% of CEV molecule), was highly variable and contained many nucleotide changes, while the P_L domain of 49 nucleotides (14% of CEV molecule) contained fewer changes. There were several changes in the C domain and none in the T1 and T2 domains.[8]

The sequence and infectivity data indicated that the P_L and/or the P_R regions of CEV played a role in determining the severity of symptom expression. In the case of PSTV, only the P domain has been implicated in modulating symptom expression.[26] The answer with CEV was obtained by the construction of infectious chimeric viroids in which approximately one half of one sequence variant was joined through the C domain with the other half of another variant (Figure 8).[25] For example, one chimeric clone in the vector pSP64, designed M_L/S_R, contained the left-hand region of a mild variant of CEV-DE30 adjoined to the right-hand region of a severe variant of CEV-A. Infectivity results on tomato seedlings clearly showed that severe symptoms were determined only by the P_L domain, but they also indicated that the P_R domain may have an effect on the level of viroid which develops in infected plants (Section IV.A.3). There is no evidence to indicate how these two domains exert these effects.

A most important part of this work was to determine if any sequence changes had occurred

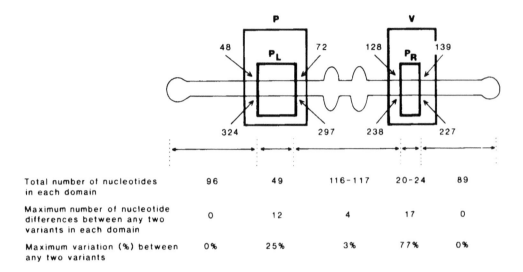

FIGURE 7. Location of nucleotide changes in the P_L and P_R domains and the surrounding regions of the proposed structures of 17 sequence variants of CEV.[8] The larger P and V domains common to all PSTV-like viroids (Figures 3 and 4) are indicated.

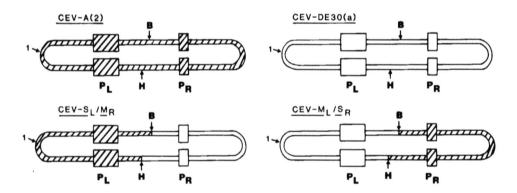

FIGURE 8. Schematic diagrams of two parental CEV viroids and two chimeric viroids in circular form. The parental viroid CEV-A(2) induces severe symptoms on tomato seedlings, while the parental viroid CEV-DE30(a) induces mild symptoms. The BamHI (B) and HindIII (H) cloning sites used to prepare the chimeric viroids are indicated. (From Visvader, J. E. and Symons, R. H., *EMBO J.*, 5, 2051, 1986. With permission.)

during the replication of the CEV variants derived from the chimeric clones. Hence, full-length cDNA clones in the vector M13mp93 were prepared[8] from the progeny viroid purified from pooled plants. The complete sequence of the progeny viroids was always the same as that of the clone used for inoculation for both the parents as well as the chimeras.

It may be of functional significance that the two classes of CEV sequence variants (A and B) have evolved so that one class of P_L domain is always linked to the same class of P_R domain. That this is not obligatory for a viable viroid was demonstrated by the infectivity of the chimeric molecules. Although these hybrid constructs were infectious, they may have a replication disadvantage in competition with the homologous species and have therefore not been detected in field isolates.[8]

3. V Domain May Affect Level of CEV

In the domain model of viroids, the V domain is the site for RNA rearrangements of the T2 domain between viroids.[46] However, in the work on infectivity of chimeric cDNA clones of CEV, the data indicated that the V domain may have an additional role in modulating the efficiency of the initial infection or the replication process. For example, although infection by the parental clone M_L/M_R and the chimeric clone M_L/S_R always give mild symptoms on tomato seedlings, inoculation with the chimeric clone S_L/M_R produced plants with either mild or severe symptoms. This effect was traced to the concentration of cDNA clone plus RNA transcript used in the inoculum; with a high concentration, all plants showed severe symptoms.[25] Further, plants with severe symptoms gave extracts with a concentration of viroid 9 to 20 times higher than in extracts from plants with mild symptoms. Hence, at least in the case of the chimeric viroid, S_L/M_R, symptom expression was dependent on the level of viroid which accumulated in the plant.

4. Potential Translation Products of CEV Sequence Variants

It is feasible that analysis of potential polypeptide products of the plus and minus strands of a large number of sequence variants of CEV may indicate conserved sequences with a possible functional role in vivo. Four out of 17 sequence variants contain one AUG in the plus strand, while all 17 variants contain one AUG in the minus strand.[8] There are no fully conserved, or even partially conserved, polypeptides encoded by AUG-initiated open reading frames in the minus strand. Only one potential GUG-initiated polypeptide product of 15 amino acids is conserved between the CEV variants; this initiates just to the left of the P_L domain on the bottom strand of the T1 domain of the plus strand. Since a 5′ proximal AUG initiation codon appears essential for the initiation of translation of eukaryotic RNA,[50] it seems highly unlikely the CEV encodes any functional polypeptide products. These findings[8] are consistent with the lack of conservation between possible translation products of both the plus and minus strands of the closely related viroids, CEV, PSTV, and CSV.[3,6] This aspect is discussed more fully in References 21 and 22.

5. How Can There Be So Many Sequence Variants in a CEV Isolate?

The two CEV isolates which contain many sequence variants, CEV-DE30 and CEV-J, were obtained from two orange trees on *Poncirus trifoliata* rootstock in orchards over 400 km apart in New South Wales, Australia. Because of the low level of CEV in infected orange trees, each isolate was propagated on tomato plants prior to purification of the CEV for analysis of sequence variants. It is feasible that the relative concentration of sequence variants in each isolate could be significantly affected by this change of host, from a woody species in the family Rutaceae to an herbaceous species in the family Solanaceae. In addition, it is not known how many of these sequence variants are independently viable or require complementation since only one cloned sequence variant of CEV-DE30 has been assayed and shown to be infectious.[25] Hence, it will be necessary to systematically prepare BamHI cDNA clones in the BamHI site of the vector pSP64[24] to determine the complete sequence of each clone and its infectivity on tomato seedlings for as many sequence variants as possible. The data so obtained will be essential for a much better understanding of the origin and epidemiology of the large number of sequence variants of CEV.

It is feasible that sequence variants could arise by either (1) a high copy error rate of an RNA polymerase[51] replicating a single RNA species, or (2) infection of one plant by several sequence variants during propagation of citrus varieties by grafting or during regular practices such as pruning.[17] The long potential life of citrus trees in the field (over 60 years) would allow the accumulation of sequence variants by either route. In the case of the production of mutants during replication, it is highly likely that most random mutants would be nonviable in view of the severe restrictions on which parts of the viroid molecule can be modified by in vitro site-directed mutagenesis.[8,25,31]

In contrast to the large number of sequence variants of CEV in field citrus trees, the analysis of nine coconut palms infected with CCCV from nine separate locations in the

Philippines only showed two sequence variants which differed by the addition of one residue at a fixed position.[9] The sequence analysis in this work was done by the direct RNA sequencing method which would not have picked up a population of one or minor components in the presence of one or two dominant species. The possibility that other sequence variants exist can only be assessed by the systematic sequencing of a large number of full-length cDNA clones prepared from a number of CCCV isolates.

B. Sequence Variants of PSTV

Sänger and colleagues have investigated the correlation between structure and pathogenicity of PSTV.[26] Eight field isolates of PSTV which gave symptoms varying from mild to lethal when inoculated on to tomato seedlings were purified and sequenced by the use of three oligonucleotide primers for the reverse transcriptase synthesis of cDNA. The primers were 5' ^{32}P-labeled, and the cDNA was sequenced by the Maxam-Gilbert procedure.[52] The sequence variants were placed in four groups: mild, intermediate, severe, and lethal, and the nucleotide changes, insertions, and deletions compared to the mild reference sequence (Figure 9). The intermediate and severe isolates showed changes at only three or one sites, respectively, whereas the two lethal isolates showed changes at either six or seven sites. Most of the changes occurred at the left-hand edge of the P domain, as defined above, while the U to AA change occurred at the left-hand edge of the V domain.

The nucleotide changes in the P domain were considered to be part of a smaller region called the virulence modulating (VM) region.[26] Increasing pathogenicity was correlated with an increasing thermodynamic instability of the VM as calculated theoretically. It was considered that this decrease in stability may lead to the increased binding of an unknown host factor which aggravates severity of symptoms.

This work was carried out with total viroid purified from a field isolate. In view of the increasing awareness that field isolates of viroids are likely to contain two or more sequence variants which could have varying effects on symptom expression unrelated to the concentration of each variant,[8,26] it is necessary to confirm the above results with infectious, single cDNA clones of each of the sequence variants.

C. Sequence Variants of Two Viroids from Coconut Palms

1. Sequence Variation Between the P Domains of CCCV and CTiV

The cadang-cadang disease of coconuts in the Philippines is a well-characterized, slow-developing disease that eventually results in the death of the host.[53,54] The causative agent is the smallest viroid known of 246 nucleotides (Figure 10).[9,54] More recently, viroid etiology has been implicated for tinangaja, another disease of coconut palms located on the island of Guam (Figure 10).[10] The RNA associated with this disease, CTiV, is 254 nucleotides long and shares 64% overall sequence homology with CCCV (Figure 2).[11]

Tinangaja exhibits similar pathology to cadang-cadang disease except for the effect on nut production. Whereas the coconuts from palms infected with cadang-cadang become more spherical and scarified, palms afflicted with tinangaja produce mummified nuts with no kernel present.[10] By analogy with sequence variants of CEV and PSTV, the P domains of CCCV and CTiV share the least sequence homology amongst the five domains (Table 4). Nevertheless, the P domain of CTiV possesses an oligo-(A) sequence in the lower strand of the P domain, similar to that found for CCCV, but unlike other PSTV-like viroids which all have an oligo-(U) sequence in the same relative position (Figures 1 and 10).

2. Association of Sequence Variants of CCCV with Disease Progression

The sequencing of CCCV isolated from individual palms and from fronds of different age within a single palm has established the existence of at least three types of sequence variant in addition to the basic 246 nucleotide species. These include a C insertion at position 198

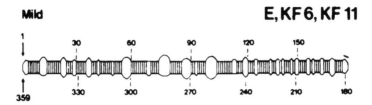

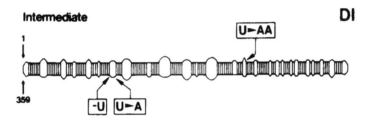

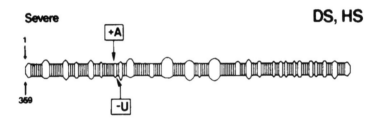

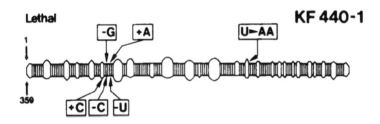

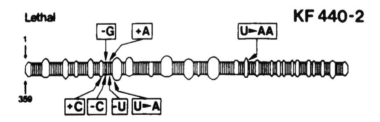

FIGURE 9. Location of nucleotide changes in the secondary structure model of PSTV causing mild, intermediate, severe, or lethal disease symptoms in tomato (*Lycopersicon esculentum* cv. Rutgers). Changes relative to the mild isolate are indicated. The names of the eight field isolates that provided the six pathogenic groups are given at the right-hand end above each structure. (From Schnölzer, M., Haas, B., Ramm, K., Hofmann, H., and Sänger, H. L., *EMBO J.*, 4, 2181, 1985. With permission.)

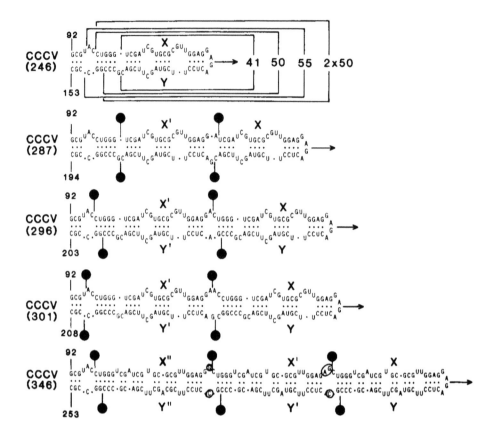

FIGURE 10.   Partial sequence duplications of CCCV (246). Two adjacent sequences, X and Y of either 41, 50, or 55 nucleotides, are duplicated either once in CCCV (287), CCCV (296), and CCCV (301), or twice in CCCV (346). The arrows indicate the boundaries of the X and Y sequences, while the filled circles mark the boundaries of the duplicated sequences. The circled nucleotides are sites of mutations in sequence variants. Data obtained from Haseloff et al.[9] and Keese et al.[11a]

of the C domain to give a 247-nucleotide sequence variant,[9] duplications of the V and T2 domains of 41, 50, 55, or 100 nucleotides (Figure 11), and mutations of bases adjacent to the boundaries of some of the partial duplications (Figure 11).

The mechanism(s) of CCCV transmission from palm to palm remains unknown, so that it is uncertain as to when and how these different-sized variants of CCCV arise. Screening of individual fronds from a single palm showed changes from the 246-nucleotide CCCV in older fronds, to a larger sequence variant in newly developing fronds.[55] Therefore, the larger variant appeared only in older palms after the initial appearance of the 246-nucleotide sequence variant. It has been postulated that the larger sequence variants of either 287, 296, or 301 nucleotides arose *de novo* in each palm from the 246-nucleotide variant which is the only variant found early in infection. In addition, only one of the 287-, 296-, or 301-nucleotide sequence variants is found in individual palms.[9,55,56] Recent evidence supporting this hypothesis is provided by mechanical inoculations of the primary shoot of germinating coconut seeds with partially purified 246-nucleotide CCCV that resulted in the appearance of the 246-nucleotide variant after 6 months, and the 296-nucleotide variant after 10 months.[57] Final confirmation would be provided by inoculation with infectious monomeric cDNA clones of the 246-nucleotide variant.

Consistent with this progression from the smallest CCCV variant to larger variants with time of infection is the observation that the 346-nucleotide variant (Figure 11) was only

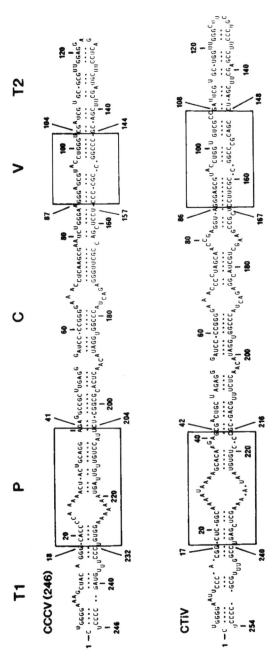

FIGURE 11.   Comparative sequences of CTIV[11] and CCCV (246).[9] The original two-dimensional structure of CCCV (246)[9] has been modified slightly to demonstrate increased structural homology with CCCV (246). The positions of the five domains are indicated. (Modified from Keese, P. and Symons, R. H., in *The Viroids*, Diener, T. O., Ed., Plenum Press, New York, 1987, in press. With permission.)

observed in the last dying stages of the cadang-cadang disease.[58] Hence, it apparently arose in the order CCCV (246), CCCV (296), CCCV (346).

Similar findings have been reported for the single nucleotide C additions with disease progression.[59] The initial infecting form appears to be the 246-nucleotide CCCV variant which is eventually succeeded by a 247-nucleotide CCCV with C addition at nucleotide 198. The same phenomenon occurs with the larger sequence variants in which the appearance of the 296-nucleotide CCCV variant is eventually replaced by a 297-nucleotide form.

Although the functional implications of this single base addition are unknown, there is appreciable change in the secondary and tertiary structure of the native molecule as the 246- and 247-nucleotide variants migrate with markedly different mobilities when electrophoresed through nondenaturing polyacrylamide gels.[59] Since this C addition occurs in the C domain, it may modify the postulated switching event predicted to occur during replication in which the native structure in the C domain is converted into a stem loop structure (Figure 5A).

3. T2 Domain Duplications of CCCV May Provide Multiple Host Component Binding Sites

The increase in size of CCCV as a result of duplications of the T2 domain during the infected life of the coconut palm prior to death raises two intriguing questions. First, why does the smallest 246-nucleotide variant appear to be the main, if not the sole, form of the viroid that initiates infection? This is puzzling since the larger 296-nucleotide variant is also readily infectious when mechanically inoculated.[59] Second, and more important in terms of its functional implications, why do the sequence variants of CCCV with partial duplications appear to have a selection advantage over the CCCV (246) sequence variant? Once a larger sequence variant of CCCV becomes detectable in different palms, it eventually becomes the dominant form. Presumably, it has some replication advantage, in spite of its larger size. Furthermore, it is unlikely that the longer sequence variants offer novel biochemical or functional possibilities since the duplications retain essentially the same sequence and structural arrangement found in CCCV (246). The only result in effect is to provide two T2 domains in the case of CCCV (287), CCCV (296), and CCCV (301), and three T2 domains in the case of CCCV (346) (Figure 11).

The functional difference between the small and large sequence variants may be ascribed to competition for binding to some host component important for replication, but which is in limited supply. For example, purified DNA-dependent RNA polymerase II from tomato binds to the T1 and T2 domains of PSTV.[49] It is feasible that binding of such a host component could provide a replication advantage.

4. Mutations Adjacent to Boundaries of CCCV Partial Duplications

The boundaries of the partial duplications of CCCV (346) are not precise due to mutations proximal to the border of the duplicated segment (Figure 11). Similar mutations have also been found for CCCV (296).[11] It has been suggested that the partial duplication of the T2 domain of CCCV occurs by discontinuous transcription by an RNA polymerase switching or jumping from one template to another.[46] Mutations could therefore occur at the site of switching. The model for CCCV is analogous to the jumping polymerase model proposed for the generation of DI RNAs of influenza virus.[60]

Given that mutations at the boundaries of the duplication arise during template switching, it is feasible that a number of single or double base mutations could occur to give a population of molecules, one or more of which could then have a replication advantage. This potentially complex situation can only be resolved by the preparation of full-length cDNA clones to the purified longer sequence variant, and the determination of the sequence of a large number of individual clones.[8]

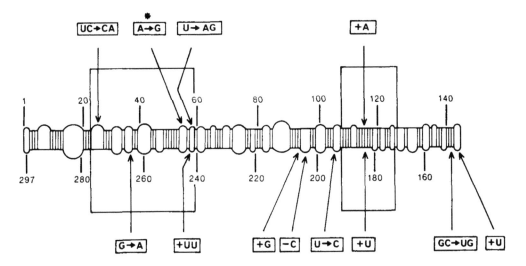

FIGURE 12. Schematic diagram of HSV to indicate the changes necessary to convert HSV to CPMV.[13,14] The positions of the P and V domains are indicated. The A → G change marked with an asterisk is also the single base change found in a grapevine isolate of HSV.[61]

V. SEQUENCE VARIANTS OF HSV, CPFV, AND CSV

A. Sequence Variants of HSV and CPFV

Although CPFV is six residues longer than HSV, it is considered to be a sequence variant of HSV because the two viroids have greater than 90% homology. The changes in HSV necessary to convert HSV to CPFV are indicated in the schematic diagram of Figure 12 together with the location of the P and V domains (boxed). The only other sequence variant characterized so far is one isolated from grapevines in Japan,[61] where there is a single base change at residue 54; this is also one of the changes found in CPFV (starred box in Figure 12).

The nucleotide differences between HSV and CPFV are scattered throughout the molecule with the greatest concentration in the P domain (Figure 12). The sequences of many more sequence variants will be needed to see if HSV conforms to the pattern set by CEV, where most changes occur in the P and V domains (Section IV.A.1).

B. Sequence Variants of CSV

Only two sequence variants of CSV have been sequenced, an Australian isolate of 356 nucleotides[3] and a European isolate of 354 nucleotides.[4] The sequence changes in the European isolate that are different to the Australian isolate are given in Figure 13 and show that about one half of the changes occur in the P plus V domains. As in the case of HSV, many more sequence variants need to be characterized to determine the variable and conserved regions.

VI. VIROID MUTANTS PREPARED IN VITRO

The ability to prepare infectious cDNA clones of viroids[24,25,28,30-32,36,67] allows the ready manipulation of sequences in vitro by mutagenesis and selection at the level of DNA (see also Section II.B.2 and B.3). This has included site-directed mutagenesis involving small deletions and additions or base substitutions, as well as construction of viroid chimeras, followed by bioassay to investigate structure/function relationships. However, all site-directed sequence changes in CEV, PSTV, and HSV have so far been located in the more

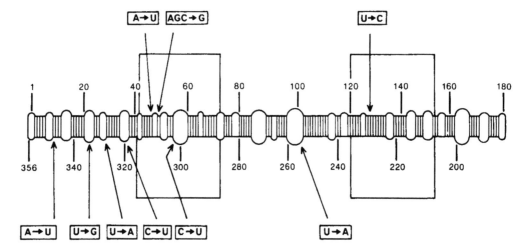

FIGURE 13. Schematic diagram of an Adelaide isolate of CSV[3] with the nucleotide changes found in a European isolate.[4] The positions of the P and V domains are indicated.

conserved T1, T2, and C domains (Figure 3) and all appear to be lethal.[24,25,31,36] These results indicated that only a small proportion of all potential sequence changes may result in a viable viroid, in spite of the large number of sequence differences that occur between viroid species (Figure 2, Table 4).

The most obvious explanation for any loss of viability as a consequence of site-directed mutation is that even a single base change can affect the local secondary structure and, perhaps, also tertiary interactions. Such changes could then affect the normal functioning of that region. For example, a G to U transition at nucleotide 351 of CEV-A in the T1 domain would give the loss of a G.C base-pair at the end of a base-paired helix[25] which is strictly conserved in CEV-A, PSTV, TPMV, and TASV and CSV (Figure 1). Although the end loop sequence of the T2 domain is variable between closely related viroids, the deletion of four nucleotides, AGCU, which form part of the end loop in CEV-A, eliminated infectivity. This deletion would greatly affect the secondary structure in this region.

The C domain is the most highly conserved region in PSTV-like viroids, especially the bulged helix (Figure 3). It is therefore not surprising that base substitutions in this conserved helix are lethal in the case of CEV[8] and PSTV.[31] For CEV-A, a G to U conversion at nucleotide 96 leads to the loss of a base-pair. For PSTV, a C to U conversion at nucleotide 93 would allow significant rearrangement of base-pairing in the neighboring nucleotides.

The successful preparation of chimeric viroids with junctions within the C domain of two sequence variants of CEV is described in Section I.A.2. Similar constructs between different viroid species, PTSV and TASV, were not infectious.[31] This may be a consequence of the incompatible nature of the C domains of these two viroids that does not allow complementation for viability.

On the basis of all the unsuccessful in vitro mutagenesis studies so far, it appears that the P and V domains offer the best regions for future experiments. The many changes found in these domains for 17 sequence variants of CEV (Section IV.A) would indicate a high probability for success. In addition, the P domain offers the possibility of a detailed examination of the sequence and structure of this domain in relation to symptom expression.[8,25]

VII. STRUCTURE-FUNCTION CORRELATIONS OF SEQUENCE VARIANTS

Although viroids consist of only 246 to 375 nucleotides, they contain sufficient information to regulate such functions as binding of a host RNA polymerase, specific processing

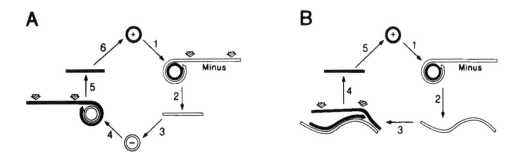

FIGURE 14.   Two rolling circle models for the replication of viroids.[33,34] (A) A circular input plus strand is copied by a host RNA polymerase to give an oligomeric minus strand which is processed at specific sites to give a full-length linear minus strand. This is circularized and copied to give an oligomeric plus strand which is also processed to full-length linear plus strands which are circularized to give the progeny viroid. (B) Similar to mechanism in A except that the oligomeric minus strand is not processed, but copied to give an oligomeric plus strand. (From Forster, A. C. and Symons, R. H., *Cell,* 49, 211, 1987. With permission.)

of replicative intermediates, the ratio of plus to minus strands in each cell, cell-to-cell movement, symptom expression, and host range. All evidence indicates that this information is in the primary, secondary, and tertiary structure of each viroid. What has been postulated so far on the basis of sequence data, secondary structural analyses, and limited functional information is the modular nature of viroids in which different regions or domains of the rod-shaped molecules are responsible for different functions.[46] This domain model may well apply to ASBV which, at present, lacks suitable viroid relatives for comparison.

A. Domains Involved in the Processing of Viroid Precursors

Many aspects of the correlation of domains with particular functions have been considered in Sections III and IV. Another important function that can be attributed to particular domains is the specific processing of viroid precursors.

Viroid infections result in the production of longer-than-unit-length plus and minus viroid RNAs. These are presumed to be replicative intermediates that are cleaved specifically to generate the monomeric form which is then circularized by a plant RNA ligase.[29,33,34] It is generally accepted that replication occurs by a rolling circle mechanism in which the invading plus circular viroid is copied by a yet-to-be-identified host RNA polymerase.

Two related pathways are likely (Figure 14). In the first (Figure 14A), the invading circular viroid is copied by a host RNA polymerase by continual transcription to give a long minus strand. This is then cleaved at specific sites, either by a host enzyme or by a self-cleavage reaction,[35] to produce linear, minus strand monomers which are then circularized by a host RNA ligase. These circular minus strand species then act as a template for the rolling circle synthesis of plus strand RNA by the same or a different RNA polymerase. Specific cleavage produces linear plus strand monomers which are circularized to produce the progeny viroid which accumulate in vivo. In the second pathway (Figure 14B), the linear oligomeric minus strand is not cleaved, but acts as a template for the synthesis of the oligomeric plus strand which is specifically cleaved to give the linear plus monomer. The pathway in Figure 14A is believed to operate in the case of ASBV.[34,35]

1. Self-Cleavage Domains of ASBV

In vitro synthesized dimeric RNA transcripts prepared from tandem dimeric cDNA clones of ASBV in the plasmid vector pSP64 using the phage SP6 RNA polymerase are capable of autocatalytic cleavage at a unique site to generate full-length monomers.[35] The self-cleavage reaction occurs in both plus and minus transcripts and generates 5'-hydroxyl and 2',3'-cyclic phosphate termini. Only Mg^{2+} and a pH of 7 to 8 appear to be necessary for the reaction. Highly conserved sequence and secondary hammerhead-shaped structures are feasible around the plus and minus cleavage sites (Figure 15A) which occur in neighbouring parts of the ASBV molecule (Figure 15B). Hence, it appears that about one third of the ASBV molecule may have a function in providing the self-cleavage domains for the rolling circle replication of this viroid.

ASBV has little sequence homology with the PSTV-like viroids (Figures 1 and 2). However, it does share the ability to self-cleave in vitro RNA transcripts of cDNA clones with the virusoid of lucerne transient streak virus (vLTSV).[62,63] vLTSV is a 324-nucleotide, single-stranded circular satellite RNA, dependent on a helper virus for replication.[64,65] RNA transcripts of cDNA clones of only part of the LTSV virusoid specifically self-cleave in both the plus and minus species, although only one cleavage site is present in each transcript. As in the case of ASBV, plus and minus cleavage sites occur in neighboring parts of the virusoid molecule. In addition, sequence and secondary hammerhead-shaped structures around the plus and minus cleavage sites are highly conserved with those of ASBV, and about one third of the molecule is involved in the provision of the self-cleavage domains.[62,63]

2. Processing of Replication Intermediates of PSTV-Like Viroids

In contrast to the extensive self-cleavage of dimeric transcripts prepared from cDNA clones of ASBV,[35] no detectable self-cleavage of dimeric transcripts of cDNA clones of CEV was observed.[24] Robertson et al.[66] have recently reported the self-processing of in vitro-synthesized dimeric PSTV transcripts. However, only 1 to 5% of the dimeric transcripts were cleaved, and this occurred between nucleotides 250 to 270, on the bottom strand of the C domain. It is feasible that the cleavage observed was not specific, but random chemical hydrolysis in a susceptible region of the molecule, especially since the site of processing determined by CEV on genetic evidence is on the top strand of the C domain.[24]

In the case of CEV,[24] a full-length monomeric BamHI cDNA clone (single BamHI site between nucleotides 89 and 90 of CEV-A) in the BamHI site of the plasmid vector pSP64 was infectious when inoculated with its RNA transcripts on tomato seedlings. Only a clone in the plus orientation (giving plus strand RNA transcripts), but not the minus, was infectious. The excised monomeric BamHI cDNA insert was also infectious. Site-directed mutagenesis of nucleotide G97 to U or A (Figure 16) resulted in an infectious clone, but a noninfectious excised insert. Surprisingly, the infectious progeny CEV from the infectious mutated cDNA clone was the wild-type sequence. The result was explained by the in vivo processing of the RNA transcripts, generated in vitro or in vivo, at one of the three phosphodiester linkages (marked by 1, 2, or 3 in Figure 16) on the 3' side of the point mutation. Because of the remarkable coincidence that a 11 nucleotide sequence on the 3' side of the BamHI site in the viroid sequence is repeated in the cloning site of the particular vector used, it is possible to generate the full-length wild-type sequence. The correlation of the same 11-nucleotide repeat with infectivity of monomeric PSTV and HSV cDNA clones has been reported by Tabler and Sänger[32] and Meshi et al.[67]

Equivalent secondary structures to those proposed for the self-cleavage of plus and minus ASBV (Figure 15A) have not been found in the region of the putative processing site of CEV (Figure 16). Hence, it is feasible that a different type of self-cleavage reaction or enzymatic processing event occurs during the rolling circle replication of the PSTV-like viroids. In addition, the presence of the putative CEV processing site in the C domain is consistent with highly conserved features of this domain and the central roles it may play in viroid replication (Section III.C.1).

ACKNOWLEDGMENTS

The work from this laboratory described in this review was supported by the Australian

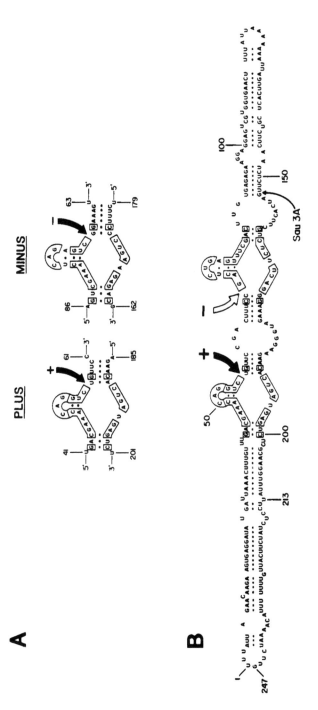

FIGURE 15. (A) Secondary structure proposed around the plus and minus self-cleavage sites of ASBV.[35] Residues conserved between the plus and minus structures are boxed and numbered according to structure in Figure 1.[2] The site of self-cleavage is indicated by an arrow. (B) Native structure of ASBV[2] modified to incorporate the plus self-cleavage structure and the complement of the minus self-cleavage structure.[35] (From Hutchins, C. J., Rathjen, D. D., Forster, A. C., and Symons, R. H., *Nucleic Acids Res.* 14, 3627, 1986. With permission.)

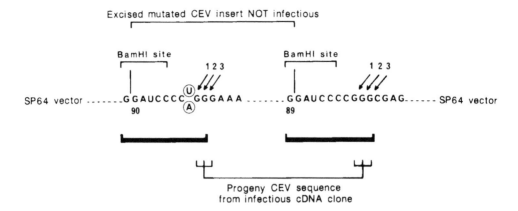

FIGURE 16.   Potential in vivo processing sites of RNA transcripts derived from two-point mutant cDNA clones of CEV-A.[24] CEV-A was cloned at its single BamHI site (nucleotides 90 to 89) into the BamHI site of the plasmid vector pSP64. Both the intact clone and the BamHI-excised insert were infectious when inoculated on to tomato seedlings.[24] The circled nucleotides represent the point mutations introduced into CEV-A at position 97 by the conversion of G97 to A97 or U97. The intact, mutated clones were infectious on tomato seedlings, whereas the BamHI-excised inserts were not.[24] The 11-nucleotide repeat sequence that correlates with infectivity and the proposed sites of processing at positions 1, 2, or 3 are indicated. The origin of the wild-type progeny sequence derived from the infectious, mutated cDNA clone is shown. (Modified from Visvader, J. E., Forster, A. C., and Symons, R. H., *Nucleic Acids Res.*, 13, 5843, 1985. With permission.)

Research Grants Scheme and by a Commonwealth Government Grant to the Adelaide University Centre for Gene Technology in the Department of Biochemistry. We thank all members of our reseach group for access to unpublished data and for discussions, Dr. Adrian Gibbs for advice and help with the computer analyses of sequence homology, Jennifer Cassady and Tammy Edmonds for assistance with the figures, and Leanne Goodwin for typing the manuscript.

REFERENCES

1. **Palukaitis, P., Hatta, T., Alexander, D. M., and Symons, R. H.**, Characterization of a viroid associated with avocado sunblotch disease, *Virology*, 99, 145, 1979.
2. **Symons, R. H.**, Avocado sunblotch viroid: primary sequence and proposed secondary structure, *Nucleic Acids Res.*, 91, 6527, 1981.
3. **Haseloff, J. and Symons, R. H.**, Chrysanthemum stunt viroid: primary sequence and secondary structure, *Nucleic Acids Res.*, 9, 2741, 1981.
4. **Gross, H. J., Krupp, G., Domdey, H., Raba, M., Jank, P., Lossow, C., Alberty, H., Ramm, K., and Sänger, H. L.**, Nucleotide sequence and secondary structure of citrus exocortis and chrysanthemum stunt viroid, *Eur. J. Biochem.*, 121, 249, 1982.
5. **Schlemmer, A., Roistacher, C. N., and Semancik, J. S.**, Detection of a unique, infectious RNA from citrus showing typical symptoms of citrus exocortis disease, *Phytopathology*, 75, 946, 1985.
6. **Visvader, J. E., Gould, A. R., Bruening, G. E., and Symons, R. H.**, Citrus exocortis viroid: nucleotide sequence and secondary structure of an Australian isolate, *FEBS Lett.*, 137, 288, 1982.
7. **Visvader, J. E. and Symons, R. H.**, Comparative sequence and structure of different isolates of citrus exocortis viroid, *Virology*, 130, 232, 1983.
8. **Visvader, J. E. and Symons, R. H.**, Eleven new sequence variants of citrus exocortis viroid and the correlation of sequence with pathogenicity, *Nucleic Acids Res.*, 13, 2907, 1985.
9. **Haseloff, J., Mohamed, N. A., and Symons, R. H.**, Viroid RNAs of cadang-cadang disease of coconuts, *Nature (London)*, 299, 316, 1982.
10. **Boccardo, G., Beaver, R. G., Randles, J. W., and Imperial, J. S.**, Tinanagaja and bristle top, coconut diseases of uncertain etiology in Guam, and their relationship to cadang-cadang disease of coconut in the Philippines, *Phytopathology*, 71, 1104, 1981.

11. **Keese, P., Keese, M. E., and Symons, R. H.**, Coconut tinangaja viroid: sequence homology with coconut cadang-cadang viroid and other potato spindle tuber viroid related RNAs, *Virology*, 162, 1988, in press.

11a. **Keese, P., Keese, M. E., and Symons, R. H.**, unpublished data.

12. **Owens, R. A., Smith, R. D., and Diener, T. O.**, Measurement of viroid sequence homology by hybridization with complementary DNA prepared *in vitro*, *Virology*, 89, 388, 1978.

13. **Sano, T., Uyeda, I., Shikata, E., Ohno, T., and Okada, Y.**, Nucleotide sequence of cucumber pale fruit viroid: homology to hop stunt viroid, *Nucleic Acids Res.*, 12, 3427, 1984.

14. **Ohno, T., Takamatsu, N., Meshi, T., and Okada, Y.**, Hop stunt viroid: molecular cloning and nucleotide sequence of the complete cDNA copy, *Nucleic Acids Res.*, 11, 6185, 1983.

15. **Gross, H. J., Domdey, H., Lossow, C., Jank, P., Rabba, M., and Alberty, H.**, Nucleotide sequence and secondary structure of potato spindle tuber viroid, *Nature (London)*, 273, 203, 1978.

16. **Kiefer, M. C., Owens, R. A., and Diener, T. O.**, Structural similarities between viroids and transposable genetic elements, *Proc. Natl. Acad. Sci. U.S.A.*, 80, 6234, 1983.

17. **Diener, T. O.**, *Viroids and Viroid Diseases*, John Wiley & Sons, New York, 1979.

18. **Sänger, H. L.**, Minimal infectious agents: the viroids, in *The Microbe 1984: Viruses, Part 1, Soc. General Microbiology Symp. Ser. No. 36*, Mahy, B. W. J. and Pattison, J. R., Eds., Cambridge University Press, New York, 1984, 281.

19. **Diener, T. O.**, Viroids, *Adv. Virus Res.*, 28, 241, 1983.

20. **Riesner, D. and Gross, H. J.**, Viroids, *Annu. Rev. Biochem.*, 54, 531, 1985.

21. **Keese, P. and Symons, R. H.**, The structure of viroids and virusoids, in *Viroids and Viroid-Like Pathogens*, Semancik, J. S., Ed., CRC Press, Boca Raton, Florida, 1987, 1.

22. **Keese, P. and Symons, R. H.**, Molecular structure (primary and secondary) of viroids, in *The Viroids*, Diener, T. O., Ed., Plenum Press, New York, 1987, in press.

23. **Wilbur, W. J. and Lipman, D. J.**, Rapid similarity searches of nucleic acid and protein data banks, *Proc. Natl. Acad. Sci. U.S.A.*, 80, 726, 1983.

24. **Visvader, J. E., Forster, A. C., and Symons, R. H.**, Infectivity and *in vitro* mutagenesis of monomeric cDNA clones of citrus exocortis viroid indicates the site of processing of viroid precursors, *Nucleic Acids Res.*, 13, 5843, 1985.

25. **Visvader, J. E. and Symons, R. H.**, Replication of *in vitro*-constructed viroid mutants: location of the pathogenicity-modulating domain of citrus exocortis viroid, *EMBO J.*, 5, 2051, 1986.

26. **Schnölzer, M., Haas, B., Ramm, K., Hofmann, H., and Sänger, H. L.**, Correlation between structure and pathogenicity of potato spindle tuber viroid (PSTV), *EMBO J.*, 4, 2181, 1985.

27. **Duran-Vila, N., Flores, R., and Semancik, J. S.**, Characterization of viroid-like RNAs associated with the citrus exocortis syndrome, *Virology*, 150, 75, 1985.

28. **Cress, D. E., Kiefer, M. C., and Owens, R. A.**, Construction of infectious potato spindle tuber viroid cDNA clones, *Nucleic Acids Res.*, 11, 6821, 1983.

29. **Ishikawa, M., Meshi, T., Ohno, T., Okada, Y., Sano, T., Ueda, I., and Shikata, E.**, A revised replication cycle for viroids: the role of longer than unit length RNA in viroid replication, *Mol. Gen. Genet.*, 196, 421, 1984.

30. **Meshi, T., Ishikawa, M., Ohno, T., Okada, Y., Sano, T., Ueda, I., and Shikata, E.**, Double-stranded cDNAs of hop stunt viroid are infectious, *J. Biochem.*, 95, 1521, 1984.

31. **Owens, R. A., Hammond, R. W., Gardner, R. C., Kiefer, M. C., Thompson, S. M., and Cress, D. E.**, Site-specific mutagenesis of potato spindle tuber viroid cDNA: alterations within premelting region 2 that abolish infectivity, *Plant Mol. Biol.*, 6, 179, 1986.

32. **Tabler, M. and Sanger, H. L.**, Cloned single- and double-stranded DNA copies of potato spindle tuber viroid (PSTV) RNA and co-inoculated subgenomic DNA fragments are infectious, *EMBO J.*, 3, 3055, 1984.

33. **Branch, A. D. and Robertson, H. D.**, A replication cycle for viroids and other small infectious RNA's, *Science*, 223, 450, 1984.

34. **Hutchins, C. J., Keese, P., Visvader, J. E., Rathjen, P. D., McInnes, J. L., and Symons, R. H.**, Comparison of multimeric plus and minus forms of viroids and virusoids, *Plant Mol. Biol.*, 4, 293, 1985.

35. **Hutchins, C. J., Rathjen, P. D., Forster, A. C., and Symons, R. H.**, Self-cleavage of plus and minus RNA transcripts of avocado sunblotch viroid, *Nucleic Acids Res.*, 14, 3627, 1986.

36. **Ishikawa, M., Meshi, T., Okada, Y., Sano, T., and Shikata, E.**, *In vitro* mutagenesis of infectious viroid cDNA clone, *J. Biochem.*, 98, 1615, 1985.

37. **Tinoco, I., Uhlenbeck, O. C., and Levine, M. D.**, Estimation of secondary structure in ribonucleic acids, *Nature (London)*, 230, 362, 1971.

38. **Riesner, D., Henco, K., Rokohl, U., Klotz, G., Kleinschmidt, A. K., Domdey, H., Jank, P., Gross, H. J., and Sänger, H. L.**, Structure and structure formation of viroids, *J. Mol. Biol.*, 133, 85, 1979.

39. **Riesner, D., Colpan, M., Goodman, T. C., Nagel, L., Schumacher, J., and Steger, G.**, Dynamics and interactions of viroids, *J. Biomol. Struct. Dyn.*, 1, 669, 1983.

40. **Steger, G., Hofmann, H., Förtsch, J., Gross, H. J., Randles, J. W., Sänger, H. L., and Riesner, D.,** Conformational transitions in viroids and virusoids: comparison of results from energy minimization algorithm and from experimental data, *J. Biomol. Struct. Dyn.,* 2, 543, 1984.

41. **Nussinov, R. and Jacobson, A. B.,** Fast algorithm for predicting the secondary structure of single-stranded RNA, *Proc. Natl. Acad. Sci. U.S.A.,* 77, 6309, 1980.

42. **Zuker, M. and Stiegler, P.,** Optimal computer folding of large RNA sequences using thermodynamics and auxiliary information, *Nucleic Acids Res.,* 9, 133, 1981.

43. **Comay, E., Nussinov, R., and Comay, O.,** An accelerated algorithm for calculating the secondary structure of single stranded RNAs, *Nucleic Acids Res.,* 12, 53, 1984.

44. **Branch, A. D., Benenfeld, B. J., and Robertson, H. D.,** Unusual properties of two branched RNA's with circular and linear compounds, *Nucleic Acids Res.,* 13, 4889, 1985.

45. **Branch, A. D., Benenfeld, B. J., and Robertson, H. D.,** Ultraviolet light-induced crosslinking reveals a unique region of local tertiary structure in potato spindle tuber viroid and HeLa 5S RNA, *Proc. Natl. Acad. Sci. U.S.A.,* 82, 6590, 1985.

46. **Keese, P. and Symons, R. H.,** Domains in viroids: evidence of intermolecular RNA rearrangements and their contribution to viroid evolution, *Proc. Natl. Acad. Sci. U.S.A.,* 82, 4582, 1985.

47. **Imperial, J. S. and Bautista, R. M.,** Transmission of the coconut cadang-cadang viroid to six species of palm by inoculation with nucleic acid extracts, *Plant Pathol.,* 34, 391, 1985.

48. **Runia, W. Th. and Peters, D.,** The response of plant species used in agriculture and horticulture to viroid infections, *Neth. J. Plant Pathol.,* 86, 135, 1980.

49. **Goodman, T. C., Nagel, L., Rappold, W., Klotz, G., and Riesner, D.,** Viroid replication: equilibrium association constant and comparative activity measurements for the viroid-polymerase interaction, *Nucleic Acids Res.,* 12, 6231, 1984.

50. **Kozak, M.,** Inability of circular mRNA to attach to eukaryotic ribosomes, *Nature (London),* 280, 82, 1979.

51. **Holland, J., Spindler, K., Horodyski, F., Grabau, E., Nichol, S., and Van de Pol, S.,** Rapid evolution of RNA genomes, *Science,* 215, 1577, 1982.

52. **Maxam, A. M. and Gilbert, W.,** Sequencing end-labelled DNA with base-specific chemical cleavages, *Methods Enzymol.,* 65, 499, 1980.

53. **Zelazny, B., Randles, J. W., Boccardo, G., and Imperial, J. S.,** The viroid nature of the cadang-cadang disease of coconut palm, *Sci. Filipinas,* 2, 45, 1982.

54. **Randles, J. W.,** Coconut cadang-cadang viroid, in *Subviral Pathogens of Plants and Animals: Viroids and Prions,* Maramorosch, K. and McKelvey, J. J., Eds., Academic Press, New York, 1985, 39.

55. **Imperial, J. S., Rodriguez, M. J. B., and Randles, J. W.,** Variation in the viroid-like RNA associated with cadang-cadang disease: evidence for an increase in molecular weight with disease progress, *J. Gen. Virol.,* 56, 77, 1981.

56. **Mohamed, N. A., Haseloff, J., Imperial, J. S., and Symons, R. H.,** Characterization of the different electrophoretic forms of the cadang-cadang viroid, *J. Gen. Virol.,* 63, 181, 1982.

57. **Keese, M. E.,** personal communication.

58. **Imperial, J. S.,** personal communication.

59. **Imperial, J. S. and Rodriguez, M. J. B.,** Variation in the coconut cadang-cadang viroid: evidence for single-base additions with disease progress, *Philipp. J. Crop Sci.,* 8, 87, 1983.

60. **Jennings, P. A., Finch, J. T., Winter, G., and Robertson, J. S.,** Does the higher order structure of the influenza virus ribonucleoprotein guide sequence rearrangements in influenza viral RNA?, *Cell,* 34, 619, 1982.

61. **Sano, T., Ohshima, K., Uyeda, I., Shikata, E., Meshi, T., and Okada, Y.,** Nucleotide sequence of grapevine viroid: a grapevine isolate of hop stunt viroid, *Proc. Jpn. Acad.,* B61, 265, 1985.

62. **Forster, A. C. and Symons, R. H.,** Self-cleavage of plus and minus RNA transcripts of a virusoid and a structural model for the active sites, *Cell,* 49, 211, 1987.

63. **Forster, A. C. and Symons, R. H.,** Self-cleavage of virusoid RNA is performed by the proposed 55-nucleotide active site, *Cell,* 50, 9, 1987.

64. **Keese, P., Bruening, G., and Symons, R. H.,** Comparative sequence and structure of circular RNAs from two isolates of lucerne transient streak virus, *FEBS Lett.,* 159, 185, 1983.

65. **Francki, R. I. B.,** Plant virus satellites, *Annu. Rev. Microbiol.,* 39, 151, 1985.

66. **Robertson, H. D., Rosen, D. L., and Branch, A. D.,** Cell-free synthesis and processing of an infectious dimeric transcript of potato spindle tuber viroid RNA, *Virology,* 142, 441, 1985.

67. **Meshi, T., Ishikawa, M., Watanabe, Y., Yamaya, J., Okada, Y., Sano, T., and Shikata, E.,** The sequence necessary for the infectivity of hop stunt viroid cDNA clones, *Mol. Gen. Genet.,* 200, 199, 1985.

Gene Reassortment and Evolution in Segmented RNA Viruses

Chapter 5

GENETIC DIVERSITY OF MAMMALIAN REOVIRUSES

Elizabeth A. Wenske and Bernard N. Fields

TABLE OF CONTENTS

I. INTRODUCTION

The mammalian reoviruses (orthoreoviruses) belong to the Reoviridae family, viruses which have a double-stranded segmented RNA genome. The Reoviridae are ubiquitous in their geographic distribution and host range;[1-4] whales and monotremes are the only mammals examined in which neither virus nor antibody has been found.[3] There are three mammalian reovirus serotypes (type 1, 2, and 3) that are identified based on neutralization and hemagglutination-inhibition tests.[5-7] RNA hybridization studies have shown that serotypes 1 and 3 are 70% related to each other, whereas serotype 2 is only 10% related to either type 1 or type 3.[8,9] Viruses from the three serotypes are morphologically identical, but the homologous genes and proteins from viruses of different serotypes differ in their mobility on polyacrylamide gels.[10,11]

Although structurally quite similar, the three serotypes interact differently with mammalian hosts producing distinct patterns of disease. Reoviruses have been studied extensively in mice because a wide variety of disease processes can be experimentally produced. This variety provides a useful model for studying the genetic basis for pathogenesis of viral disease. Spontaneously, reovirus type 3 (Dearing) causes an infection in mice consisting of diarrhea, oily hair, growth retardation, jaundice and ataxia. When inoculated into new born mice, all three reovirus serotypes can cause acute infections. The resulting clinical syndromes vary depending upon the amount of virus inoculated, the serotype, the route of infection, as well as the age and strain of the mouse.[12-15]

The theme of this chapter is genetic diversity among the mammalian reoviruses. We will consider (1) the structure of reoviruses; (2) the function of reovirus genes encoding outer capsid proteins; and (3) genetic diversity among various isolates and how it is being defined at the nucleic acid level. Because this chapter is focused on genetic diversity, the reader is referred elsewhere for in-depth reviews of reovirus pathogenesis, virus-cell interaction, structure, replication, and genetics.[16-20]

II. REOVIRUS STRUCTURE

The reovirus virion consists of a segmented, double-stranded RNA genome surrounded by a closely applied inner icosahedral protein shell (core) enclosed in a second concentric icosahedral outer protein capsid.[21,22] The ten segments of the genome are divided by molecular weight into three classes:[23-26] large (segments L1, L2, and L3), medium (segments M1, M2, and M3), and small (segments S1, S2, S3, and S4). Each double-stranded RNA segment encodes a unique messenger RNA that is translated into at least one polypeptide.[27-30] Nine of the mRNAs appear to be monocistronic, while the S1 mRNA has been shown to be bicistronic. The bicistronic nature of the S1 mRNA was suggested by the nucleotide sequences which indicate two overlapping reading frames.[31-34] A 5′ proximal AUG is followed by an open reading frame which codes 418, type 1,[32] 399, type 2,[32] and 455, type 3,[31-34] amino acids. The second AUG from the 5′ end is followed by a short open reading frame encoding 119, type 1,[32] 125, type 2,[32] and 120, type 3,[31,33-35] amino acids.

Like the genome segments that encode them, the viral proteins are grouped by molecular weight into large (lambda), medium (mu), and small (sigma) classes. The L genome segments encode lambda polypeptides 1, 2, and 3; the M segments encode mu polypeptides 1, 2, and NS (nonstructural protein); and the S segments encode sigma polypeptides 1, 2, NS, 3, and a newly described smaller nonstructural polypeptide.[29,33,35-37]

Sakar et al., have described the synthesis of a small peptide, designated σ_s, from the S1 mRNA.[35] A truncated clone of the S1 gene minus the first 5′ proximal AUG was expressed in an in vitro protein system to yield an approximate 13-kdalton polypeptide. A similar molecular weight polypeptide is present in reovirus-infected cells. Comparative trypic peptide analysis of the two polypeptides synthesized in vivo and in vitro showed them to be identical.[35] Jacobs and Samuel[36] have also reported that the S1 mRNA of reovirus serotype 1 (Lang) and 3 (Dearing) encodes both the well-characterized hemaglutinin protein (σ1) and a smaller polypeptide designated p12. The p12 polypeptide was synthesized in vitro in L-cell-free protein synthesizing systems and in vivo in L-cell cultures infected with either serotype 1 or 3 reovirus.[36] In addition, a second polypeptide designated p14 was identified by Ernst and Shatkin.[33] This polypeptide was synthesized in vitro by rabbit reticulocyte and wheat germ protein synthesizing systems in response to hybrid selected reovirus S1 mRNA. p14 is also detected in reovirus-infected L cells. The smaller peptide described by the three independent laboratories is thought to be the same polypeptide.

The outer capsid of the virus consists of three polypeptides; sigma 1, sigma 3, and mu 1C.[22,28,29,38] The mammalian reoviruses are divided into serotypes 1, 2, and 3 on the basis of hemagglutinin-inhibition and antibody neutralization tests.[5] The serotype-specific protein is the viral hemagglutinin (σ1 protein) encoded by the S1 genome segment.[38,39] The mu 1C polypeptide is derived from mu 1 by proteolytic cleavage.[28,40] Lactoperoxidase-labeling and monoclonal antibody studies have shown that the lambda 2 polypeptide in the viral core is also exposed on the outer surface of the virus.[41]

Other polypeptides in the core including the major core proteins (λ1, λ2, σ2) and the minor core proteins (λ3, μ1, μ2). λ1 to λ3 may play a role or be components of transcriptase activity.[41a,41b] λ1 and λ2 together may contain the catalytic site of the viral transcriptase.[41b] A small protein, "component viii" has also been described.[22] It is suggested that this protein is the fragment (8000 Mr) produced after cleavage of μ1-μ1c,[17] although no direct evidence is yet available.

III. FUNCTION OF OUTER CAPSID PROTEINS

The life cycle of a virus can arbitrarily be defined as beginning with release from a host, survival in the environment, and transmission to the next host. Following ingestion into the host, many viruses undergo a complex series of steps beginning with survival at the portal of entry, followed by 1° replication at or near the portal of entry, spread through the host, and targeting to selected cells and tissues (tropism). Ultimately, if the host plays a central role in the life cycle of the virus, the virus must find a portal for exit in order for the virus to infect a new host and continue its survival. This complexity mandates that viruses have the capacity to interact with, and adapt to, a wide variety of environments. For example, it must be stable to changes in temperature, pH, salts, as well as remaining infectious in the face of various nonspecific inhibitors that it encounters in the host (mucous, enzymes, fats, etc.). It is perhaps not surprising that, for the reoviruses, viruses that have an outer protein shell surrounding an inner shell (or "core"), almost all of the proteins whose functions have been found to be related to interaction with the environment and the extracellular sites in the host, are proteins found in the outer capsid. Once in the cytoplasm, the conditions are more constant and the viral components responsible for replication include core proteins. This section will emphasize the various properties of the outer capsid proteins (σ1 [viral hemagglutinin], μ1C, σ3, plus λ2) since their roles in generating diversity are more important and better understood than the core proteins.

The σ1 protein is the reovirus protein that confers serotype specificity and is the target of serotype-specific neutralizing and hemagglutination-inhibiting antibody.[42-44] The σ1 protein also plays a role in specific recognition by the immune system as a major target of cytolytic T lymphocytes (CTL) and other types of T cells (T_{DTH}, T_S).[45] In addition, σ1 causes inhibition of cellular DNA synthesis[46] and binding to cellular microtubules[47] and defines tissue tropism and virulence.[48]

The μ1C polypeptide partially defines the extent to which different isolates grow following inoculation of virus in the gastrointestinal tract.[49] Likewise, when reovirus type 3 isolates are compared, differences in yield in the nervous system and relative neurovirulence reside in the M2 gene.[50] The M2 gene product plays a role in generating suppressor T cell following peroral inoculation of virus.[51]

The σ3 protein (S4 gene) is responsible for differences in the efficiency of inhibition of host cell RNA and protein synthesis.[52] The σ3 protein also plays a role in initiation of persistent infection in cell cultures.[53-55]

The L2 dsRNA segment encodes the λ2 spike protein of reovirus.[56] The spike is composed of pentamers of λ2 that form hollow tubes stretching from the viral core to the vertices of the icosahedral outer capsid.[57] Removal of the λ2 protein eliminates the RNA transcriptase activity of isolated cores and ultrastructural studies suggest that messenger RNA is actively extruded through the spikes during RNA transcription.[41,58] The λ2 protein also serves a major structural role in the virus. Temperature-sensitive mutants of L2 show defective assembly at the nonpermissive temperature.[59] The L2 gene also governs the generation of defective virus in high passage stocks in vitro. Reassortants bearing the L2 dsRNA segment of type 3 generate deletions at high frequency in high-passage cell cultures, while those bearing the L2 genes of type 2 do not.[60]

IV. GENETIC DIVERSITY OF REOVIRUSES

The mammalian reoviruses contain a segmented dsRNA genome. Diversity has been shown to be generated in two ways: genome segment reassortment and mutation. Genome segment reassortment occurs in vitro and in vivo. Similar to other RNA viruses, mutation occurs at high frequency. Mutants have been generated in cell cultures, especially in persistent infection. The role of mutants appearing in the animal host is less well defined. At the present time there is no evidence that reovirus RNA undergoes true ("breakage and rejoining") recombination. Thus, reoviruses have served as a model system for studying reassortment and mutation. An important goal of genetic studies is to predict how genetic diversity affects virus-host interaction.

The advantages of using the reoviruses to analyze genetic diversity can be summarized as follows:

1. The mammalian reoviruses serotypes 1 (Lang) and 3 (Dearing) differ in a variety of biologic properties which are identifiable at a genomic level, i.e., (a) spread, (b) tropism, (c) transmission.
2. Reassortants of reoviruses are readily generated and have been isolated from coinfected tissue culture cells and laboratory mice.
3. A variety of mutants have been isolated and analyzed.

A. Types of Mutants and Strategy of Reversion
A variety of mutants have been identified for the reoviruses. In addition, there has been extensive analysis concerning the nature of reversion of the mutants.

1. Temperature-Sensitive Mutants
Different reovirus mutants have been identified and have been useful in investigating genetic interactions, morphogenesis, pathogenesis, and the physiology of reovirus. Conditional lethal, temperature-sensitive (*ts*) mutants of reovirus type 3 have been generated by chemical mutagenesis.[61,62] Subsequently, *ts* mutants were isolated from serial high-MOI passage stocks of wild-type virus and from "pseudorevertants" isolated from persistently infected L cells.[18,63-65] *Ts* mutants have been isolated for each of the 10 possible reassortant groups,[18,61,63-65,67,68] and the *ts* group mutations have been assigned to discrete dsRNA segments.[18,29,66,68,69]

Early experiments with *ts* mutants were directed at determining whether the dsRNA segments behave as *independently reassorting genetic units*. The generation of *ts*(+) progeny at high frequency following mixed infection of cells with certain pairs of *ts* mutants at 31°C (the permissive temperature) strongly suggested that reovirus exchange genetic material via reassortment.[61] Those mutant pairs which did not generate *ts*(+) progeny were later confirmed to contain *ts* lesions on the same gene.[65,66]

2. Other Mutant Markers

Other genetic markers have been utilized to confirm the existence of reovirus reassortment. A μ1-μ1c polypeptide complex was identified that showed an aberrant electrophoretic migration. The aberrant polypeptide complex was identified in several of the *ts* mutants.[70,71] Except for the *ts* A201 lesion, the aberrant polypeptide was independent of the *ts* mutations with which it was associated. Therefore, it proved useful as an unselected marker in three factor crosses. Through the use of three factor genetic crosses, convincing evidence was provided to support the concept that reovirus RNA segments underwent "reassortment" during viral replication.

3. Deletion Mutants

In order for viral multiplication to occur, the infecting reovirus particle requires a full complement of ten genome segments. However, deletion mutants have been generated upon serial passage in tissue culture[72] that lack one or more genome segments. Both *ts* and small-plaque/low-yield mutants have also been rescued from deletion-mutant stocks. Pure populations of deletion mutants are difficult to obtain because they require the presence of a helper virus for propagation. However, pure stocks are obtained by complementation with a *ts* helper virus.[73] In addition, Ahmed and Fields found that the denser fractions of a gradient were enriched for deletion mutants.[74] Thus, they were able to generate a relatively pure population of deletion mutants through several cycles of density-gradient centrifugation.

Studies using deletion mutants are useful for defining gene product function. Using deletion mutants, the S4 gene has been shown to play a critical role in the establishment of persistent infection in L cells.[53] After coinfecting L cells with a wild-type helper virus and a defective virus, analysis of the genomic double-stranded RNA pattern of the virus population selected during persistent infection revealed that the S1 gene was derived from the defective parent in three independently established lines, suggesting that the S4 gene plays an essential role in the establishment of persistent infection. The S4 gene derived from the defective virus is selected in lytic infections, but selection of reassortant containing the mutant S4 gene is specific for persistent infection. Thus, although defective virions may contain multiple defects, the viral components responsible for initiating persistence may be highly specific.[53,63]

4. Extragenic Suppression

Mutations carried on separate RNA segments may suppress each other such that reversion of a mutant virus to a wild-type phenotype (pseudorevertant) may result (extragenic suppression). If a *ts*(−) mutant is suppressed by a mutation on another segment, the mutant phenotype can be restored by segregating the segment from the suppressor mutant gene. This is accomplished by backcrossing the revertant with wild-type virus and selecting for the original mutant.[75]

Of 28 *ts* to *ts*+ revertants examined, 25 were suppressed pseudorevertants, indicating that extragenic suppression is the major reversion pathway in reovirus *ts* mutants.[75,76] Thus, extragenic suppression in the segmented RNA viruses provides a means to overcome the effects of deleterious mutations, and since the reassortment mechanism is efficient, the suppression mutations can quickly spread throughout the population.

In summary, studies have shown that reovirus can expand their genetic diversity through exchange of genetic material (reassortment) and/or mutations. Genetic diversity can be increased further by the interaction of more than one mutation located on discrete gene fragments.

B. Diversity in Nature and in Animal Hosts

Most genetic analyses have utilized viral mutants and reassortants generated in vitro (cell culture) and started with prototype strains. Such strains as the reovirus T3 Dearing have been passaged multiple times and are thus significantly altered by passage. The following sections summarize approaches to analyze diversity among isolates that have not been as extensively passaged as well as studies that have concentrated on the impact of viral genetic changes on virus-host interactions.

1. Natural Isolates

In order to examine genetic diversity among reoviruses found in nature, advantage has been made of a series of 94 isolates originally isolated in the late 1950s from humans, cattle, and mice. These isolates were eventually serologically classified by either hemagglutination inhibition or neutralization.[5,77] These isolates have provided a useful collection for studies on genetic diversity among natural strains. The isolates showed extensive variability in the patterns of migration of the ten double-stranded RNA genome segments ("electropherotypes") (Figure 1).[77] Isolates showing the highest degree of RNA segment heterogeneity often were isolated from widely different geographic regions or different mammalian hosts, whereas in some cases, samples from a single species from the same area over a period of time showed more limited variations. However, there were often multiple genetic variants of a single serotype present in a population. In addition, the gene products exhibited a high degree of electrophoretic mobility both among and within serotypes.[77]

As has been discussed above, reovirus has available different mechanisms for altering its genetic material. Determining the basis for heterogeneity is complicated by the fact that only the S1 gene product has an identifiable serotype specific marker. Since serologic markers for the polypeptides encoded by the other nine viral genes have not been found, alternative approaches have been needed to determine the extent that reassortment, point mutations, deletions, or insertions in viral RNA are responsible for the heterogeneity of reovirus.

One of the approaches has been to examine selected outer capsid proteins by analyzing tryptic peptide digests of polypeptides purified from the laboratory strains and natural isolates.[78,79] The results show that the outer capsid polypeptides of reoviruses isolated from different mammalian species have both conserved and unique methionine-containing tryptic peptides. For example, tryptic peptides from μ1C polypeptides of 5 mammalian type 3 isolates and 3 mammalian type 1 isolates were highly conserved, as only one tryptic peptide pattern was observed. However, the μ1C polypeptide of the type 3 Dearing strain contained one tryptic peptide not found in any other reovirus isolate examined. Although the majority of the tryptic peptides were conserved,[79] the reason for the distinct differences in μ1C-mediated biological properties[50] or the differences in electrophoretic migration rates in acrylamide gels, despite the extreme conservation of tryptic peptide structure, is not known. It is suggested that amino acid substitutions resulting from base changes could cause differences in electrophoretic migration or biological function and not be detected by tryptic peptide patterns. Correlation of M2 gene sequence with biological function of the μ1C polypeptide will help to determine how base changes may cause differences between the μ1C polypeptide functions.

In contrast to the high degree of conservation of the μ1C polypeptide, σ3 polypeptides from human and bovine strains contain a significant number of conserved, as well as unique, tryptic peptides. The methionine-containing tryptic peptides of the σ3 polypeptides from bovine isolates fall into two patterns.[79] One type 3 clone, 18, and two type 1 clones, 28 and 50, have peptide maps more closely related to each other than to their prototype strains

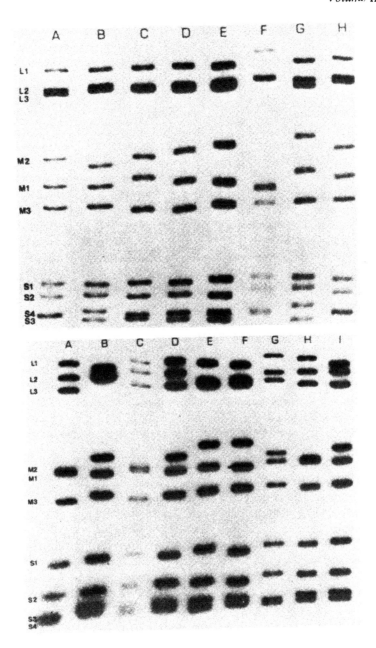

FIGURE 1. ³²P-labeled dsRNA from type 1 and type 3 reovirus isolates from humans, cattle, and mice, showing the extensive variability in patterns of migration of the ten dsRNA genome segments. Electrophoresis is from top to bottom in a single 10% acrylamide gel containing the Laemmli *tris*-glycine buffer system. Top: type 1 reovirus isolates: (A) type 1 (Lang); (B) human type HT1a, TM; (C) human type HT1b, WA; (D) human type HT1c, WA; (E) bovine type BT1a, CP; (F) bovine type BT1b, CP; (G) bovine type BT1c, JE; (H) type 1 (Lang). The RNA genome segments of type 1 (Lang) are identified. Capital-letter abbreviations denotes sites of isolation: (TM) Toluca, Mexico; (WA) Washington, D.C.; (CP) College Park herd; (JE) Jessup herd. Bottom: Type 3 reovirus isolates: (A) Type 3 (Dearing); (B) human type HT3a, TA; (C) human type HT3b, WA; (D) human type HT3c, WA; (E) bovine BT3a, CP; (F) bovine type BT3b, CP; (G) murine type MT3a, France; (H) Type 3 (Dearing); (I) bovine type BT3c, JE. RNA genome segments of type 3 (Dearing) are identified. (TA) Tahiti. (From Hardy, D. B., Rosen, L., and Fields, B. N., *J. Virol.*, 31, 104, 1979. With permission.)

type 1 Lang and type 3 Dearing. However, the σ3 tryptic peptide digests of another type 3 isolate, 31, are similar to that of the laboratory strain type 1 Lang (Table 1). Thus, it is likely that reassortment in nature is occurring, allowing for similar polypeptides to be found in virus particles irrespective of serotype.

2. In Vivo Reassortment

In light of the data demonstrating the heterogeneity of genomic RNA and viral proteins of reoviruses in nature, and the experimental data demonstrating reassortment in tissue culture cells, an experiment was designed to verify that reassortment could occur in laboratory mice coinfected with 2 reovirus serotypes. The results indicate reassortant viruses are recovered from animals coinfected by the oral route with both type 1 and type 3 reoviruses.[80] The RNAs of 1276 plaque-purified clones were examined from samples collected on days 1 through 13 postinfection. Before day 7, type 1 was the only virus isolated from the four tissues examined (ileum, spleen, liver, and brain). At later times (days 9, 11, and 13), type 3 virus was also isolated from spleen or liver samples. Beginning on day 7, reassortants were isolated, and 121 reassortant clones were isolated from the total sample of 1276 clones examined. The 121 reassortant clones comprised five different electropherotypes isolated from mouse organs (designated E1 through E5).[80]

The reassortants isolated do not represent the varied spectrum of electrophenotypes typically found when reoviruses reassort in vitro (cell cultures). Instead, the majority of the isolates have one of two electrophenotypes (E1 or E2, Figure 2). In both reassortants the majority of the RNA segments are derived from the type 1 parent. However, the M2 and S1 segments of E1 and the S1 and S2 segments of E2 are derived from type 3.

Thus, the finding of in vivo reassortment under laboratory conditions gives more evidence to suggest that reassortment is one explanation for the heterogeneity of reoviruses in nature, contributing to variability of pathogenicity among different reovirus isolates.

3. Determination of Genetic Diversity by Analyses of Gene Sequences

In order to fully understand the ramifications of genetic diversity it will be advantageous to compare sequence data of the genes from a large number of natural isolates and correlate the nucleotide sequence with the amino acid sequence of the gene product. It is important to know if nucleotide alterations, as reflected in amino acid sequences, will allow us to correlate sequence changes with alternations in function. Sequence data are being accumulated especially with regards to the S1 genes of all three serotypes.[31,32,34,35,79,81,82] The results of Cashdollar et al.[32] indicate that the serotype 1 and 2 S1 genes are more related (28% homology) to each other than to the serotype 3 S1 gene. The serotype 3 S1 gene is related to type 1 and type 2 S1 by 5 and 9%, respectively.[32]

Correlation of RNA sequence to functions of the viral proteins is being examined directly by comparing sequence and gene function of different type 3 S1 mutants with the prototype 3 (Dearing) S1 wild-type gene. A single base change in the S1 gene correlates with a change in neurovirulence.[83] This will be discussed in greater detail in a later section. In addition, Fields et al. (unpublished data) are currently sequencing S1 genes of natural isolates in order to correlate sequence relatedness and function in natural populations.

4. Genetics of Host-to-Host Transmission

Reovirus differ in their transmissibility among laboratory mouse littermates. Serotype 1 (Lang) and 3 (Dearing), standard laboratory strains, show marked differences in transmission. When a dose of 10^5 plaque-forming units of type 1 is given to two members of a litter (five litters tested) 57 ± 10% of the littermates are infected.[84] Under the same conditions, only one instance of transmission (3.7 ± 6%) is observed in type 3 (Dearing)-infected mice. To test whether transmission is dependent on serotype, a murine reovirus type 3 isolate (clone

Table 1
COMPARISON OF THE CONSERVED AND UNIQUE METHIONINE-CONTAINING PEPTIDES IN THE σ3 POLYPEPTIDES OF SEVERAL HUMAN, BOVINE, AND MURINE STRAINS OF REOVIRUS SEROTYPES 1 AND 3

Serotype	Origin	Presence[a] (+) or absence (−) of prototype peptide[b]													New σ3 peptide				
		1	2	3	4	5	6	7	8	9	10	12	14	15	A	B	C	D[c]	E
3 (Dearing)	Human	+	+	−	−	−	−	+	+	+	+	+	+	+	−	−	−	+	+
3 (clone 18)	Bovine	+	+	+	+	−	−	?	+	+	+	−	−	−	+	+	+	+	+
3 (clone 31)	Bovine	+	+	+	+	+	+	?	+	+	+	−	−	−	−	−	−	+	+
1 (Lang)	Human	+	+	+	+	+	+	+	+	+	+	−	−	−	−	−	−	+	+
1 (clone 28)	Bovine	+	+	+	+	−	−	?	−	+[d]	+	−	−	−	+	+	+	+	+
1 (clone 50)	Bovine	+	+	+	+	−	−	?	+	+	+	−	−	−	+	+	+	+	+

[a] σ3 peptides 1, 2, and 8 to 10 were shared among all strains examined.

[b] The peptide numbers were adapted from a previous report from this laboratory,[78] using type 3 Dearing and type 1 Lang as reference strains.

[c] Peptides D and E were faint in our original preparations[78] and were not numbered.

[d] Spots 9 and 10 of clones 28 and 50 showed a slight migration difference.

From Gentsch, J. R. and Fields, B. N., *J. Virol.*, 49, 641, 1984. With permission.

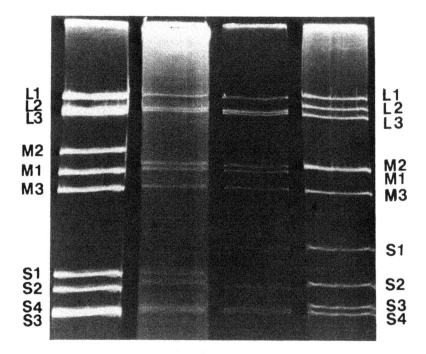

FIGURE 2.   Electropherotypes of parental viruses and reassortants. Extracts of cells infected with type 1, type 3, 4F10 (E2) or 8B18 (E1) were electrophoresed in 10% polyacrylamide gels and the dsRNA was stained with ethidium bromide as previously described.[77] RNA segments of parental type 1 (Lang) and type 3 (Dearing) are identified on the left and right sides of the gels, respectively.

9) was tested. In contrast to serotype 3 (Dearing), clone 9 is highly transmissible (Figure 3). Thus the property of transmission examined is not serotype specific.[84]

Using reassortants between reovirus serotypes type 1 (Lang) and type 3 (Dearing), the L2 gene has been identified as the gene that is responsible for these differences in transmission between mice and for the enhanced secretion into feces.

In the studies that determined the genetic basis for differences in transmission, mice were inoculated with reassortants composed of a mixture of genome segments derived from type 1 (Lang) and type 3 (Dearing). The reassortants were grouped into two categories, those that were capable of being transmitted between littermates and those that were not. In each case of positive transmission, the L2 dsRNA segment of the reassortant virus was derived from the type 1 parent. Those reassortants containing the L2 dsRNA segment of type 3 failed to transmit, regardless of the presence of any other type 1 dsRNA (Table 2). Furthermore, the analysis of reassortants demonstrated that high shedding ($>10^5$ p.f.u.) from the gastrointestinal tract correlated with the presence of the L2 segment of type 1. If the L2 was derived from type 3, the reassortant was shed in lower amounts ($<10^5$ p.f.u.). One reassortant containing the L2 gene of type 1 was not shed in titers above 10^6 indicating some type 3 dsRNA segments may attenuate the effect of the type 1 L2.[84]

In summary, differences in transmission between mice correlate with the L2 gene segment. Furthermore, those viruses shed in large amounts are transmitted at a higher frequency.

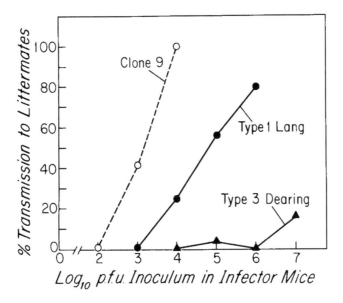

FIGURE 3. Frequency of reovirus transmission among members of a litter as a function of inoculum given to infected mice. In standard transmission experiments (five replications), one litter of eight to ten suckling mice of the NIH/Swiss strain was placed in each cage. At age 1 to 2 days, two members of the litter were infected by oralgastric intubation with a dose of virus in 30 $\mu\ell$ of gelatin saline, as previously described.[49] These "infector mice" were designated by clipping the tail and were replaced in the litter. At 10 days, the littermates were killed and their intestines were removed (the site of primary replication for reovirus).[86] Intestines were placed in 1 mℓ of gelatin saline, frozen and thawed three times, and sonicated. The sonicate was diluted from 10^{-1} to 10^{-5}, and titers of virus were determined in the plaque assay previously described.[71] Because of the toxicity of the intestinal contents to the L cells used in the assay, the 10^{-1} dilution wells were difficult to interpret. However, the assay detected the presence of virus $\geq$ 1000 PFU per intestine. Presence of virus in the intestines of the littermates was scored as (+) transmission and its absence as (−) transmission. From reference 84, by permission of Science.

Transmission in this case is a property associated with amount of virus present in the stool, rather than susceptibility of the host.

5. Viral Entry and Spread Within the Host

The upper alimentary tract is the natural portal for reoviruses; however, once they have entered their hosts, reoviruses differ as to their mode of spread. Reovirus type 1 penetrates the intestinal epithelium through M cells, a population of specialized epithelial cells that overlies the lymphoid tissue of Peyer's patches. Soon after penetration, reovirus type 1 appears within Peyer's patches.[85,86] In neonatal mice inoculated perorally, type 1 reovirus grows well in intestinal tissue, whereas the titer of type 3 progressively falls.

In both neonatal and adult mice, type 1 reovirus is initially found in Peyer's patches and then spreads sequentially to mesenteric lymph nodes and spleen. Type 3 reovirus also enters Peyer's patches in adult and neonatal mice, but rapidly loses infectivity and spreads only in young mice, reaching just the mesenteric lymph node.[85] The S1 gene, which encodes the viral hemagglutinin, determines the pathway of spread of reovirus from Peyer's patches to mesenteric lymph nodes in young and adult mice, probably for reovirus type 1 by the lymphatic route.[85] Recent studies suggest that lymphocytes may transport a large proportion of type 1 reovirus to mesenteric lymph nodes.

Table 2
PATTERN OF TRANSMISSION AND PATTERN OF SPREAD TO THE CNS OF T1, T3 AND REOVIRUS REASSORTANTS

| Virus | Origin of genome segment | | | | | | | | | | Transmission[a,b] (littermates infected/total) | Pattern of spread[c] |
| | Capsid polypeptides | | | Core polypeptides | | | | | Nonstructural polypeptides | | | |
	S1	M2	S4	L1	L2	L3	M1	S2	M3	S3		
T1	1	1	1	1	1	1	1	1	1	1	17/30 (57%)	Hematogenous
T3	3	3	3	3	3	3	3	3	3	3	1/27 (3.7%)	Neural
Reassortants 121,258,143	3	1	1	3	1	1	1	1	1	1	10/22 (45%)	Neural
144	1	3	1	1	1	1	1	1	3	3	7/19 (37%)	Hematogenous
H24	1	1	3	3	3	1	1	1	1	1	6/20 (30%)	Hematogenous
126	1	3	1	3	3	3	3	3	1	3	0/18	Hematogenous
145	1	3	3	3	3	3	3	3	1	3	0/6	Hematogenous
G2	3	1	1	1	3	1	1	3	1	1	0/13	Neural

[a]   Standard transmission experiments were conducted with one litter per cage of eight to ten suckling mice of the NIH/Swiss strain. 10^5 p.f.u. of virus in 30 $\mu\ell$ of gelatin saline was given to two members of the litter at 1 to 2 days after birth (the infectors). All mice were killed at 7 days (eight experiments) or 10 days (45 experiments). The day of death did not affect the result. Titers of virus were determined for the intestine, and transmission was scored.

[b]   Data from Reference 84.

[c]   Data from Reference 87.

Both reovirus type 1 and type 3 can infect the central nervous system (CNS), but they seek out different cell types (described below). Although the Dearing strain of reovirus type 3 is avirulent after peroral inoculations, it is highly neurovirulent if administered parenterally or intracerebrally and can cause acute and fatal encephalitis in newborn mice. Viruses with reassorted genome segments that contain a protease-resistant M2 gene derived from type 1 and an S1 gene from type 3 are neurovirulent after oral inoculation. Presumably, the M2 gene from type 1 permits virus to grow and spread, and the S1 gene from type 3 permits neural entry.[50,51]

Recent studies indicate that reovirus serotypes 1 and 3 spread by different pathways to reach the CNS after peripheral inoculation into newborn mice. In order to differentiate the pathways of spread, advantage was taken of the fact that motor and sensory neurons innervating the hindlimb and forelimb are located in completely different regions of the spinal cord. Thus, if type 3 spread via nerves, it should appear first in the spinal cord in the region containing the neurons innervating the skin and musculature at the site of viral inoculation. Conversely, if type 1 spreads through the bloodstream, it should appear in all regions of the spinal cord with similar kinetics.

Three approaches were used to look at the spread of reovirus to the CNS.[87] First, the animals were inoculated peripherally in either the forelimb or hindlimb and sections of the spinal cord titered for amount of virus present at various times post inoculation. Second, animals were inoculated in their hindlimb after the sciatic nerve had been cut to determine if this affected the spread of type 1 and/or type 3. Third, animals were treated with colchicine, known to interrupt fast axonal transport, prior to inoculation. It was postulated that if the virus spread via fast axonal transport, the colchicine would slow down or inhibit spread. The results of the first experiment are shown in Figures 4 and 5. After hindlimb or forelimb inoculation, type 3 appeared first in the region of the spinal cord innervating the injected limb; whereas after either hindlimb or forelimb inoculation, type 1 appeared at the same time and in equivalent titer in all regions of the spinal cord. The spread of type 3 and type 1 from the hindlimb to the spinal cord was next studied in neonatal mice following section of the sciatic nerve. Nerve section contralateral to the injected limb had no effect on the spread of type 3; however, sciatic nerve section completely inhibited spread of type 3 to the spinal cord if the inoculation was in the limb sectioned. The spread of type 1 was not affected by sciatic nerve section. Thus, these experiments support the hypothesis that type 3 spreads to the CNS through nerves, and that type 1 spreads through nonneural hematogenous pathways. The rapidity with which type 3 spread to the spinal cord suggested that the virus spread via the fast axonal transport (Figure 4). Colchicine was therefore used to identify the transport mechanism involved for the spread of type 3. Colchicine interrupts fast axonal transport by causing reversible dissociation of microtubules into their component tubulin monomers.[88] In order to assess the effect of colchicine, animals were pretreated with the drug prior to inoculation. In colchicine-treated mice there was an inhibition of the spread of type 3 to the spinal cord, which lasted for at least three days (Figure 6). However, multiple doses of colchicine produced a more profound inhibition of spread of type 3. Thus, the inhibition of spread in colchicine-treated animals supported the hypothesis that type 3 uses the fast axonal transport mechanism for neural spread. Furthermore, colchicine does not affect the growth of type 3 in the hindlimb site of inoculation of treated animals as compared to untreated controls. Also, colchicine has no effect on the spread of type 1. These experiments indicate that reovirus type 3 and type 1 use distinct pathways to spread to the spinal cord following footpad inoculations in neonatal mice.

By analyzing eleven reassortant viruses, it was possible to determine that the S1 gene segment is responsible for the pattern of spread. All reassortant viruses containing a type 3 S1 dsRNA segment show a neural pattern of spread to the spinal cord, whereas all reassortants containing a type 1 S1 dsRNA segment show a hematogenous pattern of spread to the spinal

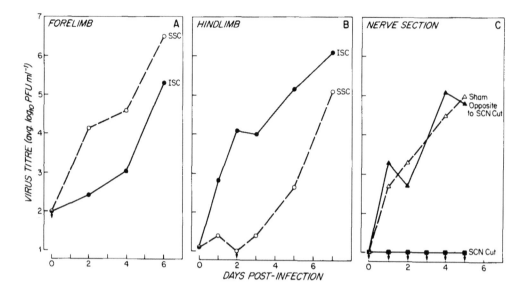

FIGURE 4. Pattern of spread of reovirus type 3 to the spinal cord of neonatal mice after inoculation of the virus into either the forelimb (A) or hindlimb (B) footpad. (O) Superior spinal cord (SSC); (●) inferior spinal cord (ISC). (C) Sciatic nerve section in limb ipsilateral (■) or contralateral (▲) to virus inoculation. (△) Sham operation. A downward-pointing arrow indicates that no virus was detected at the lowest dilution shown. (A-C) A Hamilton syringe equipped with a 30-gauge needle was used to inoculate 1-day-old NIH Swiss mice with 2.2×10^5 to 4.5×10^5 PFU of virus in a volume of 0.010 mℓ. Spinal cord blocks were removed by microdissection with the aid of a stereomicroscope. Superior spinal cord blocks (SSC) included the cervical cord segments, and inferior spinal cord blocks (ISC) included the thoracic and lumbosacral cord segments. Virus titers were determined by plaque assay on mouse L cells.[71] Specimens were titered in duplicate at serial tenfold dilutions. Plotted points represent the average of the $\log_{10}$ titer (measured in plaque-forming units per milliliter) of 9 to 14 specimens (days 0, 1, 3, and 5) or of 4 to 7 specimens (days 2 and 7). The standard error or the mean (SEM) is less than 0.3 for ISC and less than 0.5 for SSC at all time points. (C) Animals, method, and dose of virus and procedure for titration of tissue blocks were as described above. Sciatic nerve section was performed on anesthetized (methoxyflurane, Pitman-Moore) 1-day-old mice by microsurgical techniques. Adequacy of section was demonstrated by the post-operative finding of hindlimb analgesia and paresis. Virus was inoculated 24 hr after nerve section of sham operation. Sham operation was identical to the operation for nerve section, except the nerve was not cut. Each plotted point represents the $\log_{10}$ titer of duplicate assays of the ISC tissue block from a single animal. The SEM was less than 0.5 for controls at all time points. (From Tyler, K. L., McPhee, D. A., and Fields, B. N., *Science*, 233, 770, 1986. With permission.)

cord (Table 2). The parental origin of the other 9 segments does not effect the pattern of spread. Thus, these results indicate that the reovirus S1 dsRNA segment determines the capacity of reoviruses to spread to the CNS via distinct pathways.

6. Tissue Tropism and Virus-Receptor Interactions

Once a virus reaches a target tissue there must be a mechanism for recognition and uptake. Viral tropism, the specificity of a virus for a particular host tissue, is determined in part by the interaction of viral surface structures with cell-surface receptors on host cells. These virus-receptor interactions enable virus to penetrate and damage the host cell and are a major determinant of virulence. The cell and tissue tropism of reovirus is determined by a single, viral surface protein, the sigma 1 polypeptide.

The sigma 1 polypeptide was first implicated in reoviral erythrocyte-surface interactions when it was identified as the viral hemagglutinin.[38] Like many other animal viruses, reovirus agglutinates erythrocytes by binding to an erythrocyte-surface receptor. Reoviral hemagglutination is type specific; type 1 agglutinates human erythrocytes and type 3 agglutinates bovine erythrocytes.[89]

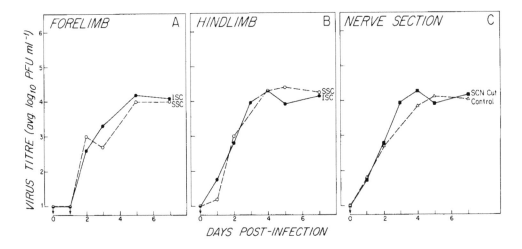

FIGURE 5. Pattern of spread of reovirus type 1 to the spinal cord of neonatal mice after inoculation of the virus into the forelimb (A) or hindlimb (B) footpad. (O) Superior spinal cord (SSC); (●) inferior spinal cord (ISC). Each plotted point represents the average $\log_{10}$ titer of five to seven specimens. The SEM is less than 0.6 for SSC and ISC at all time points. (C) Spread of type 1 to the ISC after sectioning of the sciatic nerve (■) is compared with spread in control animals with intact nerves (△). A downward-pointing arrow below a data point indicates that no virus was detected at the lowest dilution. Dose of virus, method of inoculation, animals, and details of procedure are identical to those described for Figure 4. (From Tyler, K. L., McPhee, D. A., and Fields, B. N., *Science*, 233, 770, 1986. With permission.)

Studies of reovirus tropism in the CNS further substantiated the role of the sigma 1 polypeptide in virus-receptor interactions.[48] Type 1 reovirus appears primarily in ependymal cells, where it produces a nonlethal infection that can lead to hydrocephalus,[90] whereas type 3 localizes in neuronal tissue and causes acute and fatal encephalitis in newborn mice, accompanied by destruction of neuronal cells without damage to ependymal cells.[91,92] When animals are injected intracerebrally with viruses possessing reassorted genomes, these differences in tropism in the CNS are found to be determined by the S1 genome segment[38,93] which codes for the sigma 1 polypeptide.

In addition, reovirus type 3 sigma 1 antigenic variants that are resistant to neutralization are 10,000 to 100,000 times less neurovirulent in vivo (as measured by plaque-forming units)[59,94] than the type 3 strain from which they were derived. When these variants are injected intracerebrally in mice, they infect only a subset of the neurons that are infected by the parental virus.[95] In addition, sigma 1 variants are less able to spread to the brain from peripheral sites.[93] Thus, alteration of the neutralization domain can decrease the extent and severity of viral injury to the brain.

Tropism within the pituitary gland is also attributed to the S1 genome segment.[96] Only type 1 reovirus can infect the growth hormone-producing cells of the anterior pituitary and cause a runting syndrome. In SJL/3 mice, type 1 reovirus infection also leads to production of antibodies against insulin, growth hormone, normal pancreas, and pituitary and gastric mucosa.[12,96,97] Studies with viruses possessing reassorted genome segments show that a type 1 S1 segment is required for induction of autoantibodies to growth hormone. All these data suggest that the sigma 1 protein is involved in recognizing and binding cell-surface receptors.

Analysis of the σ1 protein by using type 3 σ1-specific monoclonal antibodies indicates that there are at least three distinct domains[93,95] on the σ1 protein. One domain, termed the neutralization site, is primarily defined by two monoclonal antibodies, A2 and G5, that show high titers of neutralizing activity. The G5 monoclone was used to generate viral variants (designated A, F, and K) that resist neutralization.[95] These variants are markedly less virulent than parental type 3 virus from which they were selected. In addition, examination of brain

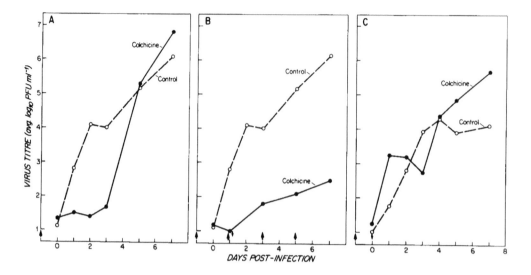

FIGURE 6.   Pattern of spread of type 3 (A and B) or type 1 (C) to the inferior spinal cord in colchicine-treated (●) and untreated control (○) mice. Upward-pointing arrows above the time scale indicate time of colchicine administration. Colchicine (J. T. Baker)-treated mice received 2×10^{-7} g of colchicine per gram of body weight. For studies of the effect of a single dose of colchicine (A and C), 1-day-old mice were given 0.010 mℓ of a $5 \times 10^{-5} M$ (20×10^{-6} g/mℓ) solution of colchicine into a hindlimb muscle by means of a 30-gauge needle and a Hamilton microsyringe. In (B) animals receive 2×10^{-7} g of colchicine per gram of body weight; either a $5 \times 10^{-5} M$ colchicine solution (ages 1 and 3 days) or a 1×10^{-4} M solution (ages 5 and 7 days) was used. Only mice showing evidence of hindlimb analgesia and paresis were used for the colchicine multidose experiments. Virus was inoculated in a 0.010- to 0.015-mℓ volume into the hindlimb footpad 24 hr after the first colchicine dose or at age 2 days (controls). The dose of virus was 3.8×10^5 to 4.5×10^5 PFU. Tissue specimens were collected and assayed as described in Figure 4. For colchicine-treated animals, each plotted point represents the average of the $\log_{10}$ titer (plaque-forming units per milliliter) of ISC specimens from two to four animals (A and C) or five to seven animals (B). Controls for (A) and (B) are identical to those in Figure 4. Each plotted point in (C) represents the average of two to four specimens. For colchicine-treated animals, the SEM was less than 0.3 at all time points in (A) and less than 0.5 for all time points in (B) and (C). (From Tyler, K. L., McPhee, D. A., and Fields, B. N., *Science*, 233, 770, 1986. With permission.)

sections of suckling mice injected intracerebrally with the variants show that the variant viruses cause tissue injury in restricted regions of the mouse brain. The results indicate that mutations in the reovirus type 3 σ1 protein cause marked alterations in viral virulence and tropism.[93] The sequences of the S1 gene and the S1 gene of the variants have been determined and their sequences compared.[83] The variants were selected by growing reovirus type 3 in the presence of either monoclonal antibody A2 or G5. Resistant clones to A2 or G5 were then selected. All five variants (resistant lanes A, F, K, AI4, and AI7) are attenuated in terms of neurovirulence. A is at least 10,000 times less virulent, and F, K, AI4, and AI17 are at least 100,000 times less virulent than parental reovirus type 3.[94]

To identify the altered site(s) of the σ1 proteins of all five variants, the entire S1 dsRNA segment (1416 nucleotide bases) from each of the variants was sequenced. A single nucleotide base substitution is found per variant. The nucleotide base changes are shown in Figure 7 and summarized in Table 3. Variants F, K, AI4, and AI7 all show base substitutions in codon 419 that alter amino acid 419 of the σ1 protein. Variant A has one nucleotide base substitution at amino acid 340 and no change at codon 419. This results in a change in the σ1 protein at amino acid 340 (Figure 7, Table 3). The specific amino acid substitutions associated with these attenuated variants, glutamic acid replaced by lysine (Glu → Lys), or Glu → Ala, or Glu → Gly, or Asp → Val, have the potential to significantly alter both protein structure and function.[83]

A.

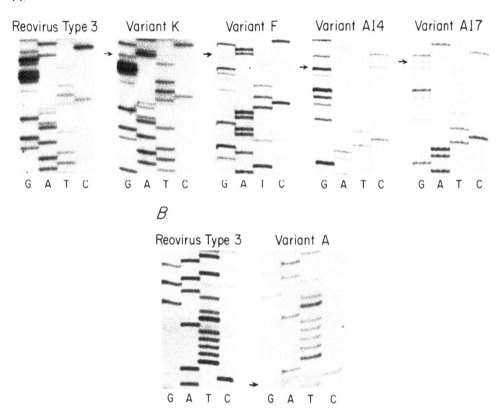

FIGURE 7. Nucleotide sequence analysis of the S1 gene from reovirus type 3 and the viral variants. Shown are gels of ddGTP-, ddATP-, ddTTP-, and ddCTP-terminated reactions (lanes G, A, T, and C, respectively) of reverse transcriptase. Arrowheads indicate the position of the mutation. (A) The sequences shown are in the region of codon 419 of the S1 gene of reovirus type 3 and variants K, F, A14, and A17. (B) The sequences shown are in the region of codon 340 of the S1 gene of reovirus type 3 and variant A. (From Bassel-Duby, R., Spriggs, D. R., Tyler, K. L., and Fields, B. N., *J. Virol.*, 60, 64, 1986. With permission.)

Table 3
LOCATION OF ATTENUATING MUTATIONS ON THE REOVIRUS TYPE 3 S1 dsRNA

Virus	Antibody	LD_{50} (p.f.u.)	Codon 419	Amino acid 419	Codon 340	Amino acid 340
Reovirus type 3	None	10	GAG	Glutamic acid	GAC	Aspartic acid
Variant A	G5	10^5	GAG	Glutamic acid	GTC	Valine
Variant F	G5	$>10^7$	AAG	Lysine	GAC	Aspartic acid
Variant K	G5	$>10^7$	AAG	Lysine	GAC	Aspartic acid
Variant A14	A2	$10^{6.9}$	GCG	Alanine	GAC	Aspartic acid
Variant A17	A2	$10^{6.6}$	GGG	Glycine	BAC	Aspartic acid

From Bassel-Duby, R., Spriggs, D. R., Tyler, K. L., and Fields, B. N., *J. Virol.*, 60, 64, 1986. With permission.

The biological alterations of variant A (attenuation and resistance to neutralization) are not as dramatic as that of the other four variants. This might indicate that the altered amino acid, amino acid 340, is situated in a less critical position in terms of structural conformation of the antigenic site.[31]

V. SUMMARY

This chapter has emphasized the diversity that is found both within reovirus serotypes and among the three serotypes. This diversity was presented in terms of the different phenotypic characteristics and how these characteristics are now being defined at the genomic level. This latter point has been aided by both the diversity of reoviruses and their genetic property of reassortment. Thus, the assignment of specific properties to discrete gene segments is made easier by the ability of reoviruses to reassort their genome. In conclusion, reoviruses are a model system that affords researchers the advantage of investigating a wide variety of viral-host-environment interactions. The question of how the reovirus diversity occurs will be resolved by examining and comparing natural isolates in greater detail. Future investigations will continue to unravel the role that evolution and reassortment have made to generating diversity among the reoviruses.

ACKNOWLEDGMENTS

We gratefully acknowledge Dr. Rhonda Bassel-Duby for her insightful comments regarding this review. We also thank Mary Quinn for her outstanding assistance and patience in the preparation of this review. Support for work in the authors' laboratories came from NINCDS program project grant 5 PO1 NS16998 and NIAID Grant 5 RO1 AI 13178. Additional support came from the Shipley Institute of Medicine.

REFERENCES

1. **Jackson, G. G. and Muldoon, R. L.,** Viruses causing respiratory infection in man. IV. Reoviruses and adenoviruses, *J. Infect. Dis.*, 128, 811, 1973.
2. **Simpson, D. I. H., Haddow, A. J., Woodall, J. P., Williams, M. C., and Bell, T. M.,** Attempts to transmit reovirus type 3 by the bite of Aedes (stegomyia) aegypti Linnaeus, *E. Afr. Med. J.*, 42, 708, 1965.
3. **Stanley, N. F.,** Reoviruses, *Br. Med. Bull.*, 23, 150, 1967.
4. **Sturm, R. T., Lang, G. H., and Mitchell, W. R.,** Prevalence of reovirus 1, 2 and 3 antibodies in Ontario racehorses, *Can. Vet. J.*, 21, 206, 1980.
5. **Rosen, L.,** Serologic groupings of reovirus by hemagglutination-inhibition, *Am. J. Hyg.*, 71, 242, 1960.
6. **Rosen, L.,** Reoviruses in animals other than man, *Ann. N.Y. Acad. Sci.*, 101, 461, 1962.
7. **Sabin, A. B.,** Reoviruses, *Science*, 130, 1387, 1959.
8. **Gaillard, R. K. and Joklik, W. K.,** Quantitation of the relatedness of the genomes of reovirus serotypes 1, 2, and 3 at the gene level, *Virology*, 123, 152, 1982.
9. **Martinson, H. G., and Lewandowski, L. J.,** Sequence homology studies between double-stranded RNA genomes of cytoplasmic polyhedrosis virus, wound tumor virus and reovirus strains 1, 2, and 3, *Intervirology*, 4, 91, 1975.
10. **Ramig, R. F., Cross, R. K., and Fields, B. N.,** Genome RNAs and polypeptides of reovirus serotypes 1, 2, and 3, *J. Virol.*, 22, 726, 1977.
11. **Sharpe, A. H., Ramig, R. F., Mustoe, T. A., and Fields, B. N.,** A genetic map of reovirus. I. Correlation of genome RNAs between serotypes 1, 2, and 3, *Virology*, 84, 63, 1978.
12. **Onodera, T., Ray, U. R., Melez, K. A., Suzuki, H., Toniolo, A., and Notkins, A. L.,** Virus-induced diabetes mellitus: autoimmunity and polyendocrine disease prevented by immunosuppression, *Nature (London)*, 297, 66, 1982.
13. **Stanley, N. F.,** Diagnosis of reovirus infections: comparative aspects, in *Comparative Diagnosis of Viral Diseases*, Kurstak, E. and Kurstak, K., Eds., Academic Press, New York, 1974, 385.
14. **Stanley, N. F. and Joske, R. A.,** Animal model: chronic murine hepatitis induced by reovirus type 3, *Am. J. Pathol.*, 80, 181, 1975.
15. **Stanley, N. F. and Joske, R. A.,** Animal model: chronic biliary obstruction caused by reovirus type 3, *Am. J. Pathol.*, 80, 185, 1975.
16. **Fields, B. N. and Greene, M. I.,** Genetic and molecular mechanisms of viral pathogenesis: implications for prevention and treatment, *Nature (London)*, 300, 19, 1982.

17. **Joklik, W. K.,** The reovirus particle, in *The Reoviridae,* Jollik, W. K., Ed., Plenum Press, New York, 1983, 9.
18. **Ramig, R. F. and Fields, B. N.,** Genetics of reovirus, in *The Reoviridae,* Joklik, W., Ed., Plenum Press, New York, 1983, 197.
19. **Sharpe, A. H. and Fields, B. N.,** Pathogenesis of reovirus infection, in *The Reoviridae,* Joklik, W. K., Ed., Plenum Press, New York, 1983, 229.
20. **Zarbl, H. and Milward, S.,** The reovirus multiplication cycle, in *The Reoviridae,* Joklik, W. K., Ed., Plenum Press, New York, 1983, 107.
21. **Gomatos, P. J., Tamm, I., Dales, S., and Franklin, R. M.,** Reovirus type 3: physical characteristics and interactions with L cells, *Virology,* 17, 441, 1962.
22. **Smith, R. E., Zweerink, H. J., and Joklik, W. K.,** Polypeptide components of virions, top component and cores of reovirus 3, *Virology,* 39, 791, 1969.
23. **Bellamy, A. R., Shapiro, L., August, J. T., and Joklik, W. K.,** Studies on reovirus RNA. I. Characterization of reovirus genome RNA, *J. Mol. Biol.,* 29, 1, 1967.
24. **Furuichi, Y., Morgan, M., Muthukrishnan, S., and Shatkin, A. J.,** Reovirus messenger RNA contains a methylated blocked 5'-terminal structure: M⁷G(5')ppp(5')GᵐpCp⁻, *Proc. Natl. Acad. Sci. U.S.A.,* 72, 362, 1975.
25. **Milward, S. and Graham, A. F.,** Structural studies on reovirus: discontinuities in the genome, *Proc. Natl. Acad. Sci. U.S.A.,* 65, 422, 1970.
26. **Shatkin, A. J., Sipe, J. D., and Loh, P. C.,** Separation of 10 reovirus genome segments by polyacrylamide gel electrophoresis, *J. Virol.,* 1, 986, 1968.
27. **Both, G. W., Banerjee, A. K., and Shatkin, A. J.,** Methylation-dependent translation of viral messenger RNAs in vitro, *Proc. Natl. Acad. Sci. U.S.A.,* 72, 1189, 1975.
28. **McCrae, M. A. and Joklik, W. K.,** The nature of the polypeptide encoded by each of the ten double-stranded RNA segments of reovirus type 3, *Virology,* 89, 578, 1978.
29. **Mustoe, T. A., Ramig, R. F., Sharpe, A. H., and Fields, B. N.,** Genetics of reovirus: identification of the dsRNA segments encoding the polypeptides of the μ and σ size classes, *Virology,* 89, 594, 1978.
30. **Shatkin, A. J. and Rada, B.,** Reovirus-directed ribonucleic acid synthesis in infected L cells, *J. Virol.,* 1, 24, 1967.
31. **Bassel-Duby, R., Jayasuriya, A., Chattijee, D., Sonenberg, N., Maizel, J. V., Jr., and Fields, B. N.,** Sequence of reovirus haemagglutinin (cell attachment protein) predicts a coiled-coil structure, *Nature (London),* 315, 421, 1985.
32. **Cashdollar, L. W., Chmelo, R. A., Wiener, J. R., and Joklik, W. K.,** Sequences of the S1 genes of the three serotypes of reoviruses, *Proc. Natl. Acad. Sci. U.S.A.,* 82, 24, 1985.
33. **Ernst, H. and Shatkin, A. J.,** Reovirus haemagglutinin mRNA codes for two polypeptides in overlapping reading frames, *Proc. Natl. Acad. Sci. U.S.A.,* 82, 48, 1985.
34. **Nagata, L. E. S., Masri, S. A., Mah, D. C. S., and Lee, P. W. K.,** Molecular cloning and sequencing of the reovirus (serotype 3) S1 gene which encodes the viral cell attachment protein σ1, *Nucleic Acids Res.,* 12, 8699, 1981.
35. **Sakar, G., Pelletier, J., Bassel-Duby, R., Jayasuriya, A., Fields, B. N., and Sonenberg, N.,** Identification of a new polypeptide coded by reovirus gene S1, *J. Virol.,* 54, 720, 1985.
36. **Jacobs, B. L. and Samuel, C. E.,** Biosynthesis of reovirus-specified polypeptides, *Virology,* 143, 63, 1985.
37. **Zweerink, H. J., McDowell, M. J., and Joklik, W. K.,** Essential and non-essential non-capsid reovirus proteins, *Virology,* 45, 716, 1971.
38. **Weiner, H. L., Ramig, R. F., Mustoe, T. A., and Fields, B. N.,** Identification of the gene coding for the hemagglutinin of reovirus, *Virology,* 86, 581, 1978.
39. **Masri, S. A., Nagata, L., Mah, D. C. W., and Lee, P. K. W.,** Functional expression in *Escherichia Coli* of cloned reovirus S1 gene encoding the viral cell attachment protein σ1, *Virology,* 149, 83, 1986.
40. **Zweerink, H. J. and Joklik, W. K.,** Studies on the intracellular synthesis of reovirus-specified proteins, *Virology,* 41, 501, 1970.
41. **Hayes, E. C., Lee, P. W. K., Miller, S. E., and Joklik, W. K.,** The interaction of a series of hybridoma IgGs with reovirus particles: demonstration that the core protein λ2 is exposed on the particle surface, *Virology,* 108, 147, 1981.
41a. **Drayna, D. and Fields, B. N.,** Activation and characterization of the reovirus transcriptase: genetic analysis, *J. Virol.,* 41, 110, 1982.
41b. **Morgan, E. M. and Kingsbury, D. W.,** Reovirus enzymes that modify messenger RNA are inhibited by perturbation of the lambda proteins, *Virology,* 113, 565, 1981.
42. **Gaillard, R. K. and Joklik, W. K.,** The antigenic determinants of most proteins coded by the three serotypes of reovirus are highly conserved during evolution, *Virology,* 107, 533, 1980.
43. **Lee, P. W. K., Hayes, E. C., and Joklik, W. K.,** Characterization of anti-reovirus immunoglobulins secreted by cloned hybridoma cell lines, *Virology,* 108, 134, 1981.

44. **Weiner, H. L. and Fields, B. N.**, Neutralization of reovirus: the gene responsible for the neutralization antigen, *J. Exp. Med.*, 146, 1305, 1977.

45. **Finberg, R., Weiner, H. L., Fields, B. N., Benacerraf, B., and Burakoff, S. J.**, Generation of cytolytic T lymphocytes after reovirus infection: role of S1 gene, *Proc. Natl. Acad. Sci. U.S.A.*, 76, 442, 1979.

46. **Sharpe, A. H. and Fields, B. N.**, Reovirus inhibition of cellular DNA synthesis: role of the S1 gene, *J. Virol.*, 38, 389, 1981.

47. **Babiss, L. E., Luftig, R. B., Weatherbee, J. A., Weihing, R. R., Ray, U. R., and Fields, B. N.**, Reovirus serotypes 1 and 3 differ in their *in vitro* association with microtubules, *J. Virol.*, 30, 863, 1979.

48. **Weiner, H. L., Drayna, D., Averill, D. R., Jr., and Fields, B. N.**, Molecular basis of reovirus virulence: role of the S1 gene, *Proc. Natl. Acad. Sci. U.S.A.*, 74, 5744, 1977.

49. **Rubin, D. H. and Fields, B. N.**, Molecular basis of reovirus virulence: role of the M2 gene, *J. Exp. Med.*, 152, 853, 1980.

50. **Hardy, D. B., Rubin, D. N., and Fields, B. N.**, Molecular basis of reovirus neurovirulence: role of the M2 gene in avirulence, *Proc. Natl. Acad. Sci. U.S.A.*, 79, 1298, 1982.

51. **Rubin, D., Weiner, H. L., Fields, B. N., and Greene, M. I.**, Immunologic tolerance after oral administration of reovirus: requirement for two viral gene products for tolerance induction, *J. Immunol.*, 127, 1697, 1981.

52. **Sharpe, A. H. and Fields, B. N.**, Reovirus inhibition of cellular RNA and protein synthesis: role of the S4 gene, *Virology*, 122, 381, 1982.

53. **Ahmed, R. and Fields, B. N.**, Role of the S4 gene in the establishment of persistent reovirus infection in L cells, *Cell*, 28, 605, 1982.

54. **Ahmed, R. and Graham, A. F.**, Persistent infections in L cells with temperature-sensitive mutants of reovirus, *J. Virol.*, 23, 250, 1977.

55. **Ahmed, R., Canning, W. M., Kauffman, R. S., Sharpe, A. H., Hallum, J. V., and Fields, B. N.**, Role of the host cell in persistent viral infection: coevolution of L cells and reovirus during persistent infection, *Cell*, 25, 325, 1981.

56. **White, C. K. and Zweerink, H. J.**, Studies on the structure of reovirus cores: selective removal of polypeptide λ2, *Virology*, 70, 171, 1976.

57. **Ralph, S. J., Harvey, J. D., and Bellamy, A. R.**, Subunit structure of the reovirus spike, *J. Virol.*, 36, 894, 1980.

58. **Bartlett, N. M., Gillies, S. C., Bullivant, S., and Bellamy, A. R.**, Electron microscope study of reovirus reaction cores, *J. Virol.*, 14, 315, 1974.

59. **Fields, B. N., Raine, C. S., and Baum, S. G.**, Temperature-sensitive mutants of reovirus type 3: defects in viral maturation as studied by immunofluorescence and electron microscopy, *Virology*, 43, 569, 1971.

60. **Brown, E. G., Nibert, M. L., and Fields, B. N.**, The L2 gene of reovirus serotype 3 controls the capacity to interfere, accumulate deletions and establish persistent infection, in *Double-Stranded RNA Viruses*, Compans, R. W. and Bishop, D. H. L., Eds., Elsevier, New York, 1983.

61. **Fields, B. N. and Joklik, W. K.**, Isolation and preliminary genetic and biochemical characterization of temperature-sensitive mutants of reovirus, *Virology*, 37, 335, 1969.

62. **Ikegami, N. and Gomatos, P. J.**, Temperature-sensitive conditional lethal mutants of reovirus 3, *Virology*, 36, 447, 1968.

63. **Ahmed, R., Chakraborty, P. R., and Fields, B. N.**, Genetic variation during lytic virus infection: high passage stocks of wild-type reovirus contain temperature-sensitive mutants, *J. Virol.*, 34, 285, 1980.

64. **Ahmed, R., Chakraborty, P. R., Graham, A. F., Ramig, R. F., and Fields, B. N.**, Genetic variation during persistent reovirus infection: presence of extragenically suppressed temperature-sensitive lesions in wild-type virus isolated from persistently infected cells, *J. Virol.*, 34, 383, 1980.

65. **Ramig, R. F., Ahmed, R., and Fields, B. N.**, A genetic map of reovirus: assignment of the newly defined mutant groups H, I and J to genome segments, *Virology*, 125, 299, 1983.

66. **Ramig, R. F., Mustoe, T. A., Sharpe, A. H., and Fields, B. N.**, A genetic map of reovirus. II. Assignment of the double-stranded RNA-negative mutant groups C, D, and E genome segments, *Virology*, 85, 531, 1978.

67. **Cross, R. K. and Fields, B. N.**, Reovirus-specific polypeptides: analysis using discontinuous gel electrophoresis, *J. Virol.*, 19, 162, 1976.

68. **Cross, R. K. and Fields, B. N.**, Temperature-sensitive mutants of reovirus type 3: studies on the synthesis of viral RNA, *Virology*, 50, 799, 1972.

69. **Mustoe, T. A., Ramig, R. F., Sharpe, A. H., and Fields, B. N.**, A genetic map of reovirus. III. Assignment of the double-stranded RNA mutant groups A, B, and G to genome segments, *Virology*, 85, 545, 1978.

70. **Cross, R. K. and Fields, B. N.**, Temperature-sensitive mutants of reovirus type 3: evidence for aberrant μ1 and μ2 polypeptide species, *J. Virol.*, 19, 174, 1976.

71. **Cross, R. K. and Fields, B. N.**, Use of an aberrant polypeptide as a marker in three-factor crosses: further evidence for independent reassortment as the mechanism of recombination between temperature-sensitive mutants of reovirus type 3, *Virology*, 74, 345, 1976.

72. **Nonoyama, M., Watanabe, Y., and Graham, A. F.**, Defective virions of reovirus, *J. Virol.*, 6, 226, 1970.

73. **Spandidos, D. A. and Graham, A. F.**, Complementation between temperature-sensitive and deletion mutants of reovirus, *J. Virol.*, 16, 144, 1975.

74. **Ahmed, R. and Fields, B. N.**, Reassortment of genome segments between reovirus defective interfering particles and infectious virus: construction of temperature-sensitive and attenuated viruses by rescue of mutations from DI particles, *Virology*, 111, 351, 1981.

75. **Ramig, R. F. and Fields, B. N.**, Revertants of temperature-sensitive mutants of reovirus: evidence for frequent extragenic suppression, *Virology*, 92, 155, 1979.

76. **Ramig, R. F. and Fields, B. N.**, Reoviruses, in *The Molecular Biology of Animal Virus*, Vol. I, Nayak, D. P., Ed., Marcel Dekker, New York, 1977.

77. **Hrdy, D. B., Rosen, L., and Fields, B. N.**, Polymorphism of the migration of double-stranded RNA segments of reovirus isolates from humans, cattle, and mice, *J. Virol.*, 31, 104, 1979.

78. **Gentsch, J. R. and Fields, B. N.**, Tryptic peptide analysis of outer capside polypeptides of mammalian reovirus serotypes 1, 2, and 3, *J. Virol.*, 38, 208, 1981.

79. **Gentsch, J. R. and Fields, B. N.**, Genetic diversity in natural populations of mammalian reoviruses: tryptic peptide analysis of outer capsid polypeptides of murine, bovine, and human type 1 and 3 reovirus strains, *J. Virol.*, 49, 641, 1984.

80. **Wenske, E. A., Chanock, S. J., Krata, L., and Fields, B. N.**, Genetic reassortment of mammalian reovirus in mice, *J. Virol.*, 56, 613, 1985.

81. **Giantini, M., Seliger, L. S., Furuchi, Y., and Shatkin, A.**, Reovirus type 3 genome segment S4: nucleotide sequence of the gene encoding a major virion surface protein, *J. Virol.*, 52, 984, 1984.

82. **Richardson, M. A. and Furuichi, Y.**, Nucleotide sequence of reovirus genome segment S3, encoding non-structural protein sigma NS, *Nucleic Acids Res.*, 11, 6399, 1983.

83. **Bassel-Duby, R., Spriggs, D. R., Tyler, K. L., and Fields, B. N.**, Identification of attenuating mutations on the Reovirus type 3 S1 double-stranded RNA segment with a rapid sequencing technique, *J. Virol.*, 60, 64, 1986.

84. **Keroack, M. and Fields, B. N.**, Viral shedding and transmission between hosts determined by reovirus L2 gene, *Science*, 232, 1635, 1986.

85. **Kauffman, R. S., Wolf, J. L., Finberg, R., Trier, J. S., and Fields, B. N.**, The sigma 1 protein determines the extent of spread of reovirus from the gastrointestinal tract of mice, *Virology*, 124, 403, 1983.

86. **Wolf, J. L., Rubin, D. H., Finberg, R., Kauffman, R. S., Sharpe, A. H., Trier, J. S., and Fields, B. N.**, Intestinal M cells: a pathway for entry of reovirus into the host, *Science*, 212, 471, 1981.

87. **Tyler, K. L., McPhee, D. A., and Fields, B. N.**, Distinct pathways of viral spread in the host determined by reovirus S1 gene segment, *Science*, 233, 770, 1986.

88. **Paulson, J. C. and McClure, W. O.**, Inhibition of axoplasmic transport by colchicine, podophyllotoxin, and vinblastine: an effect on microtubules, *Ann. N.Y. Acad. Sci.*, 253, 517, 1975.

89. **Eggers, H. J., Gomatos, P. J., and Tamm, I.**, Agglutination of bovine erythrocytes: a general characteristic of reovirus type 3, *Proc. Soc. Exp. Biol. Med.*, 110, 879, 1962.

90. **Kilham, L. and Margolis, S.**, Hydrocephalus in hamsters, ferrets, rats and mice following inoculations with reovirus type 1, *Lab. Invest.*, 21, 183, 1969.

91. **Margolis, G., Kilham, L., and Gonatos, N.**, Reovirus type III encephalitis: observations of virus-cell interactions in neural tissues. I. Light microscopy studies, *Lab. Invest.*, 24, 91, 1971.

92. **Raine, C. S. and Fields, B. N.**, Ultrastructural features of reovirus type 3 encephalitis, *J. Neuropathol. Exp. Neurol.*, 32, 19, 1973.

93. **Spriggs, D. R., Bronson, R. T., and Fields, B. N.**, Hemagglutinin variants of reovirus type 3 have altered central nervous system tropism, *Science*, 220, 505, 1983.

94. **Kaye, K. M., Spriggs, D. R., Bassel-Duby, R., Fields, B. N., and Tyler, K. L.**, Genetic basis for altered pathogenesis of an immune-selected antigenic variant of reovirus type 3 (Dearing), *J. Virol.*, 59, 90, 1986.

95. **Spriggs, D. R. and Fields, B. N.**, Attenuated reovirus type 3 strains generated by selection of hemagglutinin antigenic variants, *Nature (London)*, 197, 68, 1982.

96. **Onodera, T., Toniolo, A., Ray, U. R., Jensen, A. B., Kanzek, R. A., and Notkins, A. L.**, Virus-induced diabetes mellitus. XX. Polyendocrinopathy and autoimmunity, *J. Exp. Med.*, 153, 1457, 1981.

97. **Haspel, M. V., Onodera, T., Prabhakar, B. S., Horita, M., Suzuki, H., and Notkins, A. L.**, Virus-induced autoimmunity: monoclonal antibodies that react with endocrine tissues, *Science*, 220, 6, 1983.

Chapter 6

INFLUENZA VIRUSES: HIGH RATE OF MUTATION AND EVOLUTION

Frances I. Smith and Peter Palese

TABLE OF CONTENTS

I. INTRODUCTION

Influenza viruses can be classified into three types: A, B, and C. These viruses are similar in morphology, composition, and in their replication processes and are clearly evolutionarily related. Nevertheless, they exhibit major differences in their epidemiology, genetic variability, and evolution. The most dramatic differences are seen on a comparison of influenza A and C viruses. This chapter will review briefly the epidemiology and genetic variability of these two influenza virus types, contrast their rates and patterns of evolution, and consider mechanisms which could account for the differences.

II. INFLUENZA A AND C VIRUSES ARE EVOLUTIONARILY RELATED

Influenza A virions contain eight single-stranded RNA segments of negative sense, and it has been determined which RNA segment codes for which viral protein (Figure 1; for reviews, see References 1 to 3). In contrast, influenza C viruses contain only seven RNA species, and only three of these — the hemagglutinin esterase (HE), nucleoprotein (NP), and nonstructural protein (NS) genes — have been sequenced and assigned as coding for a given viral protein(s).[4-7a]

The fourth longest RNA segment of both virus types codes for a receptor-binding protein, but it is difficult to detect nucleotide and amino acid sequence homologies between these genes and their products.[4] Furthermore, there is also an acetyl esterase activity associated with the C virus glycoprotein not found with the A virus hemagglutinin (HA).[5,7a] On the other hand, there is significant structural homology between the two surface protein molecules as a consequence of conserved hydrophobic domains and cysteine residues. Also, in analogy with influenza A viruses, cell fusion is promoted by influenza C viruses, and this activity is most likely mediated by the HE, as is agglutination of erythrocytes. Thus, there are structural and functional similarities between the HA and HE of the influenza A and C viruses which suggest a common evolutionary ancestor for these genes.

RNA segment 5 of influenza A and C viruses codes for the NP, which is the major structural protein component of the nucleocapsid. A comparison of these genes reveals little homology over the entire coding sequence.[6] However, a significant degree of homology (25%) between A and C NPs is found within the central coding region of the molecule. These results suggest that these residues may be conserved because of functional requirements in protein-protein and/or protein-nucleic acid interactions.

In both types of influenza virus, the shortest RNA segment codes for two NS proteins, NS1 and NS2. The amino acid NH_2-terminal coding region of NS1 and NS2 are shared, but the COOH-terminal coding region of the NS2 is generated by a splicing event which results in a frameshift on the mRNA molecule (for review, see References 2 and 7). There is very little homology between influenza A and C viruses at either the nucleotide or amino acid sequence level for either the NS1 or NS2 coding regions. However, at the structural level, a hydrophilicity plot of the NS2 proteins does reveal many homologous regions. This data suggests a possible common function for these nonstructural proteins. In conclusion, influenza A and C viruses appear to derive from a common ancestor, with great divergence in nucleotide sequences having occurred, but many structural and functional features being conserved.

III. EPIDEMIOLOGY

The epidemiology of influenza A viruses has been well characterized due to the association of these viruses with pandemic outbreaks of disease in man. The pandemics usually result from the appearance of a new subtype strain (antigenic shift) containing a novel HA and/

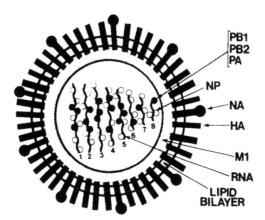

FIGURE 1. Schematic diagram of the structure of an influenza virus particle. Proteins PB1, PB2, PA, and NP comprise the virion core and are associated with the genome RNAs. M1 is in close contact with the lipid bilayer. The HA and NA spikes are surface glycoproteins.

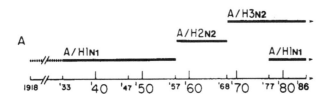

FIGURE 2. Epidemiology of influenza A viruses in nature. Broken lines indicate that virus isolates are not available from this period. The periods when influenza A viruses of a certain subtype (H1N1, H2N2, H3N2) were prevalent are designated by the thick lines. (From Palese, P. and Young, J., in *Genetics of Influenza Viruses,* Palese, P. and Kingsbury, D. W., Eds., Springer-Verlag, Vienna, 1983, 321. With permission.)

or neuraminidase (NA) that is immunologically different from that of previous circulating isolates. During this century, there appear to have been at least three occasions when new subtypes have been introduced in the human population (Figure 2). Viruses of the H1N1 subtype (HA subtype H1; NA subtype N1) circulated for many years early this century until they were displaced by H2N2 viruses which were, in turn, displaced by H3N2 viruses. During their circulation, strains within each subtype accumulated more subtle changes in antigenic character (antigenic drift).

In 1977, a unique event occurred with respect to influenza A virus epidemiology. Viruses containing surface antigens of a previous subtype, H1, reappeared and these viruses cocirculated with the H3 strains prevalent at the time. The "new" H1 strains were shown to be serologically and genetically similar to H1 strains which had circulated around 1950.[8-10] It is unusual that the new H1 viruses did not replace the H3 strains introduced in 1968, but continue to coexist with them.

The epidemiology of influenza C viruses differs markedly from that of influenza A viruses. Influenza C viruses generally cause only infrequent outbreaks of mild respiratory illness.[11] Also, although serological studies have shown that antigenic variation can occur among isolates, influenza C viruses appear more antigenically stable than A viruses, and hemagglutinin subtypes of C viruses have not been observed.[12,13]

In contrast to influenza A viruses, which have natural reservoirs in pigs, horses, and birds, the C viruses are mainly isolated from man. A report of influenza C virus isolations from pigs in Beijing, China, suggests, however, that pigs may serve as an animal reservoir for these viruses.[13]

IV. MECHANISMS OF ANTIGENIC SHIFT IN INFLUENZA A VIRUSES

A. Reassortants

Sequence comparisons between the H1, H2, and H3 HA genes,[14,15] and between the N1 and N2 NA genes,[16-17] reveal such low levels of nucleotide homology that it is highly unlikely that these different HAs or NAs arose from one another by accumulation of point mutations during a brief period of time. Rather, it is believed that the appearance of novel surface antigens or human influenza viruses are the result of reassortment (recombination) events between previously circulating human viruses and influenza viruses of animal origin. There is ample evidence supporting reassortment between human viruses and avian strains in vivo,[18] between avian strains in nature,[19] and among human viruses.[20-22] In addition to the three human HA subtypes (H1, H2, H3) and two NA subtypes (N1, N2), there are ten antigenically distinct HAs and seven NAs which have been identified in viruses isolated from horses, pigs, or birds. It is conceivable that any of these foreign HA and NA genes could be incorporated into a human virus via reassortment, thereby resulting in a new human antigenic subtype.

The best evidence for reassortment generating new human subtypes involves the origin of the H3 HA. This virus contains the NA and all other genes from an Asian (H2N2) strain, and an HA which is antigenically related to that of the A/duck/Ukraine/63 (H3N8) and A/equine/2/Miami/63 (H3N8) viruses.[23-27] The amino acid sequence homology between the HAs of the A/duck/Ukraine/63 and A/Aichi/2/68 (H3N2) viruses is 96%. This finding suggests that the Hong Kong H3N2 strain of influenza has been derived by reassortment between an H2N2 virus and a virus possessing an HA closely related to that of earlier animal strains.

B. Reemergence of Previously Circulating Strains

A second mechanism leading to the introduction of a new subtype of influenza A virus involves the reappearance of a subtype which had circulated at an earlier time. A well-documented example of this phenomenon is the reemergence of H1 strains in 1977. Genetic analysis of the new H1 strains[8,28-32] revealed that the entire genomes of the reemergent H1 strains are very similar to those of H1 viruses isolated in 1950. Several mechanisms have been postulated to explain how the new H1 strains remained relatively unchanged during a 27-year period. It is possible that influenza viruses are capable of latent or persistent infection in man in conditions in which the genetic information of the virus is highly conserved, although there is no direct evidence to support this idea. The genetic information of the virus could also have been preserved by sequential passage in an animal reservoir in which influenza viruses replicate without rapid genetic change, although, again all evidence so far collected suggests that genetic drift also occurs in animals. A possible trivial explanation is that the 1950 virus was frozen as a laboratory strain, and accidentally reintroduced into man.

V. GENETIC DRIFT IN INFLUENZA A VIRUSES

Following the emergence of a new subtype, viruses of a single subtype show minor changes in antigenic character (antigenic drift) due to an accumulation of amino acid changes caused by nucleotide substitutions in the genes encoding the HA and NA proteins. In addition to variation in the surface protein coding genes, variation among strains also occurs in the genes encoding nonsurface proteins of the virus. The two genes which have been most extensively studied with respect to genetic variation are the HA and NS genes of influenza viruses.

A. HA Gene

Information on variation of the HA gene within a subtype has been generated using nucleotide sequencing analysis of HA genes of viruses isolated over time. Early studies on the H3 HA[33] showed that most sequence changes occur in the HA1 portion of the HA molecule, which contains the antigenic sites. The resulting amino acid changes alter the antigenic sites such that antibodies generated against an initial viral infection may not be protective against infection with a later emerging virus. A comparison of nucleotide sequences of over 14 H3 subtype viruses, isolated between 1968 and 1980, showed that new H3 antigenic variants emerged by accumulating sequential amino acid changes within antigenic regions of the HA molecules.[34] These results indicate that all H3 isolates evolved along a common pathway, which consisted of a series of dominant strains, each succeeding another from which it was directly derived.

The same result was found on analysis of H1 HA variants. Raymond et al.[31] compared both nucleotide and amino acid changes in the HA1 domain of the HA molecules of five H1 viruses of the 1950 to 1957 epidemic period and 14 isolates of the 1977 to 1983 H1 epidemic period. They concluded that the HAs of isolates within both periods showed sequential changes, but that the earlier and later H1 periods have followed two different evolutionary pathways. The observations on the evolution of the H1 subtype demonstrate the extreme flexibility of the virus in its capacity to change and what may be an unlimited potential of the influenza virus to evolve successfully in many directions and escape host immunity. The results suggest the improbability that one can predict the molecular make-up of future epidemiologically important influenza A viruses based on information on the nature of epidemic variants of the past.

B. NS Gene

The high level of genetic homology observed among NS genes of viruses belonging to different subtypes[10,35] indicates that the NS gene was conserved during the emergence of new subtype strains in this century. Early work based on partial NS sequence data from three human strains[35] suggested that, as was seen for the HA gene, sequence changes in the NS gene also accumulate over time. Recently, a comparison of the complete nucleotide sequences of the NS genes of 15 influenza A viruses isolated over 53 years (1933 to 1985) confirms these observations.[32]

Buonagurio et al.[32] analyzed this sequence information by the maximum parsimony procedure[36] to determine the phylogenetic tree of minimum length (Figure 3A). This analysis allows the calculation of the evolutionary rate for this gene. The number of nucleotide substitutions between the origin of the best tree (representing the common ancestral virus) and the tip of each branch (representing each sequenced isolate) is plotted against the date of isolation of the virus whose NS gene is represented by that tip (Figure 3B). The major line, derived by linear regression analysis, shows that these sequences are evolving at the steady rate of 1.7 ± 0.1 nucleotide substitutions per year, or $1.9 \pm 0.1 \times 10^{-3}$ substitutions per nucleotide site per year. This near constant rate of evolution is consistent with the molecular clock model of evolution (for review, see Reference 37). Similar results are obtained on analysis of the HA gene, although the rate of evolution is significantly higher (see next section).

Based on a constant rate of evolution, the WSN/33 and PR/34 strains appear to have more substitutions per year than expected (Figure 3B). Since these strains were isolated before

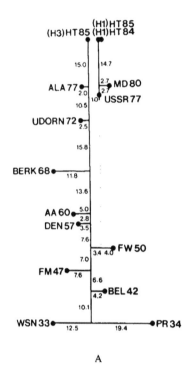

A

FIGURE 3. (A) Most parsimonious evolutionary tree for 15 influenza A virus NS genes. The nucleotide sequences[32] were analyzed by the method of Fitch.[36] The length of the trunk and side branches of the evolutionary tree are proportional to the minimal number of substitutions required to account for the differences in sequence. Nonintegral numbers arise from averaging over all possible minimal solutions. The broken line represents the predicted number of additional substitutions between the NS genes of FW/50 and USSR/77 based on the calculated evolutionary rate.[32] (B) Linearity with time of number of substitutions in the NS genes of influenza A viruses. The abscissa represents the year of isolation of the influenza A viruses used in the analysis. The ordinate indicates the number of substitutions observed in their NS genes between the first branching point formed by the WSN/33 and PR/34 sequences in Figure 3A and the tips of all branches of the evolutionary tree. A line, generated by linear regression analysis, is drawn through the points. The slope of the line is 1.7 ± 0.1 substitutions per year. In addition to the sequences found on the trunk of the evolutionary tree (solid circles), the NS genes of the four new H1 viruses are also represented in this graph (solid squares). (From Buonagurio, D. A., Nakada, S., Parvin, J. D., Krystal, M., Palese, P., and Fitch, W. M., *Science*, 232, 980, 1986. With permission. Copyright 1986 by the AAAS.)

refrigeration became available in the laboratory, we believe that continuous passaging in animal hosts and in embryonated eggs (particularly in the first 10 to 15 years after isolation of the strains) may have introduced additional mutations not present in the original isolates. Figure 3 also shows that the group of H1 subtype strains, which reemerged in the human population in 1977 after a 27-year absence, is evolving at the same rate.

VI. GENETIC DRIFT IN INFLUENZA C VIRUSES

Early evidence suggested that influenza C viruses are antigenically more stable over time than A type viruses.[12,38] Furthermore, oligonucleotide mapping analysis of the genomes of different influenza C viruses[38,39] also suggested less variation than was observed for influenza A viruses. Nucleotide sequence analysis of the HE genes of eight human influenza C virus strains isolated over approximately four decades (1947 to 1983) revealed that both on the nucleotide and amino acid levels changes can be observed, but they do not appear to accumulate with time.[40] For example, strains isolated 31 years apart may possess almost identical HE genes, whereas strains isolated only 1 or 2 years apart may differ by many changes. Similar results have been reported for the NS gene.[41]

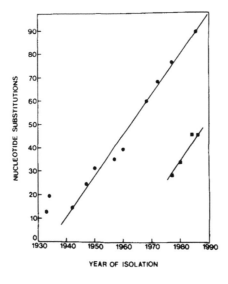

FIGURE 3B.

These results suggest that the strains of influenza C viruses examined do not directly share the same evolutionary lineage. Rather, it appears that multiple evolutionary pathways exist and C virus variants of different lineages may cocirculate in nature for extended periods of time.

VII. EVOLUTION OF INFLUENZA VIRUSES

Mutations arise randomly during viral replication, but only a small fraction of these become fixed during evolution. The neutralist theory[42] asserts that most of the evolutionary changes occurring at the molecular level are neither beneficial nor deleterious, but neutral in terms of fitness of the organism. Evolution, therefore, proceeds by the random fixation of selectively neutral mutations. An alternative view is that Darwinian natural selection of advantageous alleles is the dominant driving force in evolution. If neutral changes are randomly fixed in influenza viral genes, we would expect to see many variants cocirculating. The prediction is satisfied for both influenza A and C viruses. However, an additional mechanism is required to account for the observation that influenza A variants in general belong to a single lineage. Therefore, it appears that the evolution of influenza A virus genes is also shaped by selectional forces. The quasispecies concept[43,44] suggests that the high mutation rate of RNA viruses[45] results in populations consisting of multiple genetic variants. In the face of a strong selection pressure, probably due to the vigorous immune response triggered by influenza A virus infection, a rare antigenic variant within this population may be successful in replication and transmission. This variant would thus be strongly, positively selected.

This explanation accounts neatly for the unusual pattern of influenza A virus HA gene evolution. However, there is no evidence for immune surveillance of the NS gene products, nor have other selectional forces on the NS gene been identified. It may be that only one influenza virus gene (most likely the HA) is subject to selection. In the brief time before immunity develops to a new (antigenic) variant, that strain may spread through the population, carrying with it whatever variant of the NS gene which happens by chance to be present. In this way, the phylogeny of the NS gene may be linked (hitchhiking) to that of another gene undergoing extensive positive selection. The fixation of substitutions in the NS genes is not, therefore, simply the result of random genetic drift. The observation that the evo-

Table 1
EVOLUTIONARY
RATES
(SUBSTITUTIONS/
SITE/YEAR)[a]

Influenza A virus:	
NS	1.94×10^{-3}
H3 HA[b]	6.72×10^{-3}
N2 NA	3.17×10^{-3}
Hemoglobin:	
Alpha chain	4.50×10^{-9}
Beta chain	3.83×10^{-9}
Interferon:	
Alpha-1	4.94×10^{-9}
Beta-1	8.09×10^{-9}
Gamma	11.39×10^{-9}
Histone:	
H4	6.16×10^{-9}

[a] Mammalian gene values from Li et al.[46] Viral data from Buonagurio.[57]
[b] HA1 domain.

lutionary rate of the HA gene is significantly higher than that for the NS gene (Table 1) is also consistent with this hypothesis.

The evolutionary model postulated for influenza C viruses is quite different from that suggested for influenza A. Antigenically dominant A virus variants emerge with time, and successive variants accumulate mutations found in the genes of variants circulating in earlier years (see above). In contrast, influenza C viruses do not appear to rapidly accumulate changes with time, and strains of different lineages cocirculate. Figure 4 shows a diagrammatic representation of this bush-tree model highlighting the differences between the strains.

What could account for these differences? First, we may speculate that influenza A viruses induce a strong immune response which favors selection of antigenic variants, whereas influenza C viruses do not. Also, fewer people may be infected at any one time, thus insuring a larger immunologically naive population for strain propagation. However, certain intrinsic properties of the viral replication system may also contribute to this situation. For example, the polymerase of influenza A viruses may be more error-prone than that of influenza C viruses, and/or the influenza C virus genes may be under tighter functional constraint. Both situations could lead to a higher mutation rate and thus more genetic variability in the influenza A virus genome. Data reviewed in the next section indicates that mutation rates may indeed differ significantly between different RNA viruses.

VIII. MUTATION RATES OF RNA VIRUSES

RNA viruses provide the opportunity to study evolutionary processes which are manifested over a relatively short time frame (decades). In contrast, the eukaryotic chromosomes of the host cell take millions of years to achieve the same extent of molecular variation that viral genes can attain in just a few human generations. Evolutionary rates for mammalian genes are on the order of 10^{-9} substitutions per site per year,[46] which is one millionfold

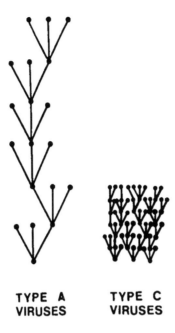

TYPE A
VIRUSES

TYPE C
VIRUSES

FIGURE 4. Evolutionary (bush-tree) model for the propagation of influenza A and influenza C viruses. The length of the branches indicates relative
genetic distances. Dots lying on a horizontal line represent influenza virus
isolates obtained in the same season (year). The left part of the diagram
shows the emergence of influenza A virus variants lying on the same
evolutionary tree. Dominant variants appear to emerge which show an
accumulation of changes. The right part of the diagram depicts the cocir-
culation of influenza C virus variants derived from multiple evolutionary
pathways. Variation appears to be slower than in the case of influenza A
viruses, as demonstrated by the shorter length of the branches on the
different evolutionary trees. For both influenza A and influenza C viruses
an arbitrary number of seven seasonal cycles is shown on the diagram.
(From Buonagurio, D. A., Nakada, S., Desselberger, U., Krystal, M.,
and Palese, P., *Virology*, 146, 221, 1985. With permission.)

lower than that of influenza A viruses (Table 1). One obvious reason for the difference in
evolutionary rates is the vast difference in generation time between viruses and mammals.
However, another important factor is that RNA polymerases apparently exhibit less fidelity
of genome replication than DNA polymerases, probably due in part to a lack of proofreading
enzymes. The proofreading exonuclease activity of DNA polymerase is able to remove
misincorporated bases from newly synthesized DNA strands so that errors in DNA replication
average as low as 10^{-8} to 10^{-11} per incorporated nucleotide. Studies on RNA viruses suggest
error rates that are orders of magnitude higher than this.[45]

However, attempts to calculate and compare mutation rates for RNA viruses have been
hampered by the lack of a reliable experimental system. One technique to assess mutation
rates of viruses is to measure the frequency of antigenic variants resistant to neutralization
by a selecting monoclonal antibody. It has been reported that the frequency of resistant
mutants is similar for influenza virus, Sendai virus, vesicular stomatitis virus, poliovirus,
and Coxsackie virus and is in the range of 10^{-4} to 10^{-5}.[47-51] This approach, however, is
limited in that only a small region of the genome can be probed for changes using monoclonal
antibody selection. The target size for mutant selection is frequently unknown and may often
include more than one gene. Additionally, differences in mutation frequencies generated
from resistant mutant selection experiments may be explained by differences in avidity of
the discriminating monoclonal antibodies.

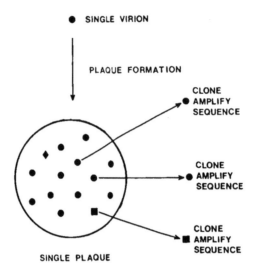

FIGURE 5. Experimental design for measuring mutation rates in viral genes. A single virion (the parental virus) formed a plaque after sufficient time for about five infectious cycles. The many virions contained within the plaque carried genes with the parental progeny sequence (represented by filled circles), and a fraction of the virions present carried genes with a point mutation (represented by square and diamond). The individual virions were cloned by plaquing and amplified under conditions that minimize the effect of new mutations on the consensus sequence of hte gene in the clone. RNA obtained from purified virus was sequenced using the dideoxy chain termination procedure. The mutation rate was determined by dividing the number of observed point mutations by the number of nucleotides analyzed and by the number of infectious cycles. (From Parvin, J. D., Moscona, A., Pan, W. T., Leider, J. M., and Palese, P., *J. Virol.*, 59, 377, 1986. With permission.)

Analysis of variation at the nucleotide level has also been used in an attempt to determine mutation rates for several different RNA viruses. Studies on the bacteriophage Qβ[52] have shown that the error level per genome doubling at given base positions in the RNA genome is between 10^{-3} and 10^{-4}. Using a similar approach, Coffin et al.[53] calculated a mutation frequency of 10^{-3} to 10^{-4} for a particualr point mutation in the Rous sarcoma virus RNA genome. These values are very high; they are both derived from observing the appearance of mutants over several passages and are mathematically based on the assumption that viral replication can be considered as a series of doubling steps. This approach is thus quite complex. More recently, Steinhauer and Holland[54] have quantitated polymerase error frequencies for vesicular stomatitis virus at one highly conserved nucleotide site using both in vivo and in vitro assays. The extremely high frequency of base misincorporation is approximately 10^{-4} substitutions per base incorporated at the site in both assays. One difficulty of this approach is that the substitution frequency at a single nucleotide site may not be representative of all the nucleotide positions of the genome RNA.

Another approach has been taken by Parvin et al.,[55] who assayed the mutation rate in tissue culture for the NS gene of influenza A virus and for the VP1 gene of type 1 poliovirus. Each gene was directly sequenced in over 100 randomly selected viral clones which had arisen from a single virion in one plaque generation (see Figure 5 for diagrammatic representation of this experimental system). The VP1 gene encodes one of the structural proteins comprising the capsid of the poliovirus. Seven mutants of the influenza A virus NS gene were detected, whereas no VP1 gene poliovirus mutant was observed. Thus, the calculated mutation rates are 1.5×10^{-5} and less than 2.1×10^{-6} mutations per nucleotide per infectious cycle for the influenza A virus and the poliovirus 1, respectively. These values represent "neutral" mutations rates since the viral mutants were not compromised in their replication. Thus, the mutation rate of influenza A virus was found to be significantly higher than that of poliovirus 1. A similar mutation rate for poliovirus was found by Sedivy et al.[55a] who assayed for amber mutation revertants.

Durbin and Stollar[56] also present evidence that another RNA virus — Sindbis virus — has a much lower mutation rate than does influenza A virus. These authors took the clonal progeny of a given virus (grown under conditions not prejudicial to a certain host type revertant) and plaqued it on cells permissive for the revertant, but nonpermissive for wild type. Because they knew from previous sequencing studies which base changes can reverse the phenotype, they were able to calculate a mutation rate. They report that Sindbis virus RNA polymerase has a mutation rate of less than 10^{-6} errors per base incorporation.

In conclusion, it is difficult to compare the above data on variation in different RNA viruses. Different experimental systems were used and different mathematical approaches were taken. Nevertheless, it is clear that although mutation rates may be high for all RNA viruses, there may still be significant differences in mutation rates between these viruses. These differences may prove to have important epidemiological consequences.

ACKNOWLEDGMENTS

The work in the authors' laboratories was supported by NIH grants AI-11823 and AI-18998 to P. Palese and DK 38381 to F. Smith. We thank Jeffrey D. Parvin for stimulating discussions.

REFERENCES

1. **Palese, P.,** The genes of influenza virus, *Cell*, 10, 1, 1977.
2. **Lamb, R. A.,** The influenza virus RNA segments and their encoded proteins, in *Genetics of Influenza Viruses*, Palese, P. and Kingsbury, D. W., Eds., Springer-Verlag, New York, 1983, 21.
3. **Air, G. M. and Compans, R. W.,** Influenza B and influenza C viruses, in *Genetics of Influenza Viruses*, Palese, P. and Kingsbury, D. W., Eds., Springer-Verlag, New York, 1983, 280.
4. **Nakada, S., Creager, R. S., Krystal, M., Aaronson, R. P., and Palese, P.,** Influenza C virus hemagglutinin: comparison with influenza A and B virus hemagglutinins, *J. Virol.*, 50, 118, 1984.
5. **Herrler, G., Rott, R., Klenk, H.-D., Muller, H.-P., Shukla, A. K., and Schauer, R.,** The receptor-destroying enzyme of influenza C virus is neuraminate-*O*-acetylesterase, *EMBO J.*, 4, 1503, 1985.
6. **Nakada, S., Creager, R. S., Krystal, M., and Palese, P.,** Complete nucleotide sequence of the influenza C/California/78 virus nucleoprotein gene, *Virus Res.*, 1, 433, 1984.
7. **Nakada, S., Graves, P. N., and Palese, P.,** The influenza C virus NS gene: evidence for a spliced mRNA and a second NS gene product (NS2 protein), *Virus Res.*, 4, 263, 1986.
7a. **Vlasak, R., Krystal, M., Nacht, M., and Palese, P.,** The influenza C virus glycoprotein (HE) exhibits receptor-binding (hemagglutinin) and receptor-destroying (esterase) activities, *Virology*, 160, 419, 1987.
8. **Nakajima, K., Desselberger, U., and Palese, P.,** Recent human influenza A (H1N1) viruses are closely related genetically to strains isolated in 1950, *Nature (London)*, 274, 334, 1978.
9. **Kendal, A. P., Noble, G. R., Skehel, J. J., and Dowdle, W. R.,** Antigenic similarity of influenza A (H1N1) viruses from epidemics in 1977—1978 to "Scandinavian" strains isolated in epidemics of 1950—1951, *Virology*, 89, 632, 1978.
10. **Scholtissek, C., Rohde, W., Von Hoyningen, V., and Rott, R.,** On the origin of the human influenza virus subtypes H2N2 and H3N2, *Virology*, 87, 13, 1978.
11. **Katagiri, S., Ohizumi, A., and Homma, M.,** An outbreak of type C influenza in a children's home, *J. Infect. Dis.*, 148, 51, 1983.
12. **Chakraverty, P.,** Antigenic relationship between influenza C viruses, *Arch. Virol.*, 58, 341, 1978.

13. **Guo, Y. J., Jin, F. G., Wang, P., Wang, M., and Zhu, J. M.,** Isolation of influenza C virus from pigs and experimental infection of pigs with influenza C virus, *J. Gen. Virol.,* 64, 177, 1983.

14. **Winter, G., Fields, S., and Brownlee, G. G.,** Nucleotide sequence of the haemagglutinin of a human influenza virus H1 subtype, *Nature (London),* 292, 72, 1981.

15. **Krystal, M., Elliott, R. M., Benz, E. W., Young, J. F., and Palese, P.,** Evolution of influenza A and B viruses: conservation of structural features in the hemagglutinin gene, *Proc. Natl. Acad. Sci. U.S.A.,* 79, 4800, 1982.

16. **Markoff, L. and Lai, C.-J.,** Sequence of the influenza A/Udorn/72 (H3N2) virus neuraminidase gene as determined from cloned full-length DNA, *Virology,* 119, 288, 1982.

17. **Colman, P. M. and Ward, C. W.,** Structure and diversity of influenza virus neuraminidase, *Curr. Top. Microbiol. Immunol.,* 114, 177, 1985.

18. **Webster, R. G., Campbell, C. H., and Granoff, A.,** The *in vivo* production of "new" influenza viruses. I. Genetic recombination between avian and mammalian influenza viruses, *Virology,* 44, 317, 1971.

19. **Desselberger, U., Nakajima, K., Alfino, P., Pedersen, F. S., Haseltine, W. A., Hannoun, C., and Palese, P.,** Biochemical evidence that "new" influenza virus strains in nature may arise by recombination (reassortment), *Proc. Natl. Acad. Sci. U.S.A.,* 76, 3341, 1978.

20. **Young, J. F. and Palese, P.,** Evolution of human influenza A viruses in nature: recombination contributes to genetic variation of H1N1 strains, *Proc. Natl. Acad. Sci. U.S.A.,* 76, 6547, 1979.

21. **Bean, W. J., Cox, N. J., and Kendal, A. P.,** Recombination of human influenza A viruses in nature, *Nature (London),* 284, 638—640, 1980.

22. **Cox, N. J., Bai, Z. S., and Kendal, A. P.,** Laboratory-based surveillance of influenza A (H1N1) and A (H3N2) viruses in 1980—81: antigenic and genomic analyses, *Bull. WHO,* 61, 143, 1983.

23. **Laver, W. G. and Webster, R. G.,** Studies on the origin of pandemic influenza. III. Evidence implicating duck and equine influenza viruses as possible progenitors of the Hong Kong strain of human influenza, *Virology,* 51, 383, 1973.

24. **Schulman, J. L. and Kilbourne, E. D.,** Independent variation in nature of hemagglutinin and neuraminidase antigens of influenza virus: distinctiveness of hemagglutinin antigen of Hong Kong/68 virus, *Proc. Natl. Acad. Sci. U.S.A.,* 63, 326, 1969.

25. **Fang, R., Min Jou, W., Huylebroeck, D., Devos, R., and Fiers, W.,** Complete structure of A/duck/Ukraine/63 influenza hemagglutinin gene: animal virus as progenitor of human H3 Hong Kong 1968 influenza hemagglutinin, *Cell,* 25, 315, 1981.

26. **Ward, C. W. and Dopheide, T. A.,** Amino acid sequence and oligosaccharide distribution of the hemagglutinin from an early Hong Kong influenza virus variant A/Aichi/2/68 (X-31), *Biochem. J.,* 193, 953, 1981.

27. **Ward, C. W. and Dopheide, T. A.,** Evolution of the Hong Kong influenza A subtype, *Biochem. J.,* 195, 337, 1981.

28. **Young, J. F., Desselberger, U., and Palese, P.,** Evolution of human influenza A viruses in nature: sequential mutations in the genomes of new H1N1 isolates, *Cell,* 18, 73, 1979.

29. **Scholtissek, C., Von Hoyningen, V., and Rott, R.,** Genetic relatedness between the new 1977 epidemic strains (H1N1) of influenza and human influenza strains isolated between 1947 and 1957 (H1N1), *Virology,* 89, 613, 1978.

30. **Concannon, P., Cummings, I. W., and Salser, W. A.,** Nucleotide sequence of the influenza virus A/USSR/90/77 hemagglutinin gene, *J. Virol.,* 49, 276, 1984.

31. **Raymond, R. L., Caton, A. J., Cox, N. J., Kendal, A. P., and Brownlee, G. G.,** The antigenicity and evolution of influenza H1 hemagglutinin, from 1950—1957 and 1977—1983: two pathways from one gene, *Virology,* 148, 275, 1986.

32. **Buonagurio, D. A., Nakada, S., Parvin, J. D., Krystal, M., Palese, P., and Fitch, W. M.,** Evolution of human influenza A viruses over 50 years: rapid, uniform rate of change in NS gene, *Science,* 232, 980, 1986.

33. **Both, G. W. and Sleigh, M. J.,** Conservation and variation in the hemagglutinins of Hong Kong subtype influenza viruses during antigenic drift, *J. Virol.,* 39, 663, 1981.

34. **Both, G. W., Sleigh, M. J., Cox, N. J., and Kendal, A. P.,** Antigenic drift in influenza virus H3 hemagglutinin from 1968—1980: multiple evolutionary pathways and sequential amino acid changes at key antigenic sites, *J. Virol.,* 48, 52, 1983.

35. **Hall, R. M. and Air, G. M.,** Variation in nucleotide sequences coding for the N-terminal regions of the matrix and nonstructural proteins of influenza A viruses, *Virology,* 38, 1, 1981.

36. **Fitch, W. M.,** Toward defining the course of evolution: minimum change for a specific tree topology, *Syst. Zool.,* 20, 406, 1971.

37. **Wilson, A. C., Carlson, S. S., White, T.J.,** Biochemical evolution, *Annu. Rev. Biochem.,* 46, 573, 1977.

38. **Meier-Ewert, H., Petri, T., and Bishop, D. H. L.,** Oligonucleotide fingerprint analyses of influenza C virion RNA recovered from five different isolates, *Arch. Virol.,* 67, 141, 1981.

39. **Guo, Y. J. and Desselberger, U.**, Genome analysis of influenza C viruses isolated in 1981/1982 from pigs in China, *J. Gen. Virol.*, 65, 1857, 1984.

40. **Buonagurio, D. A., Nakada, S., Desselberger, U., Krystal, M., and Palese, P.**, Noncumulative sequence changes in the hemagglutinin genes of influenza C virus isolates, *Virology*, 146, 221, 1985.

41. **Buonagurio, D. A., Nakada, S., Fitch, W. M., and Palese, P.**, Epidemiology of influenza C virus in man: multiple evolutionary lineages and low rate of change, *Virology*, 153, 12, 1986.

42. **Kimura, M.**, The neutral theory of molecular evolution, in *Evolution of Genes and Proteins*, Nei, M. and Koehn, R. K., Eds., Sinauer Associates, Sunderland, Massachusetts, 1983, 208.

43. **Eigen, M. and Schuster, P.**, The hypercycle — a principle of natural self-organization, *Naturwissenschaften*, 64, 541, 1977.

44. **Domingo, E., Martinez-Salas, E., Sobrino, F., de la Torre, J. C., Portela, A., Ortin, J., Lopez-Galindez, C., Perez-Brena, P., Villaneuva, N., Najera, R., VandePol, S., Steinhauer, D., DePolo, N., and Holland, J.**, The quasispecies (extremely heterogeneous) nature of viral RNA genome populations: biological relevance — a review, *Gene*, 40, 1, 1985.

45. **Holland, J., Spindler, K., Horodyski, F., Grabau, E., Nichol, S., and VandePol, S.**, Rapid evolution of RNA genomes, *Science*, 215, 1577, 1982.

46. **Li, W.-H., Wu, C.-I., and Luo, C.-C.**, A new method for estimating synonymous and nonsynonymous rates of nucleotide substitution considering the relative likelihood of nucleotide and codon changes, *Mol. Biol. Evol.*, 2, 150, 1985.

47. **Portner, A., Webster, R. G., and Bean, W. J.**, Similar frequencies of antigenic variants in Sendai, vesicular stomatitis, and influenza A viruses, *Virology*, 104, 235, 1980.

48. **Prabhakar, B. S., Haspel, M. V., McClintock, P. R., and Notkins, A. L.**, High frequency of antigenic variants among naturally occurring human Coxsackie B4 virus isolates identified by monoclonal antibodies, *Nature (London)*, 300, 374, 1982.

49. **Lubeck, M. D., Schulman, J. L., and Palese, P.**, Antigenic variants of influenza viruses: marked differences in the frequencies of variants selected with different monoclonal antibodies, *Virology*, 102, 458, 1980.

50. **Emini, E. A., Kao, S.-Y., Lewis, A. J., Crainic, R., and Wimmer, E.**, Functional basis of poliovirus neutralization determined with monospecific neutralizing antibodies, *J. Virol.*, 46, 466, 1983.

51. **Minor, P. D., Schild, G. C., Bootman, J., Evans, D. M. A., Ferguson, M., Reeve, P., Spitz, M., Stanway, G., Cann, A. J., Hauptmann, R., Clarke, L. D., Mountford, R. C., and Almond, J. W.**, Location and primary structure of a major antigenic site for poliovirus neutralization, *Nature (London)*, 310, 674, 1983.

52. **Domingo, E., Sabo, D., Taniguchi, T., and Weissmann, C.**, Nucleotide sequence heterogeneity of an RNA phage population, *Cell*, 13, 735, 1978.

53. **Coffin, J. M., Tsichlis, P. V., Barker, C. S., and Voynow, S.**, Variation in avian retrovirus genomes, *Ann. N.Y. Acad. Sci.*, 354, 410, 1980.

54. **Steinhauer, D. A. and Holland, J. J.**, Direct method for quantitation of extreme polymerase error frequencies at selected single base sites in viral RNA, *J. Virol.*, 57, 219, 1986.

55. **Parvin, J. D., Moscona, A., Pan, W. T., Leider, J. M., and Palese, P.**, Measurement of the mutation rates of animal viruses: influence A virus and poliovirus type 1, *J. Virol.*, 59, 377, 1986.

55a. **Sedivy, J. M., Capone, J. P., Rajbhandary, U. L., and Sharp, P. A.**, An inducible mammalian amber suppressor: propagation of a poliovirus mutant, *Cell*, 50, 379, 1987.

56. **Durbin, R. K. and Stollar, V.**, Sequence analysis of the E1 gene of a hyperglycosylated, host restricted mutant of Sindbis virus and estimation of mutation rate from frequency of revertants, *Virology*, 154, 135, 1986.

57. **Buonagurio, D. A.**, Evolutionary Patterns of Influenza A and C Viruses in Man, Ph.D. thesis, Mount Sinai School of Biomedical Sciences, New York, 1986.

Role of Genome Variation in Virus Evolution

Chapter 7

ANTIGENIC VARIATION IN INFLUENZA VIRUS HEMAGGLUTININS

J. J. Skehel and D. C. Wiley

TABLE OF CONTENTS

I. INTRODUCTION

Antigenic variation in influenza viruses primarily involves the hemagglutinin (HA) and neuraminidase (NA) glycoproteins of the virus membranes. Proteins enclosed by the membranes are not invariant; they can be distinguished antigenically by their partial reactions with heterologous, type-specific sera and by using monoclonal antibodies.[1,2] The genes which specify these proteins can also be differentiated by nucleic acid hybridization[3] and their nucleotide sequences have been shown to differ in a limited number of analyses by up to 8%.[4,5] By comparison, however, the amino acid sequences of the membrane glycoproteins differ more extensively, by up to 80%, and influenza A viruses are placed in subtypes on the basis of the exclusive reactions of their HAs and NAs with subtype-specific sera.[6] There are 13 antigenic subtypes of HA and nine of NA. All 13 HAs have been identified in viruses from avian species, and more restricted ranges have been obtained from horses, swine, and humans. The three subtypes of HA, H_1, H_2, and H_3, detected in humans since the first isolation of influenza viruses from humans in 1933[7] have been most extensively analyzed, and this review of antigenic variation in HAs will concentrate on them. The time periods of their prevalence were H_1, 1918 to 1957 and 1977 to the present; H_2 1957 to 1968; and H_3 1968 to the present.

II. ANTIGENIC SHIFT

The sudden appearance of "Asian" influenza viruses of the H_2 subtype in 1957 and of "Hong Kong" viruses of the H_3 subtype in 1968, and the displacement by them of viruses of the H_1 and H_2 subtypes, respectively, are commonly called antigenic shifts. Studies on the origin of these viruses and the mechanism of antigenic shift strongly suggest that they are transferred to the human population from animal or avian virus reservoirs, and that the process of transfer involves the formation of recombinant viruses during mixed infections of an appropriate host with human and animal or avian viruses. For the Hong Kong viruses in particular, antigenic analyses of avian and equine viruses isolated before 1968 revealed that the HAs of the viruses A/Duck/Ukraine/1963[8] and A/Equine/Miami/1963[9] were antigenically closely related to the HA of 1968 isolates from humans. Subsequent protein[10,11] and nucleic acid sequencing studies[12,13] confirmed these similarities and indicated the close structural relationship of the HAs from the avian and human viruses. The suggested mechanism of transfer of viruses with these antigenic properties into the human population is based on two main observations. First, because of the segmented nature of influenza virus genomes,[14,15] reassortment occurs readily during mixed infections both in vivo[16] and in vitro,[17] and second, comparative RNA hybridization analyses indicate that seven of the eight genome segments of the 1968 Hong Kong virus were closely related to equivalent segments of the Asian influenza viruses; the eighth which coded for the hemagglutinin, the antigen with which infectivity neutralizing antibodies interact, was distinct.[18] It is therefore likely that a recombinant virus with a novel HA, but otherwise similar to the viruses circulating in the human population at the time, was responsible for the Hong Kong pandemic. The site of the reassortment and the actual viruses involved are not known, but evidence from studies of the host range characteristics of avian and swine viruses suggest that swine are the most likely intermediates in the process.[19] Similar conclusions with regard to the mixed origin of the RNAs of the virus responsible for the Asian influenza pandemic in 1957 have been reported,[18] but a different situation exists with regard to the reintroduction of viruses of the H_1 subtype in 1977. Antigenically,[20] and as judged by the results of RNA hybridization[21] and sequencing analyses,[22] these viruses were closely related to H_1 subtype viruses infecting humans in 1950. Consequently, for this antigenic shift the emergence of a dormant virus from an unknown depot seems to have occurred.

Antigenic shift occurs unpredictably at intervals of about 40 (1918 to 1957), 10 (1957 to 1968), and at least 18 years (1968 to the present) from direct observation, and indirectly, from serological studies, at a similar interval in the last century (1890 to 1899).[23] Interestingly, the HAs of the viruses involved in the two late 19th century antigenic shifts may also have belonged to the H_2 and H_3 subtypes, but the significance of this apparent restriction is not known. The main property of the HA which might influence the ease of interspecies transfer is receptor binding specificity, which certainly differs between the HAs of viruses isolated from humans and those from birds and horses[24] and might limit the number of HAs which can successfully change their tissue tropism from avian intestinal epithelium cells to the respiratory epithelium of humans.

III. ANTIGENIC DRIFT

Antigenic drift describes the progressive change in antigenic properties which occurs with time after the introduction of a new subtype into the human population. Convincing evidence for the process is only available for human influenza, although there are suggestions that it may also occur in horses.[13] Routinely, the changes in HA antigenicity are analyzed in hemagglutination inhibition tests using convalescent ferret antisera which are highly strain specific and show clearly the relationships between viruses isolated in different years.[25] Recently, these analyses have been accompanied by studies of the nucleotide sequences of the RNA genes for HA of representative isolates[26-30] and the results obtained indicate that antigenic drift results from the accumulation of amino acid substitutions in one of the two glycopolypeptide components of the HA subunit. Hemagglutinins are trimers of molecular weight of about 220,000, in which each monomer consists of two disulfide-linked glyco polypeptides HA_1 and HA_2. The three-dimensional structure of the 1968 Hong Kong virus HA is known to 3 Å resolution.[31] The molecule is a 135-Å-long cylinder, approximately triangular in cross section, and varying in radius from about 15 to 40 Å. The HA_1 chain of each subunit extends from the base of the molecule at the virus membrane through a fibrous stem to a distal region rich in β-structure and composed exclusively of HA_1 residues. It then returns to the fibrous region of the molecule and terminates about 30 Å from the virus membrane (Figure 1). The most prominent features of the part of the subunit composed of HA_2 residues are two antiparallel α-helices, one, 29 Å long, which proceeds distally beginning about 30 Å from the membrane to connect with the other helix which is 76 Å long and stretches back to within 20Å of the membrane. These helices are the main components of the central fibrous stem of the molecule which supports the globular membrane distal HA_1 domain.

In the H_3 subtype, since 1968, 73 of the 328 residues of HA_1 have been observed to have changed, compared with 12 in HA_2.[26,27,32] The changes cluster in five surface regions of the distal domain labeled A to E[33] in Figure 1, and a list of the amino acid substitutions detected is also given. The substitutions involve changes in charge, e.g., 144 Gly to Asp, 156 Lys to Glu, or side chain length, e.g., 198 Ala to Val, or changes which lead to additional sites for glycosylation: 63 Asp to Asn in the sequence Asn–Cys–Thr, 124 Gly to Ser in the sequence Asn–Glu–Ser, and 248 Asn to Ser or Thr in the sequence Asn–Ser–Ser–Thr. For the first of the three last examples, a direct effect of the carbohydrate side chain on HA antigenicity has been shown,[34] and the other two extra sites for glycosylation are in positions where a carbohydrate side chain could mask antigenically important regions of the molecule. In some positions, e.g., 144 Gly to Asp to Val, repeated changes occur. The antigenic significance of changes in these five regions is confirmed by observations that single amino acid substitutions in one or other of the sites are detected in the HAs of monoclonal antibody selected antigenic variants.[35-39] Two such variants containing 146 Gly to Asp or 188 Asn to Asp substitutions have also been used to provide evidence that antibodies bind to the region

residue	change
2	D → N
3	L → F
9	N → S
10	T → K
31	D → N
34	I → T
48	T → A
50	K → R
53	N → D
54	N → S
58	I → V
62	I → K
67	I → V
78	V → G
82	E → K
83	T → K
91	S → G
92	K → R
122	T → N
124	G → S
126	T → N
126	T → I
129	G → E
132	Q → E
133	N → S
137	N → Y
137	N → S
143	P → S
144	G → D
144	D → V
145	S → N
145	N → K
145	S → R
145	S → I
146	G → S
155	T → Y
156	K → E
156	K → Q
157	S → L
158	G → E
159	S → Y
160	T → K
160	T → A
160	T → S
160	K → R
160	V → A
164	L → Q
170	N → D
171	D → G
172	D → G
173	N → K
174	F → S
182	I → V
186	S → I
188	N → D
188	D → Y
189	Q → R
189	Q → K
191	Q → P
193	N → D
193	S → N
193	N → K
197	Q → R
198	A → T
198	A → V
201	R → K
207	R → K
208	R → G
209	S → N
213	I → V
217	I → V
219	S → P
229	R → G
230	I → V
240	G → R
242	V → I
244	V → L
244	L → S
246	N → K
248	N → S
248	N → T
260	M → I
261	R → H
275	D → G
278	I → S
307	K → R
323	V → I

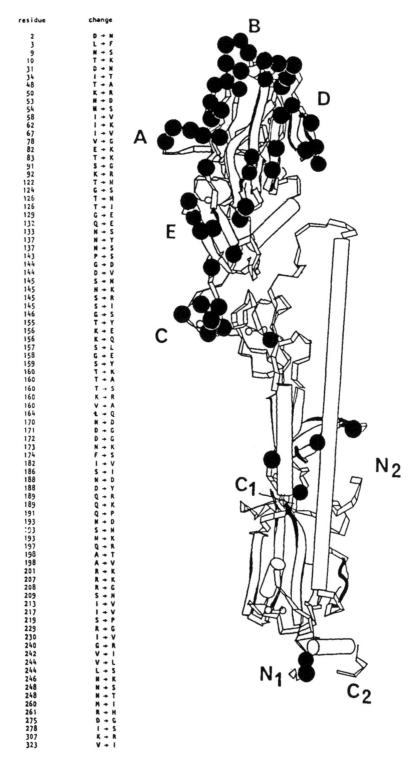

FIGURE 1. A diagram of a monomer of the A/Hong Kong/1968 hemagglutinin showing the locations of amino acid substitutions in the HA$_1$ glycopolypeptide chain and the antigenically important regions designated A, B, C, D, and E. N$_1$, C$_1$, N$_2$, and C$_2$ denote the amino and carboxyl termini of the HA$_1$ and HA$_2$ glycopolypeptide chains, respectively; both N$_1$ and C$_2$ are next to the virus membrane. The list includes all amino acid changes detected in natural variants of the H$_3$ subtype between 1968 and 1986.

of the molecule in which the amino acid substitution occurs.[40,41] In these cases, the structures of the mutant HAs were determined crystallographically and the structural differences from the wild-type HA shown to be confined to the immediate vicinity of the amino acid substitution. Such observations also indicate directly that the simple addition of an amino acid side chain is a sufficient structural alteration to prevent antibody binding. The results of electron microscopy of HA monoclonal antibody complexes for antibodies which recognize changes in sites A, B, and E[42] are consistent with the conclusion that the site of amino acid substitution is within the site of antibody binding.

The frequency at which amino acid substitutions occur in HA_1 is estimated from studies of natural isolates to be about 1.1%/year.[26,43] The 73 residue changes since 1968 correspond to about 22% of amino acid sequence difference, which compares with differences in HA_1 between H_2 and H_3 subtype HAs of 64%, and between H_1 and H_2 HAs of 42%.[44,45]

Similar studies to these have been done with viruses of the H_1 subtype isolated between 1934 and 1957 and since 1977.[28-30] Again, amino acid substitutions occur predominantly in HA_1 and appear to cluster near the five regions shown in Figure 1. Analyses of the specificities of a large number of monoclonal antibodies prepared against the HA of A/Puerto Rico/8/ 34 virus led to the designation of four antigenically important parts of the HA Sa, Sb, Ca, and Cb[46] which, by subsequent sequence analysis of the HAs of selected variants,[47] were indicated to correspond approximately to sites B, B, A + D, and E respectively. Mutants in site C, several of which were detected in the natural variants,[28,30] were not represented in the collection of monoclonal antibody selected mutants, and this observation is similar to those made for the H_3 subtype in which a single mutation at HA1 residue 54 in a monoclonal antibody selected variant[35] serves to define this site. The importance of amino acid substitutions which influence glycosylation is also suggested from these studies of the H_1 subtype in which substitutions in regions B (Sa) and E (Cb) introduce new sites for glycosylation. Amino acid substitution between 1977 and 1983 occurred at a rate of about 0.8%/year, and a higher frequency of 1.4%/year was observed in the HAs of viruses isolated between 1950 and 1957, which may reflect a high level of immunity in the population in the fourth decade of H_1 subtype prevalence. To date, however, antigenic drift in the H_3 subtype has remained approximately constant at 1.1% between 1968 and 1979,[26,43] and 1.2% between 1979 and 1986.[32]

Overall, the information obtained from these studies of antigenic drift in both H_1 and H_3 subtypes indicates that as a result of immune pressure, antigenic variants are selected which contain amino acid substitutions at the surface of the membrane distal domain of the molecule. These substitutions prevent the binding of antibodies induced following a previous infection, and as a consequence, the variants selected have the ability to reinfect. The direction of antigenic drift is dictated by specific amino acid substitutions which are shared by subsequent variants and viruses which cause major epidemics, such as A/Victoria/3/75, contain substitutions in each of the antigenically important regions. The precise mechanism of their selection is not known. Considering that the frequency at which antigenic variants are selected by monoclonal antibodies in the laboratory is between 10^{-4} and 10^{-5},[48] and that antibodies which recognize any of the five antigenic regions neutralize virus infectivity, selection of a mutant with the ability to reinfect the majority of the population would be expected to occur spontaneously at a very low frequency. Changes at each of the antigenic sites may, on the other hand, occur during reinfections by antigenic mutants changed in only one or two sites of individuals who develop only partial immunity during initial infections. Analysis of the variety of antibody specificities in postinfection human sera[49] and of the restricted ability of human sera to neutralize monoclonal antibody selected mutants[50] suggest that this is the case.

Changes in the structure of the HA which result in changes in antigenicity involve surface residues in the membrane distal domain of the molecule, and the extent to which they are accommodated (Figure 1) indicates the structural plasticity of this surface. In certain cases, however, they are not without effect on HA function. A number of reports have indicated interrelationships between changes in receptor binding properties and changes in antigenicity, particularly in sites A and B which encircle the receptor binding pocket.[51-54] The suggestion has in fact been made[55] that the accumulation of amino acid substitutions in this region of the molecule may eventually limit the viability of certain antigenic variants and the extent of variation. The relationship between receptor binding and antigenicity may also be an important consideration when assessing the significance of natural antigenic variants detected in influenza surveillance studies. The viruses used in such studies are isolated and passaged in hen eggs or in cells in tissue culture which may impose selection pressures in favor of certain receptor variants with specific antigenic properties. The heterogeneous mixture of isolates identified almost every year in such studies may, therefore, not accurately reflect the relative abundance of the viruses in nature,[56] and the potential importance of these possibilities for vaccination strategies has been noted.[57,58]

What are the properties of influenza viruses responsible for their characteristic frequency of antigenic change? Are HAs in some way less effective immunogens than, for example, the equivalent measles virus membrane glycoproteins? Probably not, considering the solid immunity of all adults born before 1957 to the reintroduced H_1 viruses in 1977. Possibly, the influenza RNA polymerases are more error prone than those of other RNA viruses, but this is unlikely since antigenic variants of rhabdo- and paramyxoviruses are selected by monoclonal antibodies at about the same frequency.[59] The answer to this basic question and to others, such as, how are viruses of one subtype displaced with the introduction of a new subtype? and what is the role of cellular immunity in antigenic variation? are not at present known. It seems likely, however, that the existence of influenza virus reservoirs in other species and the transfer of viruses from them to humans at the infrequent intervals of pandemic influenza, are major factors in disturbing the approach to equilibrium in the human population which similar but less variable viruses appear to achieve.

REFERENCES

1. **Schild, G. C., Oxford, J. S., and Newman, R. W.,** Evidence for antigenic variation in influenza A nucleoprotein, *Virology*, 93, 569, 1979.
2. **Van Wyke, K. L., Hinshaw, V. S., Bean, W. J., and Webster, R. G.,** Antigenic variation of influenza A virus nucleoprotein detected with monoclonal antibodies, *J. Virol.*, 35, 24, 1980.
3. **Bean, W. J.,** Correlation of influenza A virus nucleoprotein genes with host species, *Virology*, 133, 438, 1984.
4. **Winter, G. and Fields, S.,** The structure of the gene encoding the nucleoprotein of human influenza virus A/PR/8/34, *Virology*, 114, 423, 428, 1981.
5. **Huddleston, J. A. and Brownlees, G. G.,** The sequence of the nucleoprotein gene of human influenza A virus, strain A/NT/60/68, *Nuc. Acid Res.*, 10, 1029, 1982.
6. WHO Memorandum, A revised system of nomenclature for influenza viruses, *Bull. WHO*, 58, 585, 1980.
7. **Smith, W. C., Andrewes, C. H., and Laidlaw, P. P.,** A virus obtained from influenza patients, *Lancet*, ii, 66, 1933.
8. **Coleman, M. T., Dowdle, W. R., Pereira, H. G., Schild, G. C., and Chang, W. K.,** The Hong Kong/68 influenza A_2 variant, *Lancet*, ii, 1384, 1968.
9. **Waddell, G. H., Tiegland, M. B., and Sigel, M. M.,** A new influenza virus associated with equine respiratory disease, *J. Am. Vet. Med. Assoc.*, 143, 587, 1963.
10. **Laver, W. G. and Webster, R. G.,** Studies on the origin of pandemic influenza. III. Evidence implicating duck and equine influenza viruses as possible progenitors of the Hong Kong strain of human influenza, *Virology*, 51, 383, 1973.
11. **Ward, C. W. and Dopheide, T. A.,** Evolution of the Hong Kong influenza A subtype. Structural relationship between the haemagglutinin from A/Duck/Ukraine/63 (Hav 7) and the Hong Kong (H_3) haemagglutinins, *Biochem. J.*, 195, 337, 1981.

12. **Fang, R., Min Jou, W., Huylebroeck, D., Devos, R., and Fiers, W.,** Complete structure of A/Duck/ Ukraine/63 influenza haemagglutinin gene: animal virus as progenitor of human H_3 Hong Kong 1968 influenza haemagglutinin, *Cell,* 25, 315, 1981.
13. **Daniels, R. S., Skehel, J. J., and Wiley, D. C.,** Amino acid sequences of haemagglutinins of Influenza Viruses of the H_3 subtype isolated from horses, *J. Gen. Virol.,* 66, 457, 1985.
14. **McCauley, J. W. and Mahy, B. W. J.,** Structure and function of the influenza virus genome, *Biochem. J.,* 211, 281, 1983.
15. **Lamb, R. A. and Choppin, P. W.,** The gene structure and replication of influenza virus, *Ann. Rev. Biochem.,* 52, 467, 1983.
16. **Webster, R. G., Campbell, C. H., and Granoff, A.,** The 'in vivo' production of 'new' influenza A viruses. I. Genetic recombination between avian and mammalian influenza viruses, *Virology,* 44, 317, 1971.
17. **Tumova, B. and Pereira, H. G.,** Genetic interaction between influenza A viruses of human and animal origin, *Virology,* 27, 253, 1965.
18. **Scholtissek, C., Rohde, W., Von Hoyningen, V., and Rott, R.,** On the origin of the Human Influenza Virus subtypes H_2N_2 and H_3N_2, *Virology,* 87, 13, 1978.
19. **Scholtissek, C., Burger, H., Kistner, O., and Shortridge, K. F.,** The nucleoprotein as a possible major factor in determining host specificity of influenza H_3N_2 viruses, *Virology,* 147, 287, 1985.
20. **Kendal, A. P., Noble, G. R., Skehel, J. J., and Dowdle, W. R.,** Antigenic similarity of influenza A (H_1N_1) viruses from epidemics in 1977—1978 to 'Scandinavian' strains isolated in epidemics of 1950—1951, *Virology,* 89, 632, 1978.
21. **Scholtissek, C., Von Hoyningen, V., and Rott, R.,** Genetic relatedness between the new 1977 epidemic strains (H_1N_1) of influenza and human influenza strains isolated between 1947 and 1957, *Virology,* 89, 613, 1978.
22. **Young, J. F., Desselberger, U., and Palese, P.,** Evolution of human influenza A viruses in nature: sequential mutations in the genomes of new H_1N_1 isolates, *Cell,* 18, 73, 1979.
23. **Mulder, J. and Masurel, N.,** Pre-epidemic antibody against the 1957 strain of Asiatic influenza in the serum of older persons living in the Netherlands, *Lancet,* 810, 1958.
24. **Rogers, G. N. and Paulson, J. C.,** Receptor determinants of human and animal influenza virus isolates: differences in receptor specificity of the H_3 haemagglutinin based on species of origin, *Virology,* 127, 361, 1983.
25. **Pereira, M. S.,** Persistence of influenza in a population, in Virus Persistence Symposium 33 of the Society for General Microbiology, Ed. Mahy, B. W. J., Minson, A. C. and Darby, G. K., Eds., Cambridge University Press, 1982, 15.
26. **Both, G. W., Sleigh, M. J., Cox, N. J., and Kendal, A. P.,** Antigenic drift in influenza virus H_3 haemagglutinin from 1968 to 1980. Multiple evolutionary pathways and sequential amino acid changes at key antigenic sites, *J. Virol.,* 48, 52, 1983.
27. **Skehel, J. J., Daniels, R. S., Douglas, A. R., and Wiley, D. C.,** Antigenic and amino acid sequence variations in the haemagglutinins of type A influenza viruses recently isolated from human subjects, *Bull. WHO,* 61, 671, 1983.
28. **Daniels, R. S., Douglas, A. R., Skehel, J. J., and Wiley, D. C.,** Antigenic and amino acid sequence analyses of influenza viruses of the H_1N_1 subtype isolated between 1982 and 1984, *Bull. WHO,* 63, 273, 1985.
29. **Raymond, F. L., Caton, A. J., Cox, N. J., Kendal, A. P., and Brownlee, G. G.,** Antigenicity and evolution amongst recent influenza viruses of H_1N_1 subtypes, *Nuc. Acid Res.,* 11, 7191, 1983.
30. **Raymond, F. L., Caton, A. J., Cox, N. J., Kendal, A. P., and Brownlee, G. G.,** The antigenicity and evolution of Influenza H_1 haemagglutinin from 1950—1957 and 1977—1983: two pathways from one gene, *Virology,* 148, 275, 1986.
31. **Wilson, I. A., Skehel, J. J., and Wiley, D. C.,** Structure of the haemagglutinin membrane glycoprotein of influenza virus at 3Å resolution, *Nature,* 289, 366, 1981.
32. **Stevens, D. J., Skehel, J. J., and Wiley, D. C.,** in preparation.
33. **Wiley, D. C., Wilson, I. A., and Skehel, J. J.,** Structural identification of the antibody-binding sites of Hong Kong influenza haemagglutinin and their involvement in antigenic variation, *Nature,* 289, 373, 1981.
34. **Skehel, J. J., Stevens, D. J., Daniels, R. S., Douglas, A. R., Knossow, M., Wilson, I. A., and Wiley, D. C.,** A carbohydrate side chain on haemagglutinins of Hong Kong influenza viruses inhibits recognition by a monoclonal antibody, *Proc. Natl. Acad. Sci. U.S.A.,* 81, 1779, 1984.
35. **Laver, W. G., Air, G. M., Webster, R. G., Gerhard, W., Ward, C. W. and Dopheide, T. A.,** Antigenic drift in type A influenza virus: sequence differences in the haemagglutinin of Hong Kong (H_3N_2) variants selected with monoclonal hybridoma antibodies, *Virology,* 98, 226, 1979.
36. **Laver, W. G., Air, G. M., and Webster, R. G.,** Mechanism of antigenic drift in influenza virus. Amino acid sequence changes in an antigenically active region of Hong Kong (H_3N_2) influenza virus haemagglutinin, *J. Mol. Biol.,* 145, 339, 1981.

37. **Daniels, R. S., Douglas, A. R., Skehel, J. J., and Wiley, D. C.,** Analyses of the antigenicity of influenza haemagglutinin at the pH optimum for virus-mediated membrane fusion, *J. Gen. Virol.,* 64, 1657, 1983.
38. **Newton, S. E., Air, G. M., Webster, R. G., and Laver, W. G.,** Sequence of the haemagglutinin gene of influenza virus A/Memphis/1/71 and previously uncharacterized monoclonal antibody derived variants, *Virology,* 128, 495, 1983.
39. **Webster, R. G., Brown, L. E., and Jackson, D. C.,** Changes in the antigenicity of the haemagglutinin molecule of H_3 influenza virus at acidic pH, *Virology,* 126, 587, 1983.
40. **Knossow, M., Daniels, R. S., Douglas, A. R., Skehel, J. J., and Wiley, D. C.,** Three-dimensional structure of an antigenic mutant of the influenza virus haemagglutinin, *Nature,* 311, 678, 1984.
41. **Weiss, W. and Wiley, D. C.,** unpublished.
42. **Wrigley, N. G., Brown, E. B., Daniels, R. S., Douglas, A. R., Skehel, J. J., and Wiley, D. C.,** Electron microscopy of influenza haemagglutinin-monoclonal antibody complexes, *Virology,* 131, 308, 1983.
43. **Min-Jou, W., Verhoeyen, M., Devos, R., Saman, E., Fang, R., Huylebroeck, D., Fiers, W., Threlfall, G., Barber, C., Carey, N., and Emtage, S.,** Complete structure of the haemagglutinin gene from the human A/Victoria/3/75 (H_3N_2) strain as determined from cloned DNA, *Cell,* 19, 683, 1980.
44. **Gething, M.-J., Bye, J., Skehel, J. J., and Waterfield, M. D.,** Cloning and DNA sequence of double-stranded copies of haemagglutinin genes from H_2 and H_3 strains elucidates antigenic shift and drift in human influenza virus, *Nature,* 287, 301, 1980.
45. **Winter, G., Fields, S., and Brownlee, G. G.,** Nucleotide sequence of the haemagglutinin gene of a human influenza virus H_1 subtype, *Nature,* 292, 72, 1981.
46. **Gerhard, W., Yewdell, J., Frankel, M. E., and Webster, R. G.,** Antigenic structure of influenza virus haemagglutinin defined by hybridoma antibodies, *Nature,* 290, 713, 1981.
47. **Caton, A. J., Brownlee, G. G., Yewdell, J. W., and Gerhard, W.,** The antigenic structure of the influenza virus A/PR/8/34 haemagglutinin (H_1 subtype), *Cell,* 31, 417, 1982.
48. **Yewdell, J. W., Webster, R. G., and Gerhard, W.,** Antigenic variation in three distinct determinants of an influenza type A haemagglutinin molecule, *Nature,* 279, 246, 1979.
49. **Wang, M-L., Skehel, J. J., and Wiley, D. C.,** Comparative analyses of the specificities of anti-influenza haemagglutinin antibodies in human sera, *J. Virol.,* 57, 124, 1986.
50. **Natali, A., Oxford, J. A., and Schild, G. C.,** Frequency of naturally occurring antibody to influenza virus antigenic variants selected with monoclonal antibody, *J. Hyg. Camb.,* 87, 185, 1981.
51. **Daniels, R. S., Douglas, A. R., Skehel, J. J., Wiley, D. C., Naeve, C. W., Webster, R. G., Rogers, G. N., and Paulson, J. C.,** Antigenic analyses of influenza virus haemagglutinins with different receptor-binding specificities, *Virology,* 138, 174, 1984.
52. **Underwood, P. A.,** Receptor binding characteristics of strains of the influenza Hong Kong subtype using a periodate sensitivity test, *Arch. Virol.,* 84, 53, 1985.
53. **Underwoood, P. A., Skehel, J. J., and Wiley, D. C.,** Receptor binding characteristics of monoclonal antibody-selected antigenic variants of influenza virus, *J. Virol.,* 61, 206, 1987.
54. **Yewdell, J. W., Caton, A. J., and Gerhard, W.,** Selection of influenza A virus adsorptive mutants by growth in the presence of a mixture of monoclonal antihaemagglutinin antibodies, *J. Virol.,* 57, 623, 1986.
55. **Underwood, P. A.,** unpublished.
56. **De Jong, J. C.,** in preparation.
57. **Schild, G. C., Oxford, J. S., De Jong, J. C., and Webster, R. G.,** Evidence for host-cell selection of influenza virus antigenic variants, *Nature,* 303, 706, 1983.
58. **Robertson, J. S., Naeve, C. W., Webster, R. G., Bottman, J. S., Newman, R., and Schild, G. C.,** Alterations in the haemagglutinin associated with adaptation of influenza B virus to growth in eggs, *Virology,* 143, 166, 1985.
59. **Portner, A., Webster, R. G., and Bean, W. J.,** Similar frequencies of antigenic variants in sendai, vesicular stomatitis and influenza A viruses, *Virology,* 104, 235, 1980.

Chapter 8

VARIATION OF THE HIV GENOME: IMPLICATIONS FOR THE PATHOGENESIS AND PREVENTION OF AIDS

Flossie Wong-Staal

TABLE OF CONTENTS

I. INTRODUCTION

The disease named acquired immune deficiency syndrome (AIDS) has rapidly reached epidemic proportions in the U.S. and Europe since the first case documentation in 1981. The last few years have witnessed, on one hand, great frustration on the part of clinicians and health care workers in coping with the morbidity and mortality of the increasing number of patients, but, on the other hand, tremendous progress in the basic understanding of the infectious agent of this disease. The virus, called human T-lymphotropic virus type III (HTLV-III), lymphadenopathy virus (LAV), or, more recently, human immunodeficiency virus (HIV), is a member of the retrovirus family. Ironically, its genetic structure is by far the most complex among retroviruses (Figure 1). In addition to the three structural genes (gag, pol, and env) necessary for replication of all retroviruses, the genome of HTLV-III contains at least four additional genes (sor, tat, trs/art and 3'orf) (see References 1 and 2 for review, and References 3 and 4). Two of these (tat and trs) are necessary for virus replication and function in posttranscriptional regulation in *trans* of virus gene expression.[4-7] The trs gene, in particular, appears to regulate the accumulation of spliced mRNA for gag and env proteins.[4] The functions of the sor and 3'orf genes are still unknown, but are probably also regulatory in nature. Deletions of sor and 3'orf did not abrogate virus production, but significantly compromised or enhanced it, respectively. Basic studies on HTLV-III have already led to the development of first and second generation diagnostic reagents, so that routine blood bank screening is feasible in most developed countries and transmission of AIDS through blood products is no longer a major threat. Furthermore, our understanding of the complex, multitiered regulatory mechanism of this virus may also open up new avenues of antiviral therapy. For example, in addition to drugs that interfere with reverse transcriptase (pol) activity, those that interfere with tat or trs functions should also block virus replication. Aside from developing diagnostic and therapeutic procedures, the development of an effective vaccine has been the focus of scientific effort. In this regard, the finding of considerable heterogeneity in the early genetic analysis of various HTLV-III field isolates was discouraging. The purpose of this chapter is to summarize studies that address the question of genetic polymorphism of HTLV-III and to evaluate the implications of these studies for the pathogenesis of AIDS and the prospects of developing a broadly cross-reactive vaccine.

II. GENETIC POLYMORPHISM OF HIV ISOLATES

Heterogeneity in the genomes of different HTLV-III isolates has been assessed by several approaches: restriction enzyme analysis of uncloned DNA of infected cells or tissues from patients, restriction enzyme analysis of cloned viral genomes, heteroduplex analysis by electron microscopy, and nucleotide sequence comparisons. More than 50 virus isolates have been compared by restriction enzyme mapping.[8-12] Several conclusions are apparent from these studies. First, all isolates are related as they cross-hybridize under highly stringent conditions. Second, all isolates can be distinguished by one or more restriction enzyme sites. Some differ in as many as half of their cleavage sites. Third, there do not seem to be specific variants that correlate with tissue tropisms of the virus, at least none that can be discerned with these crude genomic analyses. For example, Shaw et al.[10] first showed that HTLV-III is enriched in brain tissues, compared to peripheral blood and lymph node tissues, of some patients with central nervous system (CNS) disorders, and that this infection of brain cells may be the primary contributing factor in the CNS disorders. However, analyses of viral DNA sequences from various tissues, including brain, from the same patients failed to reveal significant differences. Detailed comparisons of the nucleotide sequence and biological properties of isolates from different tissues would be of great interest. Fourth, regardless of

FIGURE 1. Genetic structure of HTLV-III.

the history of exposure of infected individuals, a single virus genotype or minor variant thereof was found in each individual. This observation is quite remarkable in view of the fact that many of these individuals were homosexual men from HTLV-III endemic regions who had had, in some cases, thousands of different sexual contacts. In one study, one predominant virus genotype was repeatedly isolated from such an individual over a period of 2 years.[12] This finding suggests that once infected, individuals are protected from subsequent HTLV-III infections. It is unlikely that such an effect could be the result of classical viral interference mechanism, whereby an infected cell loses its receptor for the same virus and cannot be superinfected, since only a small fraction of the T4-positive lymphocytes of chronically infected people actually harbor the virus.[13] More likely, infected people are afforded immunologic protection against low dosages of exogenous viruses.

III. GENETIC VARIATION OF SEQUENTIAL ISOLATES OF HIV

Further insight into the nature and rate of HTLV-III divergence was provided by a study of serial isolates from patients.[12] Four to six virus isolates obtained from each of three individuals over a 1- or 2-year period were analyzed. Changes throughout the viral genomes were detected in isolates from each patient, although it is equally clear that the sequential isolates were minor variants derived from a common progenitor rather than distinct isolates resulting from multiple infections. Five sequential isolates from one patient (coded WMJ), a Haitian child with AIDS who was born to an HTLV-III-positive mother, were examined in detail. The restriction maps of WMJ-1 through WMJ-5 showed that although these isolates were obtained in a temporally sequential manner over a period of 1 year, the changes were not progressive, i.e., WMJ-1 did not give rise to WMJ-2 which did not then give rise to WMJ-3, etc. That these changes were nondirectional was confirmed by comparing the envelope nucleotide sequences of WMJ-1, WMJ-2, and WMJ-3, which revealed a closer relationship of WMJ-1 to WMJ-3 than to WMJ-2 in many regions. This observation, that the sequential isolates probably arose in parallel, suggested that as variants were generated and dominated the culture, the preexisting virus population was not completely cleared and could still serve as the progenitor of new variants.

The availability of the sequential isolates also made it possible to estimate the rate of genetic evolution of HTLV-III in a natural infection. It has been documented that RNA viruses generally evolve at a much more rapid rate than DNA viruses.[14] Since patient WMJ, who had had a single perinatal exposure to the virus, was in her third year of life when isolates 1, 2, and 3 were obtained, it seemed reasonable to assume that these viruses had evolved from a common progenitor within the preceding 5 years. Using 1 to 5 years as the time of divergence, the rates of evolution for the env and gag genes of HTLV-III (WMJ) were calculated according to the equations:

$$R = D/2T$$

$$D = -(3/4) \ln[1 - (4/3)P]$$

being R = the rate of nucleotide substitutions per site per year, T = the time of divergence, and P = the proportion of nucleotides that differ between the homologous genes. On the basis of this equation, the rate of evolution (R) for the HTLV-III env gene was calculated to be between 1.58×10^{-2} and 3.17×10^{-3} nucleotide substitutions per site per year for T values of 1 and 5 years; for the gag genes, R = 1.85×10^{-3} to 3.7×10^{-4}. This rate of genetic change for HTLV-III is within the range of values derived for other RNA viruses.

It is not clear at this point whether new variants generated in vivo represented an antigenic drift to evade host immune recognition. It would be important to correlate virus change with changes in neutralizibility by analyzing sequential serum samples taken from the same patients.

IV. DIVERGENCE OF THE ENVELOPE GENES

It was not apparent from restriction enzyme studies whether variability in the HTLV-III genomes was randomly distributed or clustered in specific regions. One evidence that the envelope gene is the site of greatest divergence came from heteroduplex studies in which the cloned genomes of two divergent HTLV-III (BH-10 and HAT-3) were annealed under conditions of increasing stringency.[15] Substitution loops first appeared in the env gene region of the heteroduplex molecules. More definitive conclusions could be drawn from nucleotide sequence comparisons. The complete nucleotide sequences of three HTLV-III and related viruses and partial sequences of several additional isolates have been published.[16-20] These sequences confirmed that the various field isolates of HTLV-III, LAV, and ARV are closely related viruses with the same genomic organizations and that the env gene is one of the most divergent loci.[21] Table 1 compares the nucleotide and deduced amino acid sequences of five isolates. These data demonstrate a broad spectrum of diversity among the viruses. Overall, the differences range from 1.5% of nucleotides (14.2% amino acids) for HTLV-III (BH-10) and LAV-1a to 9.3% of nucleotides (14.2% amino acids) for HTLV-III (HAT-3) and ARV. The distribution of sequence differences was not uniform throughout the viral genomes. Changes were more prevalent in two genes, envelope and 3'orf. In the envelope gene in particular, there was as much as a 19% difference in amino acid sequence (BH-10 and HAT-3). Within the envelope gene, the signal peptide and the extracellular major envelope glycoprotein were the most highly divergent. For example, the extracellular envelope domains of HAT-3, WMJ-1, and ARV-2 differed from the same region of BH-10 in 21.4, 18.7, and 17.0% of their amino acids, respectively.

The most conserved areas in these genomes were the gag and pol genes, which differed among the viruses in less than 6% of nucleotides and less than 7% of amino acids. In these regions, nucleotide sequence changes were due almost exclusively to point mutations, in contrast to env, where clustered nucleotides changes involving in-frame deletions, insertions, and duplications were common. For example, compared to BH-10, there were 48 base-pairs of insertions and 21 base-pairs of deletions in the extracellular envelope of HAT-3, compared to only 6 base-pairs of insertions and deletions in gag, pol, and sor sequences. LAV-1a, which is more closely related to BH-10 than the other viruses analyzed, still contained a 15-base-pair insertion/duplication in the exterior envelope gene, but no insertions or deletions in gag, pol, or sor. The env gene of these viruses also differed from the gag gene in the proportion of silent, third base-pair changes. In comparison to BH-10, the point mutations in the external env genes of ARV-2, HAT-3, and WMJ-2 involved third base changes in 36, 37, and 34% of cases, leading to 32 and 42% amino acid changes, respectively. In contrast, the gag genes of these same viruses contained 63, 56, and 66% third base changes, resulting in 10, 3.5, and 0% amino acid changes, respectively. Therefore, more than half of the single nucleotide changes in env occurred in the first or second codon position, resulting in nonsilent mutations. Even the third position changes in env frequently led to

Table 1
SEQUENCE COMPARISON OF FIVE INDEPENDENT AIDS VIRUS ISOLATES

A. Nucleotide Differences Compared to BH-10 (Number of Nucleotide Changes/Number of Nucleotides Sequenced)

Clones	LTR	Leader sequence and primer binding site	gag	pol	sor	tat and adjacent sequences	Signal peptide	Envelope Extra-cellular portion	Envelope Trans-membrane portion	3' orf	Total
HAT-3	45/634 7%	13/152 8.5%	77/1285 5.9%	46/1246 3.7%	17/464 3.6%	48/584 8.2%	26/111 23.4%	202/1443 14.0%	87/1035 8.4%	69/648 10.6%	645/7489 8.6%
ARV-2	30/634 4.7%	14/152 9.2%	86/1536 5.6%	134/3045 4.4%	31/609 5.1%	49/584 8.4%	26/111 23.4%	164/1443 11.4%	66/1035 6.3%	52/648 8.0%	582/9213 6.3%
LAV-1a	10/634 1.6%	7/152 4.6%	46/1536 3.0%	59/3045 1.9%	2/609 0.3%	11/584 1.9%	2/111 1.8%	32/1443 2.2%	9/1035 0.9%	13/648 2.0%	144/9213 1.5%
WMJ-1	nd	nd	38/1162 3.3%	nd	nd	nd	27/111 24.3%	177/1443 12.3%	62/1035 6.0%	nd	nd

B. Amino Acid Differences compared to BH-10 (Number of Amino Acid Changes/Number of Amino Acids Sequenced)

Clones	LTR	Leader sequence and primer binding site	gag	pol	sor	tat and adjacent sequences	Signal peptide	Envelope Extra-cellular portion	Envelope Trans-membrane portion	3' orf	Total
HAT-3	—	—	28/428 6.5%	22/415 5.3%	10/154 6.4%	—	11/37 29.7%	103/481 21.4%	45/345 13.0%	37/216 17%	256/2039 12.5%
ARV-2	—	—	32/512 6.3%	51/1015 5.0%	20/203 9.8%	—	13/37 35.1%	82/481 17.0%	42/345 12.2%	29/216 12.2%	239/2593 9.2%
LAV-1a	—	—	16/512 3.1%	21/1015 2.1%	0/203 0%	—	2/37 5.4%	14/481 2.9%	5/345 1.4%	8/216 3.7%	58/2593 2.2%
WMJ-1	—	—	12/387 3.1%	nd	nd	—	13/37 35.1%	90/481 18.7%	29/345 8.4%	nd	nd

From Starcich et al., *Cell*, 45, 637, 1986. With permission.

amino acid changes. Conversely, the gag genes contained a lower proportion of first and second codon position changes, and the majority of the third codon position changes were silent. These findings suggest two possible interpretations which are not mutually exclusive: first, there is a stronger structure-function constraint for protein sequence conservation in gag than in env, and second, there may be positive selection for nonsilent nucleotide changes in env. Such selective pressures may be immunologic in nature. Recent evidence demonstrated that HTLV-III variants could be generated by passage of the virus in the presence of neutralizing antibodies.[21a] The new variant was resistant to the selecting antibodies. This phenomenon is reminiscent of results obtained with visna virus which is known to undergo changes in its exterior envelope protein to evade immunosurveillance of the host.[22]

V. IDENTIFICATION OF HYPERVARIABLE AND HIGHLY CONSERVED REGIONS IN THE ENVELOPED PROTEINS

Since the greatest divergence of the HTLV-III genome is localized in the env gene, and since the envelope protein is likely to be the substrate for a subunit vaccine, it was important to further scrutinize the nature of this divergence and its impact on the feasibility of vaccine development. The HTLV-III env gene encodes a precursor protein (gp160) that is heavily glycosylated. This is then cleaved into a major exterior envelope glycoprotein (gp120) and a small envelope protein (gp41) that spans across the membrane. A striking feature of the comparison was the extreme conservation of the cysteine residues. Within the extracellular envelope protein, all 18 cysteine residues were conserved, and all but one cysteine residue in the "transmembrane" small envelope protein were also conserved. This finding argues that the cysteine residues are necessary to maintain the envelope proteins in the proper functional configuration. Despite this overall structural conservation, changes were evident throughout the envelope glycoprotein. In the small envelope protein, these were comprised predominantly of single amino acid substitutions resulting from point mutations. In the extracellular major envelope protein (gp120), insertions, deletions, and duplications of oligopeptides were found in addition to isolated changes. Inspection of the aligned sequences revealed four regions of highly conserved polypeptide segments (C1 to C4) interspersed with five regions of high variability (V1 to V5) (Figure 2). The localization of these variable and constant regions was confirmed by objective determination of relative variation using computer analysis.[20,23] Potential glycosylation sites, as well as predicted secondary structures, were significantly affected in the variable regions, but well conserved in the constant regions (Table 2). The small envelope protein is well conserved overall, with a short region in the middle that exhibits some single amino acid substitutions. For convenience, it is subdivided into two hyperconserved regions (C5 and C6).

VI. PREDICTION OF ANTIGENIC SITES

To assess whether the conserved and variable regions were relevant to the immunologic properties of the env proteins, potential antigenic sites of the protein were determined using a computer program that predicts the secondary structure and calculates values for hydrophilicity, flexibility, and surface probability.[23] Continuous antigenic epitopes are mainly located in β-turn regions that have a large number of nonhydrophobic or flexible amino acid sequences and/or a high probability of surface location. Using these criteria, nine epitopes were predicted in the gp120. Five of these corresponded to the five hypervariable regions (Table 2). Three of the constant regions (C2 to C4) also contained potential antigenic sites, but only C4 is predicted to be strongly antigenic.

The small envelope protein (gp41) of HTLV-III contains three hydrophobic sequences, one immediately at the cleavage site, which is separated from the second hydrophobic

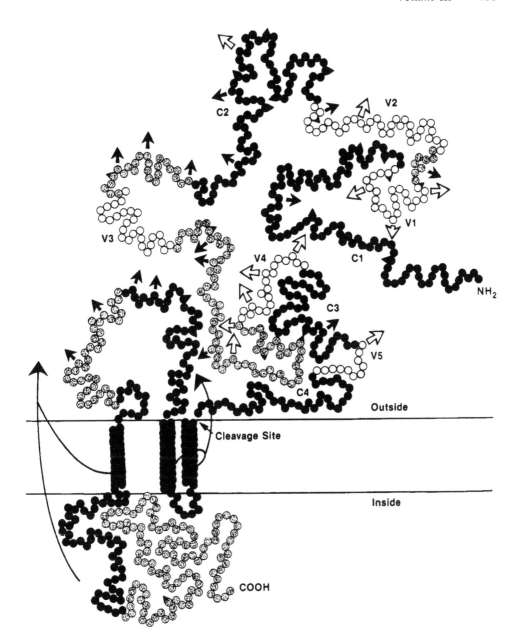

FIGURE 2. Structure of the env gene of HTLV-III showing constant and hypervariable regions (see text and Table 2).(▲) Cysteine residues, (◇) glycosylation sites.

sequence by ten amino acids. The second and third hydrophobic sequences, presumably transmembrane, are separated by a domain of 90 hydrophilic, flexible, and charged amino acids that has a likely extracellular location in the infected cells. It has been shown that peptides derived from this region are useful diagnostic reagents since they are highly antigenic and broadly cross reactive.[24] The gp41 of HTLV-III differs from the transmembrane proteins of other retroviruses in that it contains a long cytoplasmic tail. This stretch of about 100 amino acids has an additional potential antigenic determinant.

It should be emphasized that these predictions are only valid for continuous epitopes and not for conformational epitopes which are probably also important in determining the antigenicity of HTLV-III proteins.

Table 2
PARAMETERS OF PEPTIDE REGIONS OF THE HTLV-III env GLYCOPROTEIN BASED ON ANALYSIS OF SEVEN ISOLATES

Region[a]	Aminno acid deletions/insertions	% Conserved amino acids	Conserved glycosylation	Conserved β-turns	Hydrophilicity	Surface probability	Flexibility	Antigenic[b] sites
C-1 38—134	0	87%	1/1	2/4	7—10%	7—10%	8—10%	
V1 135—154	0—15	10%	0/1—3	0/0—11	13-33%	0—13%	28—82%	137—154
155—162	0	87%	1/1	0/0	0%	0%	0%	
V2 163—203	0—18	21%	1/2—5	0/2—17	14—41%	10—32%	20—46%	186—203
C2 204—279	0	89%	2/3—4	2/2—4	7—11%	0%	7—11%	232—246
280—304	0	76%	0/0—1	0/0—3	8—20%	0%	8—12%	
V3 305—330	0—5	25%	0/0—1	2/2—8	19—34%	7—22%	0—57%	300—320
331—395	0—1	72%	4/6	8/9—11	12—17%	6—8%	12—25%	358—375
V4 396—414	0—11	17%	0/1—3	0/4—11	2—11%	0—6%	10—20%	394—412
C3 415—458	0	86%	1/1	4/4—7	0—11%	0%	5—11%	445—458
V5 459—469	0—2	16%	0/1	1/1—6	1—90%	0—45%	54—100%	459—479
C4 470—510	0	87%	0/0	4/4	29—32%	43—46%	24—29%	470—483
C5 511—616	0—1	90%	2/2	6/6—8	5%	5—7%	6%	
616—653	0	44%	2/2	0/0—2	13—23%	10—13%	7—10%	611—637
C6	0	85%	0/0—1	6/6—10	34%	20—29%	32—40%	724—745

| 654—745 | | | | | | | |
| 746—856 | 0 | 69% | 0/0—2 | 4/4—7 | 15—18% | 10—12% | 7—10% |

[a] Amino acid positions based on MJ-1 isolate
[b] Numbers refer to nucleotide positions of predicted antigenic sites within each segment. Underlined are the strongly antigenic sites.

From Modrow, S., Hahn, B. H., Shaw, G. M., Gallo, R. C., Wong-Staal, F., and Wolf, H., *J. Virol.*, 1986, in press. With permission.

VII. GROUP- AND TYPE-SPECIFIC EPITOPES FOR VIRUS NEUTRALIZATION

The identification of potential antigenic sites in the hypervariable, as well as constant, regions of the env proteins of HTLV-III suggests the presence of type- and group-specific antigenic epitopes. Several laboratories have demonstrated low titer neutralizing antibodies in patients infected with HTLV-III.[25,26] Such antibodies could be adsorbed by purified gp120, suggesting that the majority of neutralizing epitopes are located in this extracellular protein (Robey, J., personal communication). A broad spectrum of HTLV-III and related viruses were usually susceptible to neutralization by these antibodies, but at different titers, again suggesting the presence of a mixture of group- and type-specific epitopes. The presence of type-specific epitopes was definitively shown in an experiment in which HTLV-III variants were generated by passage of the virus in the presence of neutralizing sera in vitro.[21a] These variants were no longer neutralized by the selected sera, but were still susceptible to other neutralizing sera. Nonetheless, the identification of highly conserved regions in the gp120 protein and of broadly cross-reactive neutralizing epitopes provides the logical basis and valuable information for developing an effective vaccine.

ACKNOWLEDGMENTS

I am grateful to all of my colleagues who contributed to the work described here, particularly Drs. George Shaw, Beatrice Hahn, Bruno Stacich, Hans Wolf, Susanna Modrow, and Robert Gallo. I also thank Elizabeth Swerda and Deborah Ann Fritzler for expert editorial assistance.

REFERENCES

1. **Wong-Staal, F. and Gallo, R. C.,** Human T-lymphotropic retroviruses, *Nature (London),* 317, 395, 1985.
2. **Wong-Staal, F.,** Molecular biology of viruses of the HTVL family, in *AIDS: Modern Concepts and Therapeutic Challenges,* Broder, S., Ed., Marcel Dekker, New York, 1987, 165.
3. **Sodroski, J. G., Groh, W. C., Rosen, C., Dayton, A., Terivilliger, E., and Haseltine, W.,** A second post-transcriptional trans-activator gene required for HTVL-III replication, *Nature (London),* 321, 412, 1986.
4. **Feinberg, M. B., Jarrett, R. F., Aldovini, A., Gallo, R. C., and Wong-Staal, F.,** HTLV-III expression and production involve complex regulation at the levels of splicing and translation of viral RNA, *Cell,* 46, 807, 1986.
5. **Fisher, A. G., Feinberg, M. B., Josephs, S. F., Harper, M. E., Marselle, L. M., Reyes, G., Gonda, M. A., Aldovini, A., Debouk, C., Gallo, R. C., and Wong-Staal, F.,** The trans-activator gene of HTLV-III is essential for virus replication, *Nature (London),* 320, 367, 1986.
6. **Dayton, A. I., Sodroski, J. G., Rosen, C. A., Goh, W. C., and Haseltine, W. A.,** The trans-activator gene of the human T-cell lymphotropic virus type III is required for replication, *Cell,* 44, 941, 1986.
7. **Rosen, C. A., Sodroski, J. G., Goh, W. C., Dayton, A. I., Lippke, J., and Haseltine, W. A.,** Post-transcriptional regulation accounts for the trans-activation of the human T-lymphotropic virus type III, *Nature (London),* 319, 555, 1986.
8. **Shaw, G. M., Hahn, B. H., Arya, S. K., Groopman, J. E., Gallo, R. C., and Wong-Staal, F.,** Molecular characterization of human T-cell leukemia (lymphotropic) virus type III in the acquired immune deficiency syndrome, *Science,* 226, 1165, 1984.
9. **Wong-Staal, F., Shaw, G. M., Hahn, B. H., Salahuddin, S., Popovic, M., Markham, P. D., Redfield, R., and Gallo, R. C.,** Genomic diversity of human T-lymphotropic virus type III, *Science,* 229, 759, 1985.
10. **Shaw, G. M., Harper, M. E., Hahn, B. H., Epstein, L. G., Gaidusek, C. D., Price, R. W., Navia, B. A., Petito, C. K., O'Hara, C. J., Cho, E. S., Oleske, J. A., Wong-Staal, F., and Gallo, R. C.,** HTLV-III infection in brains of children and adults with AIDS encephalopathy, *Science,* 227, 177, 1985.

11. **Benn, S., Rutledge, R., Folks, T., Gold, J., Baker, L., McCormick, J., Feorino, P., Piot, P., Quinn, T., and Martin, M. A.,** Genomic heterogeneity of AIDS retroviral isolates from North America and Zaire, *Science,* 230, 949, 1985.

12. **Hahn, B. H., Shaw, G. M., Taylor, M. E., Redfield, R. R., Markhan, P. D., Salahuddin, S., Wong-Staal, F., Gallo, R. C., and Parks, W. P.,** Genetic variation in HTLV-III/LAV over time in patients with AIDS or at risk for AIDS, *Science,* 232, 1548, 1986.

13. **Harper, M. E., Marselle, L. M., Gallo, R. C., and Wong-Staal, F.,** Detection of lymphocytes expressing human T-lymphotropic virus type III in lymph nodes and peripheral blood from infected individuals by in situ hybridization, *Proc. Natl. Acad. Sci. U.S.A.,* 83, 772, 1986.

14. **Holland, J., Spindler, K., Horodyski, F., Grabau, E., Nichol, S., and Vande Pol, S.,** Rapid evolution of RNA genomes, *Science,* 215, 1577, 1982.

15. **Hahn, B. H., Gonda, M. A., Shaw, G. M., Popovic, M., Hoxie, J., Gallo, R. C., and Wong-Staal, F.,** Genomic diversity of the AIDS virus HTLV-III: different viruses exhibit greatest divergence in their envelope genes, *Proc. Natl. Acad. Sci. U.S.A.,* 82, 4813, 1985.

16. **Ratner, L., Haseltine, W., Patarca, R., Livak, K. J., Starcich, B., Josephs, S. F., Doran, E. R., Rafalski, J. A., Whitwhoen, E. A., Baumeister, K., Ivanoff, L., Petteway, S. R., Pearson, M. L., Lautenberger, J. A., Papas, T. S., Ghrayeb, J., Chang, N. T., Gallo, R. C., and Wong-Staal, F.,** Complete nucleotide sequence of the AIDS virus HTLV-III, *Nature (London),* 313, 277, 1985.

17. **Wain-Hobson, S., Sonigo, P., Danos, O., Cole, S., and Alizon, M.,** Nucleotide sequence of the AIDS virus, LAV, *Cell,* 40, 9, 1985.

18. **Sanchez-Pescador, R., Power, M. D., Barr, P. J., Steimer, K. S., Stempten, M. M., Brown-Shimer, S. L., Gee, W. W., Renard, A., Randolph, A., Levy, J. A., Dina, D., and Luciw, P. A.,** Nucleotide sequence and expression of an AIDS-associated retrovirus (ARV-2), *Science,* 227, 484, 1985.

19. **Muesing, M. A., Smith, D. H., Cabradilla, C. D., Benton, C. V., Lasky, L. A., and Capon, D. J.,** Nucleic acid structures and expression of the human AIDS/lymphadenopathy retrovirus, *Nature (London),* 313, 430, 1985.

20. **Starcich, B. R., Hahn, B. H., Shaw, G. M., McNeely, P. D., Modrow, S., Wolf, H., Parks, E. S., Parks, W. P., Josephs, S. F., Gallo, R. C., and Wong-Staal, F.,** Identification and characterization of conserved and variable region in the envelope gene of HTLV-III/LAV, The retrovirus of AIDS, *Cell,* 45, 637, 1986.

21. **Ratner, L., Gallo, R. C., and Wong-Staal, F.,** HTLV-III, LAV, and ARV are variants of the same AIDS virus, *Nature (London),* 313, 636, 1985.

21a. **Robert-Guroff, M., Reitz, M. S., Jr., Robey, W. G., and Gallo, R. C.,** In vitro generation of an HTLV III variant by neutralizing antibody, *J. Immunol.,* 137, 3306, 1986.

22. **Clements, J. E., Pedersen, F. S., Narayan, O., and Haseltine, W. A.,** Genomic changes associated with antigenic variation of visna virus during persistent infection, *Proc. Natl. Acad. Sci. U.S.A.,* 77, 4454, 1980.

23. **Modrow, S., Hahn, B. H., Shaw, G. M., Gallo, R. C., Wong-Staal, F., and Wolf, H.,** Computer assisted analysis of the envelop proteins sequences of seven HTLV-III/LAV isolates: prediction of antigenic epitopes in conserved and variable regions, *J. Virol.,* 61, 570, 1987.

24. **Chang, T. W., Kato, I., McKinney, S., Chandra, P., Barone, A. D., Wong-Staal, F., Gallo, R. C., and Chang, N. T.,** Detection of anitbodies to human T-cell lymphotropic virus III (HTLV-III) with an immunoassay employing a recombinant E. coli-derived viral antigenic peptide, *Biotechnology,* 3, 905, 1985.

25. **Robert-Guroff, M., Brown, M., and Gallo, R. C.,** HTLV-neutralizing antibodies in AIDS and ARV, *Nature (London),* 316, 72, 1985.

26. **Weiss, R. A., Clapham, P. R., Cheingsong-Popov, R., Dagleish, A. G., Carne, C. A., Weller, I. V. D., and Tedder, R. S.,** Neutralizing antibodies to human T-cell lymphotropic virus type III, *Nature (London),* 316, 69, 1985.

Chapter 9

BIOLOGICAL AND GENOMIC VARIABILITY AMONG ARENAVIRUSES

P. J. Southern and M. B. A. Oldstone

TABLE OF CONTENTS

I. INTRODUCTION

The arenavirus family contains several distinct viruses with markedly different biological properties.[1-5] Each virus is associated in the wild with a particular rodent host and is usually found in well-defined geographical regions, i.e., Lassa: West Africa, Junin: Argentina, Machupo: Bolivia. In contrast, lymphocytic choriomeningitis virus (LCMV), the prototype arenavirus, is widely distributed throughout the world (see Table 1). The viruses are maintained in natural rodent populations by both vertical and horizontal transmission. Despite a life-long viremia, the rodents normally show no overt signs of disease except under conditions of extreme crowding or stress. Primary human infection occurs from contact with infected animals or their excreta. Lassa fever transmission in man has occurred via contaminated blood or syringes.

LCMV infection of laboratory mice has provided an excellent model for virus-host interactions and virus persistence in vivo.[1,2] Different strains of LCMV have been studied in different laboratories, and substantial information has been accumulated relating to differences in pathogenic potential.[6] Recently, cDNA cloning and nucleotide sequencing experiments have generated information to examine differences between arenaviruses at the molecular level.[7-11] It is now possible to predict primary amino acid sequences for the major structural proteins (nucleocapsid protein [NP] and glycoprotein precursor [GP-C]) encoded by arenavirus genomic S RNA segments. Direct comparisons of these predicted protein sequences identify both conserved regions and divergent regions which may be involved with modifications to pathogenic potential.

In this chapter, we will review variations among arenaviruses and arenavirus-induced disease, especially in experimental laboratory infections. Correlations between disease states and viral gene products have been established using reassortant (mixed genotype) viruses,[12-14] and, with sequence information now available, it is possible to reconcile dramatic differences in biological properties of viruses with relatively small numbers of amino acid changes in viral structural proteins. There is a possibility that reassortant viruses may also arise in nature and be responsible for the appearance of new arenaviruses and a wider spectrum of diseases.

II. HISTORICAL BACKGROUND

Perhaps the first record of an arenavirus infectious agent (subsequently to be named lymphocytic choriomeningitis virus) relates to a patient who had died during the 1933 epidemic of encephalitis in St. Louis. Using material obtained at autopsy, Armstrong and Lillie passed infection in monkeys and recovered a virus which, on the basis of pathological lesions in intracerebrally infected monkeys and mice, was designated the "virus of experimental lymphocytic choriomeningitis".[15] Shortly afterwards, Traub isolated a virus from an experimental mouse colony[16] and Rivers and Scott recovered viruses from two patients who had been treated for nonbacterial meningitis.[17] One of these patients was known to have worked with mice shown to be infected by Traub, but the source of infection for the second patient remained unknown. It was quickly realized that these independently isolated infectious agents were closely related, and the name lymphocytic choriomeningitis virus emerged.

There have been a number of documented cases of human disease caused by LCMV infection. For example, in the early 1970s an outbreak of LCMV occurred resulting in illness of children and adults. The source of virus was linked to persistently infected tumor cell lines and infected pet hamsters that had been obtained from a persistently infected breeding colony.[18] Most LCMV infections of adults result in subclinical, "influenza-like" illnesses and are usually resolved without further complication. The virus is, however, widely rep-

Table 1

Virus	Disease	Locality	Rodent reservoir and vector	Person to person transmission	Laboratory model of human infection
Lymphocytic choriomeningitis	Grippe, aseptic meningitis, occasional more severe forms of meningoencephalomyelitis	Probably originated in Europe, now worldwide	*Mus musclulus* natural host; colonized rodents, particularly mice and hamster; dog(?)	Never documented	Adult mouse inoculated intracerebrally
Junin	Argentinian hemorrhagic fever	Circumscribed area of Argentina; Buenos Aires to northwest	*Calomys musculinus*, possibly others	Occasional	Guinea pig
Machupo	Bolivian hemorrhagic fever	Beni region of Bolivia	*Calomys callosus*	Occasional, particularly spouses; recognized hospital outbreak	Rhesus monkey
Lassa	Lassa fever	Western Africa	*Mastomys nataliensis*	Frequent; explosive intrahospital epidemics	Squirrel monkey

resented in the U.S., as about 11 to 20% of the population show a positive serum antibody response against the virus.[4]

A number of distinct arenaviruses have been isolated and identified in the Americas and in Africa (reviewed in References 3, 4, and 19, see Table 1). In terms of human disease, Lassa virus in Africa and Junin and Machupo viruses in South America can cause severely debilitating and, on occasion, fatal infections. Individuals from outside areas face the greatest risk of serious infection, although infections are not uncommon among native populations. Clinical descriptions of human infections with Lassa, Junin, and Machupo viruses have recently been presented elsewhere[4] and will not be reviewed again here.

III. PRINCIPLES LEARNED FROM LABORATORY MODELS OF LCMV INFECTIONS

Traub first described the phenomenon of persistent LCMV infection in laboratory mice and the fundamental pathogenic consequences of acute virus infection.[20] Intracerebral infection of an adult animal results in death 7 to 9 days postinfection, whereas equivalent infection of a newborn animal within the first 24 hr of life results in a persistent infection with life-long viremia. The acute lethal infection of adult animals was later shown by Rowe[21] to be immune mediated, as infection of adult, immunosuppressed animals resulted in a persistent infection. Persistently infected animals are not tolerant to the virus but produce antibodies directed against all known viral structural proteins that react with viral antigens in the circulation to form antigen-antibody complexes.[22,23] Accumulation of these complexes results in varying degrees of immune-complex disease. Indeed, LCMV persistent infection has been the model system for studying virus immune complexes, and many findings have subsequently been extended to diverse infections of man and animals (reviewed in References 24 and 25).

The prevention of lethal, intracerebral infection in adult mice by immunosuppression and the appearance of immune complex disease in persistently infected mice established an important involvement for the host immune response in the pathogenesis of virus-induced disease. For example, study of the immune response to LCMV during acute infection led to the first description of cytotoxic T lymphocytes (CTL)[26] and the finding that such cells recognized both a specific viral determinant and a syngeneic major histocompatibility protein.[27] These observations, first recorded with LCMV infection, have been extended throughout the realm of microbiology to other animals, man, and infectious agents. In both acute and persistent infections, disease frequently follows from the immune response to the virus rather than being caused directly by virus replication. More recent work has described the ability of this nonlytic virus to replicate in differentiated cells and alter their specific differentiation product, leading to altered homeostasis and disease.[24]

IV. LCMV VARIANTS (STRAINS) AND DIFFERENT BIOLOGIC PROPERTIES

LCMV has been recognized as a manipulable and reproducible model for infection of laboratory mice, and several investigators have established independent virus isolates that have subsequently been passaged under different conditions. This has generated an extensive and sometimes conflicting literature for LCMV; however, it is now clear that discrepancies between published reports often reflect the generation of viral variants (strains) with fundamentally different properties.

The variability resulting from infections with many different combinations of animal and virus strains is summarized in Table 2. When mice are infected within the first 24 hr after birth (newborns), there is an initial period of active virus replication and release of progeny

Table 2
LCMV STRAIN AND DISEASE ASSOCIATION

LCMV-induced disease or phenomena	Disease in LCMV strains				
	ARM	E-350	WE	PASTEUR	TRAUB
Growth Hormone (GH) deficiency in persistently infected C3H/St mice					
Death	>95%	>95%	<5%	30%	<5%
Hypoglycemia	+ + + +	+ + + +	nil	+ +	nil
Poor growth	+ + + +	+ + + +	nil	+ +	nil
% GH cells containing viral antigen	>95%	>95%	<10%	40%	<10%
Hyperglycemia, abnormal glucose tolerance test in SWR/J, BALB mice, β cells in islets of Langerhans of pancreas containing virus	+ + +	+ + +	+ +	+ +	+ +
Immune complexes in persistently infected SWR/J mice	+ + + +	+ + + +	+ + +	+ + +	+
Acute death of adult guinea pigs	nil with >10⁵ PFU		+ + + + with 1 PFU		

virus particles followed by an altered pattern of viral gene expression that marks the progression from acute to persistent infection. Molecular details of this regulatory change are still being described. There is a significant reduction in the release of infectious virus particles that correlates with reduced expression of the viral glycoproteins, but viral nucleic acid sequences and viral nucleoprotein continue to accumulate.[28,29] In most circumstances, LCMV infection of newborn animals results in maintenance of low levels of virus (10^3 to 10^5 pfu per gram of tissue or per mℓ of serum) and substantial amounts of intracellular viral nucleic acid throughout the life-span. Circulating antibodies directed against viral proteins combine with viral antigens to form immune complexes that frequently complicate the infection. For example, certain mouse strains, i.e., SWR/J, are high level antibody responders to LCMV, but others, like BALB/WEHI, are low responders.[6] Responses are controlled by a number of host genes including immune response genes (Ir) located within the histocompatibility complex. Non-H2 genes also play a role. Some persistently infected animals manifest subtle alterations in specialized cell functions, e.g., hyperglycemia[30] (due to infection of beta cells of the islets of Langerhans) and decreased thyroid hormone[31] (T_3 and T_4) levels due to persistent infection of thyroid follicular cells.

A severe growth hormone deficiency disease occurs in 13- to 30-day-old C3H/St mice that have been inoculated at birth with LCMV Arm or E-350 strains. This results in approximately 95% of the animals dying from low blood sugar. These infected animals fail to grow at the same rate as uninfected littermates and, at the time of death, show about a 50% weight reduction relative to controls. Such animals have lowered growth hormone levels in the pituitary, and reconstitution experiments involving the introduction of rat pituitary cells (the GH3 cell line) that secrete growth hormone allow the infected mice to develop normally and maintain normal blood glucose levels.[32,33] This reconstitution experiment suggests that a defect in growth hormone is responsible for abnormal growth and development. LCMV ARM and E-350 replicate extensively in the growth hormone-producing cells, while the other LCMV strains, Traub and WE, that fail to induce growth hormone disease, replicate poorly in growth hormone-synthesizing cells.[34] Interestingly, these virus strains (Traub, WE)

Table 3
LCMV RNA SEGMENT AND DISEASE ASSOCIATION

LCMV-induced disease state or phenomena	LCMV strain		LCMV RNA segment causing disease
	Virulent	Avirulent	
Growth hormone deficiency in persistently infected C3H/St mice (poor growth, hypoglycemia, death)	ARM	WE	S RNA ARM
Acute death in adult guinea pigs	WE	ARM	L RNA WE
Immune complexes in persistently infected SWR/J mice	ARM	TRAUB	S RNA ARM
Interferon-induced liver necrosis and death of BALB/ WEHI mice	WE/ ARM	ARM WE ARM/WE	L RNA of WE S RNA of ARM
Induction and generation of virus specific H2 restricted cytotoxic T lymphocytes	ARM	PASTEUR	S RNA of ARM

replicate in C3H mouse liver and spleen and show a typical, widespread distribution of viral nucleic acid sequences and infectious virus in most tissues. Hence, the growth hormone disease correlates with virus replication in selected cells of the anterior pituitary. However, it is not yet clear whether the differences in LCMV strains and disease potential are exerted at the level of virus adsorption and uncoating or at the level of virus replication within the growth hormone-producing cells of the anterior pituitary. Reassortant viruses, made between an LCMV strain that causes disease (Arm) and one that does not (WE), have been used to establish that the growth hormone disease is associated with the S RNA segment of the Arm strain[13] and, by implication, genes encoded by that segment. The S RNA encodes the nucleocapsid protein and glycoproteins.[12] This suggests that the growth hormone disease may reflect a tropism of infection rather than differential replication because the viral replicase functions are encoded by the L RNA segment.[35]

V. REASSORTANT GENOTYPE LCM VIRUSES

The isolation and characterization of reassortant viruses from unique parental LCMV strains has produced considerable new information[12-14] (Table 3). Simultaneous infections with the two parental viruses allowed random interactions between input L and S genomic RNA segments, and reassortant viruses of mixed genotype were identified in the progeny virus population by screening with monoclonal antibodies and nucleic acid hybridization probes. Recovery of both pairs of potential reassortant viruses (for example, Arm L/WE S and WE L/Arm S) has allowed an unambiguous assignment of biological function to a genomic RNA segment. In this way, growth hormone disease in C3H mice has been mapped to the Arm genomic S RNA segment,[13] and a lethal infection in adult guinea pigs correlates with the presence of the WE genomic L segment.[36] Target specificity for H2-restricted CTL killing has also been mapped to the LCMV genomic S RNA segment.[14]

The frequency of recovery of reassortant viruses appears to have varied according to the pairing of parental LCMV strains. Reassortants between Arm and Pasteur were recovered at significantly greater frequency than Arm and WE reassortants.[12,14] This suggests that transcription and/or replication signals for Arm and Pasteur may be more closely related than for Arm and WE, and raises the possibility of mutational change at regulatory sites as a prerequisite for successful propagation of reassortant viruses. Experiments involving direct

RNA sequencing of virion RNA preparations are now in progress to examine the frequency of sequence change at the population level (Salvato et al., unpublished results).

VI. MOLECULAR BIOLOGY AND SEQUENCING

Recently, a number of laboratories have initiated molecular cloning experiments with arenavirus genomic RNA segments.[7-11] Different cloned cDNA sequences are now available for the following applications:

1. Evaluation of the complete genetic potential of the viruses.
2. Production of hybridization probes to monitor viral gene expression and gene regulation.
3. Comparisons of nucleotide and (predicted) protein sequences for the different arenaviruses.
4. Investigations of the molecular basis of arenavirus-induced disease.

The genomic organization of the viral S segment involves an unusual ambisense gene coding arrangement[7,9,10,35] (reviewed in Volume I, Chapter 9). Both the major viral structural proteins NP and GP-C are encoded by the S RNA segment — NP mRNA is complementary to the genome, whereas GP-C mRNA is in the sense of the genome. The NP and GP-C coding regions do not overlap and are separated by a short intergenic hairpin. The hairpin region and the ambisense gene organization are likely to be involved in regulation and discrimination between transcription and replication,[36a] but detailed schemes are not presently available.

The genetic structure of the viral L RNA segment is not as well defined. There is a very long open reading frame which is apparently involved with synthesis of a 150- to 200-kdalton viral polymerase or replicase protein. This coding region is associated with a mRNA that is complementary to the L segment.[36b]

VII. CODING ASSIGNMENTS

The major viral structural proteins NP and GP-C have been mapped to the S RNA segment by both genetic and biochemical techniques.[12,37] Definitive experiments using antisera to synthetic peptides derived from regions of the predicted protein sequences have shown that the gene order for the S segment is 3' NP, GP-2, GP-1, 5'.[10,35,38] Cleavage of the GP-C precursor, to release the mature GP-1 and GP-2 species, has been mapped to residues 262/263 in GP-C.[38] Antisera raised against synthetic peptides that converge from either side of this site recognize, respectively, GP-1 in the amino-terminal part of GP-C, and GP-2 in the carboxy-terminal part. The cleavage site, containing two adjacent basic amino acids, is conserved between LCMV Arm and WE, Pichinde, and Lassa viruses. A high-molecular-weight putative polymerase (L protein) originally assigned to the L segment by size considerations has now been detected using antibodies to L-derived synthetic peptides.[36b] Similar experiments using antipeptide antibodies will be used to evaluate additional potential protein coding regions that may be detected in genomic L cDNA clones.

VIII. TERMINAL SEQUENCE HOMOLOGIES

The 3' terminal sequences of the genomic L and S RNAs are identical for 17 of the first 19 positions, and for the S segment, the 5' terminal sequence is complementary to the 3' sequence.[39,40] There is no sequence information currently available for the genomic 5' L terminus, but, by analogy with S and other single-stranded RNA viruses, we can anticipate preservation of the complementary sequence character. These terminal sequences probably

Table 4
SIGNIFICANT AMINO ACID
CHANGES IN GP-C[a]

GP-C residue	LCMV arm	LCMV WE
110	L	P
133	T	S
173	T	S
174	F	S
177	A	P
181	Q	M
216	K	Y
240	T	R
253	S	A
265	A	S
313	A	E

[a] GP-C residues 1 to 262 = GP-1; 263 to 498
     = GP-2.

represent binding sites for the viral RNA-dependent RNA polymerase and/or a nucleation site for the binding of NP in the formation of ribonucleoprotein complexes. Auperin and Bishop[41] have noted the presence of an additional G residue at the exact 5' terminus of Pichinde and Lassa S genomic segments and have suggested that this may function to discriminate between the 5' ends of the genomic sense and genomic complementary sense RNAs. This additional G residue was not reported in the complete sequence of the WE genomic S segment,[9] so any suggested function for control of replication requires further experimental support.

IX. NUCLEOTIDE AND PROTEIN SEQUENCE CONSERVATION

The LCMV strains Armstrong (Arm) and WE represent the most homologous pair of arenaviruses for which sequence information is presently available. In the S protein coding regions there is 80 to 85% conservation of nucleotide sequence with transitions occurring much more frequently than transversions, and the sequences can be aligned without any significant insertion or deletion. Conservation of amino acid sequences is somewhat higher (90 to 95%), indicating the silent character of many of the nucleotide changes. There are only a limited number of amino acid changes that might be expected to produce significant changes in the structures of the folded proteins, and well-characterized differences in biological properties or reactivities with neutralizing monoclonal antibodies[42] may reside in single amino acid changes[10,11] (Table 4).

On the basis of protein sequence relatedness, LCMV shows somewhat more homology to Lassa than Pichinde, and Lassa and Pichinde are are no more closely related to LCMV than they are to each other.[11] The viral structural proteins show highly conserved regions which are interspersed with divergent regions. This type of arrangement was previously indicated from cross-protection studies and conserved and unique epitopes that had been defined by monoclonal antibodies.[43] The alignment of amino acid sequences for GP-C molecules indicates that the greatest diversity occurs in the region between residues approximately 120 to 240 in GP-1 (Table 5). This alignment has been made with the minimum number of gaps being introduced into the amino acid sequences and emphasizes sequence conservation among the GP-2 molecules. The mechanism of sequence evolution among arenaviruses remains to be elucidated, but there is now substantial cumulative evidence for a common ancestral virus that has diverged while becoming fixed in distinct geographical locations within specific rodent hosts.

Table 5

ARENAVIRUS GLYCOPROTEINS

```
ARM         MGQIVTMFEA LPHIIDEVIN IVIIVLIVIT GIKAVYNFAT CGIFALISFL
WE          MGQIVTMFEA LPHIIDEVIN IVIIVLIIIT SIKAVYNFAT CGILALVSFL
LA          MGQIVTFFQE VPHVIEEVMN IVLIALSVLA VLKGLYNFAT CGLVGLVTFL
PV          MGQIVTLIQS IPEVLQEVFN VALIIVSVLC IVKGFVNLMR CGLFLVTFL
CONSERVED   MGQIVT     P     EV N  I          K    N     CG   L  FL

            LLAGRSCGMY GLKGPDIYKG VYQFKSVEFD MSHLNLTMPN ACSANNSHHY
            FLAGRSCGMY GLNGPDIYKG VYQFKSVEFD MSHLNLTMPN ACSVNNSHHY
            LLCGRSCT.. ....TSLYKG VYELQTLELN METLNMTMPL SCTKNNSHHY
            ILSGRSCDSM MIDRRHNLTH VEFNLTRMFD NL......PQ SCSKNNTHHY
            L GRSC                V             P     C  NN HHY

            ISMGTS...G LELTFTNDSI ISHNFCNLTS AFNKKTFDHT LMSIVSSLHL
            ISMGSS...G LEPTFTNDSI LNHNFCNLTS ALNKKSFDHT LMSIVSSLHL
            IMVGNET..G LELTLTNTSI INHKFCNLSD AHKKNLYDHA LMSIISTFHL
            YKGPSNTTWG IELTLTNTSI ANETSGNFSN IGSLGYGNIS NCDRTREAGH
                    G   E T TN SI     N

            SIRGNSNYKA VSCDFNNG.. .......... .ITIQYNLTF SDAQSAQSQC
            SIRGNSNYKA VSCDFNNG.. .......... .ITIQYNLSS SDPQSAMSQC
            SIPNFNQYEA MSCDFNGG.. ....K..... .ISVQYNLSH SYAGDAANHC
            TLKWLLNELH FNVLHVTRHI GARCKTVEGA GVLIQYNLTV GDRGGEVGRH
                                                 QYNL

            RTFRGRVLDM F.RTAFGGKY MRSGWGWTGS DGKTTW.CSQ TSYQYLIIQN
            RTFRGRVLDM F.RTAFGGKY MRSGWGWTGS DGYTTW.CSQ TSYQYLIIQN
            GTVANGVLQT FMRMAWGGSY I......ALD SGRGNWDCIM TSYQYLIIQN
            LIASLAQIIG DPKIAWVGKC FNNCSGDTCR LTNCEGGTH. ..YNFLIIQN
                     A  G                              Y  LIIQN
                                  GP-1       GP-2
            RTWENHCTYA ..GPFGMSRI LLSQEKTKFF TRRLAGTFTW TLSDSSGVEN
            RTWENHCRYA ..GPFGMSRI LFAQEKTKFL TRRLSGTFTW TLSDSSGVEN
            TTWEDHCQFS RPSPIGYLGL LSQRTRDIYI SRRLLGTFTW TLSDSEGKDT
            TTWENHCTYT ...PMATIRM ALQRTAYSSV SRKLLGFFTW DLSDSSGQHV
            TWE HC     P                    R LG FTW    LSDS G

            PGGYCLTKWM ILAAELKCFG NTAVAKCNVN HDAEFCDMLR LIDYNKAALS
            PGGYCLTKWM ILAAELKCFG NTAVAKCNVN HDEEFCDMLR LIDYNKAALS
            PGGYCLTRWM LIEAELKCFG NTAVAKCNEK HDEEFCDMLR LFDFNKQAIQ
            PGGYCLEQWA IIWAGIKCFD NTVMAKCNKD HNEEFCDTMR LFDFNQNAIK
            PGGYCL W    A  KCF   NT   AKCN    H  EFCD  R L D N  A

            KFKEDVESAL HLFKTTVNSL ISDQLLMRNH LRDLMGVPYC NYSKFWYLEH
            KFKQDVESAL HVFKTTLNSL ISDQLLMRNH LRDLMGVPYC NYSKFWYLEH
            RLKAEAQMSI QLINKAVNAL INDQLIMKNH LRDIMGIPYC NYSKYWYLNH
            TLQLNVENSL NLFKKTINGL ISDSLVIRNS LKQLAKIPYC NYTKFWYIND
                       N L I D L    N  L     PYC NY K WY

            AKTGETSVPK CWLVTNGSYL NETHFSDQIE QEADNMITEM LRKDYIKRQG
            AKTGETSVPK CWLVTNGSYL NEIHFSDQIE QEADNMITEM LRKDYIKRQG
            TTTGRTSLPK CWLVSNGSYL NETHFSDDIE QQADNMITEM LQKEYMERQG
            TITGRHSLPQ CWLVHNGSYL NETHFKNDWL WESQNLYNEM LMKEYEERQG
                TG  S P  CWLV NGSYL NE HF        N    EM L K Y  RQG

            STPLALMDLL MFSTSAYLVS IFLHLVKIPT HRHIKGGSCP KPHRLTNKGI
            STPLALMDLL MFSTSAYLIS IFLHFVRIPT HRHIKGGSCP KPHRLTNKGI
            KTPLGLVDLF VFSTSFYLIS IFLHLVKIPT HRHIVGKSCP KPHRLNHMGI
            KTPLALTDIC FWSLVFYTIT VFLHIVGIPT HRHIIGDGCP KPHRITRNSL
             TPL L D     S    Y     FLH V IPT HRHI G  CP KPHR

            CSCGAFKVPG VKTVWKRR
            CSCGAFKVPG VKTIWKRR
            CSCGLYKQPG VPVKWKR
            CSCGYYKYQR NLTNG
            CSCG  K
```

X. CONSIDERATIONS FOR THE FUTURE

The availability of cloned arenavirus cDNA sequences should support many further advances in our understanding of virus gene regulation and expression and the mechanisms of virus pathogenesis. The isolation of a reassortant virus from a laboratory mixed infection that has a pathogenic potential possessed by neither of the parental virus strains (Table 3)[44,45] is particularly significant. This may be indicative of a mechanism that has contributed to the diversity of known arenaviruses, and may account for the appearance of new arenaviruses with new disease associations. It is now possible to examine the expression of individual viral genes to assess their relative importance for recognition and interaction with the host immune system. Also, cDNA genes can be mutated or recombined in vitro and reintroduced into cells or animals either as double-stranded DNA or RNA to identify alterations in biological properties. These approaches should define epitopes within the viral proteins that relate directly to pathogenic potential, and this may provide the key to an effective vaccine strategy for arenaviruses.

ACKNOWLEDGMENTS

This is Publication Number 4487-IMM from the Department of Immunology, Scripps Clinic and Research Foundation, La Jolla, CA. This work was supported in part by USPHS grants NS-12428, AI-09484, and AG-04342 and by USAMRIID contract C-3013. The findings in this report are not to be construed as an official Department of the Army position unless so designated by other authorized documents. We thank Kathy Nasif and Gay Schilling for secretarial assistance.

REFERENCES

1. **Buchmeier, M. J., Welsh, R. M., Dutko, F. J., and Oldstone, M. B. A.,** The virology and immunology of lymphocytic choriomeningitis virus infection, *Adv. Immunol.,* 30, 275, 1980.
2. **Lehmann-Grube, F., Martinez Peralta, L., Bruns, M., and Lohler, J.,** Persistent infection of mice with the lymphocytic choriomeningitis virus, *Compr. Virol.,* 18, 43, 1983.
3. **Lehmann-Grube, F.,** Portraits of viruses: arenaviruses, *Intervirology,* 22, 121, 1984.
4. **Peters, C. J.,** Arenaviruses, in *Textbook of Human Virology,* Belshe, R. B., Ed., John Wright-PSG, Littleton, Massachusetts, 1984, 513.
5. **Compans, R. W. and Bishop, D. H. L.,** Biochemistry of arenaviruses, *Curr. Top. Microbiol. Immunol.,* 114, 153, 1985.
6. **Dutko, F. J. and Oldstone, M. B. A.,** Genomic and biological variation among commonly used lymphocytic choriomeningitis virus strains, *J. Gen. Virol.,* 64, 1689, 1983.
7. **Auperin, D. D., Romanowski, V., Galinski, M., and Bishop, D. H. L.,** Sequencing studies of Pichinde arenavirus S RNA indicate a novel coding strategy, ambisense viral S RNA, *J. Virol.,* 52, 897, 1984.
8. **Clegg, J. C. S. and Oram, J. D.,** Molecular cloning of Lassa virus RNA: nucleotide sequence and expression of the nucleocapsid protein gene, *Virology,* 114, 363, 1985.
9. **Romanowski, V., Matsuura, Y., and Bishop, D. H. L.,** Complete sequence of the S RNA of lymphocytic choriomeningitis virus (WE strain) compared to that of Pichinde arenavirus, *Virus Res.,* 3, 101, 1985.
10. **Southern, P. J., Singh, M. K., Riviere, Y., Jacoby, D. R., Buchmeier, M. J., and Oldstone, M. B. A.,** Molecular characterization of the genomic S RNA segment from lymphocytic choriomeningitis virus, *Virology,* 157, 145, 1987.
11. **Southern, P. J. and Bishop, D. H. L.,** Sequence comparisons among arenaviruses, *Curr. Top. Microbiol. Immunol.,* 133, 19, 1987.
12. **Riviere, Y., Ahmed, R., Southern, P. J., Buchmeier, M. J., Dutko, F. J., and Oldstone, M. B. A.,** The S RNA segment of lymphocytic choriomeningitis virus codes for the nucleoprotein and glycoproteins 1 and 2, *J. Virol.,* 53, 966, 1985.
13. **Riviere, Y., Ahmed, R., Southern, P. J., and Oldstone, M. B. A.,** Perturbation of differentiated functions during viral infection *in vivo.* II. Viral reassortants map growth hormone defect to the S RNA of the lymphocytic choriomeningitis virus genome, *Virology,* 142, 175, 1985.
14. **Riviere, Y., Southern, P. J., Ahmed, R., and Oldstone, M. B. A.,** Biology of cloned cytotoxic T lymphocytes specific for lymphocytic choriomeningitis virus. Recognition is restricted to gene products encoded by the viral S RNA segment, *J. Immunol.,* 136, 304, 1986.

15. **Armstrong, C. and Lillie, R. D.,** Experimental lymphocytic choriomeningitis of monkeys and mice produced by a virus encountered in studies of the 1933 St. Louis encephalitis epidemic, *Public Health Rep. Wash.,* 49, 1019, 1934.

16. **Traub, E.,** A filterable virus recovered from white mice, *Science,* 81, 298, 1935.

17. **Rivers, T. M. and Scott, T. F. M.,** Meningitis in man caused by a filterable virus, *Science,* 81, 439, 1935.

18. **Gregg, M. B.,** Recent outbreaks of lymphocytic choriomeningitis in the United States of America, *Bull. WHO,* 52, 549, 1975.

19. **Howard, C. R.,** *Arenaviruses,* Elsevier, Amsterdam, Netherlands, 1986.

20. **Traub, E.,** Persistence of lymphocytic choriomeningitis virus in immune animals and its relation to immunity, *J. Exp. Med.,* 63, 847, 1936.

21. **Rowe, W. P.,** Studies on Pathogenesis and Immunity in Lymphocytic Choriomeningitis Infection of the Mouse. Research Rep. NM 005048. 14.01 Naval Medical Research Institute, Bethesda, Maryland, 1954.

22. **Oldstone, M. B. A. and Dixon, F. J.,** Lymphocytic choriomeningitis: production of antibody by "tolerant" infected mice, *Science,* 158, 1193, 1967.

23. **Oldstone, M. B. A. and Dixon, F. J.,** Pathogenesis of chronic disease associated with persistent lymphocytic choriomeningitis virus infection. I Relationship of antibody production to disease in neonatally infected mice, *J. Exp. Med.,* 129, 483, 1969.

24. **Oldstone, M. B. A.,** Viruses can alter cell function without causing cell pathology: disordered function leads to inbalance of homeostasis and disease, in *Concepts in Viral Pathogenesis,* Notkins, A. L. and Oldstone, M. B. A., Eds., Springer-Verlag, New York, 1984, 269.

25. **Southern, P. J. and Oldstone, M. B. A.,** Medical consequences of persistent viral infection, *N. Engl. J. Med.,* 314, 359, 1986.

26. **Cole, G. A., Nathanson, N., and Prendergast, R. A.,** Requirement for θ-bearing cells in lymphocytic choriomeningitis virus-induced central nervous system disease, *Nature (London),* 283, 335, 1972.

27. **Zinkernagel, R. M. and Doherty, P. C.,** Immunological surveillance against altered self components by sensitized T lymphocytes in lymphocytic choriomeningitis, *Nature (London),* 251, 547, 1974.

28. **Oldstone, M. B. A. and Buchmeier, M. J.,** Restricted expression of viral glycoprotein in cells of persistently infected mice, *Nature (London),* 300, 360, 1982.

29. **Southern, P. J., Blount, P., and Oldstone, M. B. A.,** Analysis of persistent virus infections by *in situ* hybridization to whole-mouse sections, *Nature (London),* 312, 555, 1984.

30. **Oldstone, M. B. A., Southern, P., Rodriguez, M., and Lampert, P.,** Virus persists in β-cells of islets of Langerhans and is associated with chemical manifestations of diabetes, *Science,* 224, 1440, 1984.

31. **Klavinskis, L. S. and Oldstone, M. B. A.,** Lymphocytic choriomeningitis virus can persistently infect thyroid epithelial cells and perturb thyroid hormone production, *J. Gen. Virol.,* 68, 1867, 1987.

32. **Oldstone, M. B. A., Sinha, Y. N., Blount, P., Tishon, A., Rodriguez, M., von Wedel, R., and Lampert, P. W.,** Virus induced alterations in homeostasis: alterations in differentiated functions of infected cells *in vivo, Science,* 218, 1125, 1982.

33. **Oldstone, M. B. A., Rodriguez, M., Daughaday, W. Y., and Lampert, P. W.,** Viral perturbation of endocrine function: disorder of cell function leading to disturbed homeostasis and disease, *Nature (London),* 307, 278, 1984.

34. **Oldstone, M. B. A., Ahmed, R., Buchmeier, M. J., Blount, P., and Tishon, A.,** Perturbation of differentiated functions during viral infection *in vivo.* I Relationship of lymphocytic choriomeningitis virus and host strains to growth hormone deficiency, *Virology,* 142, 158, 1985.

35. **Southern, P., Buchmeier, M. J., Ahmed, R., Francis, S. J., Parekh, B., Riviere, Y., Singh, M. K., and Oldstone, M. B. A.,** Molecular pathogenesis of arenavirus infections, in *Vaccines 86. New Approaches to Immunization,* Brown, F., Chanock, R. M., and Lerner, R. A., Eds., Cold Spring Harbor Laboratory, Cold Spring Harbor, New York, 1986, 239.

36. **Riviere, Y., Ahmed, R., Southern, P. J., Buchmeier, M. J., and Oldstone, M. B. A.,** Genetic mapping of lymphocytic choriomeningitis virus pathogenecity: virulence in guinea pigs is associated with the L RNA segment. *J. Virol.,* 55, 704, 1985.

36a. **Fuller-Pace, F. V. and Southern, P. J.,** Temporal analysis of transcription and replication during acute infection with lymphocytic choriomeningitis virus, *Virology,* 162, 00, 1988, in press.

36b. **Singh, M. K., Fuller-Pace, F. V., Buchmeier, M. J., and Southern, P. J.,** Analysis of the genomic L RNA segment from lymphocytic choriomeningitis virus, *Virology,* 161, 448, 1987.

37. **Harnish, D. G., Dimock, K., Bishop, D. H. L., and Rawls, W. E.,** Gene mapping in Pichinde virus: assignment of viral polypeptides to genomic L and S RNAs, *J. Virol.,* 46, 638, 1983.

38. **Buchmeier, M. J., Southern, P. J., Parekh, B. S., Wooddell, M. K., and Oldstone, M. B. A.,** Site specific antibodies define cleavage sites in a polyprotein, *J. Virol.,* 61, 982, 1987.

39. **Auperin, D., Dimock, K., Cash, P., Rawls, W. E., Leung, W. C., and Bishop, D. H. L.,** Analyses of the genomes of prototype Pichinde arenavirus and a virulent derivative Pichinde munchique: evidence for sequence conservation at the 3' termini of their viral RNA species, *Virology,* 116, 363, 1982.

40. **Auperin, D. D., Compans, R. W., and Bishop, D. H. L.,** Nucleotide sequence conservation at the 3' termini of the virion RNA species of new world and old world arenaviruses, *Virology,* 121, 200, 1982.

41. **Auperin, D. D. and Bishop, D. H. L.,** Arenavirus gene structure and organization, *Curr. Top. Microbiol. Immunol.,* 133, 5, 1987.

42. **Parekh, B. S. and Buchmeier, M. J.,** Proteins of lymphocytic choriomeningitis virus: antigenic topography of the viral glycoproteins, *Virology,* 153, 168, 1986.

43. **Buchmeier, M. J.,** Antigenic and structural studies of the glycoproteins of lymphocytic choriomeningitis virus, in *Segmented Negative Strand Viruses,* Compans, R. W. and Bishop, D. H. L., Eds., Academic Press, Orlando, Florida, 1986, 193.

44. **Riviere, Y. and Oldstone, M. B. A.,** Genetic reassortants of lymphocytic choriomeningitis virus: unexpected disease and mechanism of pathogenesis, *J. Virol.,* 59, 363, 1986.

45. **Riviere, Y.,** Mapping arenavirus genes causing virulence, *Curr. Top. Microbiol. Immunol.,* 133, 59, 1987.

Chapter 10

MODULATION OF VIRAL PLANT DISEASES BY SECONDARY RNA AGENTS *

J. M. Kaper and C. W. Collmer

TABLE OF CONTENTS

---

* This chapter was completed August 1986 and partially updated November 1987.

I. PROLOGUE

In 1972, field tomato plants in the Alsace region of France were stricken with a lethal necrotic disease of epidemic proportions.[1] The disease turned out to be of viral etiology[2,3] and probably represents the first clearcut, documented example of viral plant disease modulation by a secondary RNA agent, specifically, a viral satellite.[4] In the following pages we shall attempt to sketch how from the above events ensued a revival of interest in viral satellites, the first one of which was described in the early 1960s,[5] and several of which now known[6,7] are also capable of modifying viral plant disease expression. For those satellites on which the relevant information is available to us, but particularly for the cucumoviral satellites, we shall review the molecular data on replication, structure, and structural variability, and attempt to relate this information to the unique ways in which satellites modify disease expression of their helper virus genome.

II. PLANT DISEASE MODULATION BY VIRAL SATELLITES AND OTHER SECONDARY RNA AGENTS

For the purpose of this chapter we shall consider as secondary RNA agents, capable of modulating viral plant disease symptoms, only viral satellites and defective interfering (DI) RNAs.[8] Both types of agents have in common the following properties: (1) They depend upon the viral presence for their own replication, (2) they are capable of interfering with viral replication. However, they differ characteristically from each other in that viral satellites lack nucleotide sequence homology with their helper viruses, whereas DI RNAs are essentially deletion mutants of their helper viruses.

While thus far viral satellites have been predominantly reported in association with plant viruses and DI RNAs with animal viruses (see next chapter), notable exceptions exist. Only now the first outlines of a DI RNA or DI RNA-like association with certain plant viruses is beginning to take shape.[9,10] On the other hand, the Delta agent of hepatitis B virus (HBV) is the first reported example of an RNA-type satellite associated with an animal (albeit DNA−) virus.[11] An intermediate form or chimera of a satellite and a DI RNA has also recently been reported for a plant virus.[12] If all this seems confusing, it should not be so. In Gilbert's vision of a RNA world at the origin of life, where self-splicing introns and intron-flanked exons are moving as transposons,[13] any conceivable combination of RNA sequences containing the appropriate splicing signals can be realized in the absence of enzymes, let alone in their presence. Here, undoubtedly, other recombination events[14] possibly involving splicing/ligation or "jumping polymerases", as well as the high frequency of RNA mutation, all will add to the kaleidoscopic variability that can be observed among secondary RNA agents.

A. Cucumber Mosaic Viral Satellites and Tomato Necrosis

When tomato necrosis struck the Alsatian tomato growers in 1972, French plant pathologists were taken by surprise. It took close to a full year before they were able to conclude that the disease was caused by cucumber mosaic virus (CMV). The chronology of this investigation can be traced upon careful reading of the newspaper articles[1] and three scientific publications which appeared in 1973[15] and 1974.[2,3] The surprise is easily understandable upon examination of Figure 1 reproduced from one of the articles.[2] The newly observed syndrome of leaf epinasty, necrotic leaves, necrotic petioles, and stem necrosis leading to death of mature plants in the field can all be seen in the upper panel and was named lethal tomato necrosis by the French.[3] This disease contrasted totally with the conventional pattern of CMV infection in tomato, shown in the lower panel, which is characterized by filiform leaves and usually some form of chlorosis as an early symptom, and which is referred to as fern leaf syndrome. Even after identifying CMV as the probable causal agent of the tomato necrosis epidemic in the Alsace, much confusion remained because of the variability of symptoms obtained experimentally in tomato (*Lycopersicon esculentum*) in the greenhouse with identical samples of Alsatian necrogenic CMV. These ranged from lethal necrosis, to fern leaf syndrome, to no symptoms at all.[2] In addition, it was found that other well-established CMV strains, maintained in the French collection, induced a similar wide range of symptoms in tomato, including lethal necrosis, fern leaf syndrome, or no symptoms at all.[3] One consistent observation, in both reports, was that the incidence of tomato necrosis induced by all these CMV strains depended to an important degree upon the host in which the virus had been maintained. CMV grown in tobacco or tomato usually yielded enhanced necrotic response, whereas CMV grown in Cucurbitaceae usually yielded a reduced necrotic response. This observation received much attention in our laboratory, where since 1972 the varying presence of a fifth RNA component had been noted in preparations of CMV strain S (CMV-S). This fifth RNA occurred in large proportions when the virus was grown in tobacco (*Nicotiana tabacum* L. cv. Xanthi nc), and in much reduced proportions when the virus was increased in squash (*Cucurbita pepo* L. cv. Caserta Bush). The fifth RNA was identified to be a defective, small RNA of satellite-like nature, associated with CMV-S,[16] and was therefore designated CMV-Associated RNA 5 or CARNA 5.[4] The parallel correlations of the high or low incidence of CMV-induced tomato necrosis in the French work and the high and low proportions of CARNA 5 in CMV-S, both depending on host passage history of the virus, were strikingly obvious. With the availability of highly purified preparations of CMV-S genomic RNAs and CARNA 5 for tomato inoculations, it led to a series of experiments that conclusively identified CARNA 5 as the causal agent of lethal tomato necrosis.[4]

To our knowledge, CMV/CARNA 5-induced tomato necrosis has not been reported to occur in the field on a significant scale since 1972. Sporadic occurrences have been reported in France[17] and, more recently, in Japan.[18] From the standpoint of molecular pathology, however, the tomato necrosis syndrome is a very attractive experimental system. As will

FIGURE 1. Symptoms of cucumber mosaic virus in tomato. Upper panel shows tomato necrosis of a plant in the field; lower panel shows plant with conventional fern leaf syndrome. (From Putz, C., Kuszala, J., Kuszala, M., and Spindler, C., *Ann. Phytopathol.*, 6, 139, 1974. With permission.)

be seen later, CMV-associated and other viral satellites modulate viral disease expression in a number of different ways, but very often in the form of symptom attenuation. This can be rationalized on the basis of their presumed interference with viral replication. A problem is that certain forms of symptom modulation, perceived by the investigator as an aggrevated disease condition (for instance, intensified yellowing or chlorosis), could in reality result from a lesser viral presence in the tissues. However, with the lethal necrosis syndrome in tomato there is no such ambiguity, since it is hard to conceive how a lesser virus presence could result in death of the cells. Thus, it seems that the tomato necrogenic character is somehow encoded in the nucleotide sequence of certain CMV satellites, but not others.

Nucleotide sequences for a number of CARNA 5 variants, necrogenic as well as non-

FIGURE 2.   Modulation of cucumber mosaic virus disease symptoms by its satellite CARNA 5. Upper row of plants was infected by virus alone; lower row of plants by virus plus CARNA 5. In tabasco pepper plants on the left, CARNA 5 attenuates disease symptoms; in tomato plants on the right, CARNA 5 induces lethal necrosis.

necrogenic, isolated from CMV strains originating in different parts of the world, have now been reported.[19-23b] In addition, the sequences of several others are known (see Section III.B.2). This knowledge should facilitate both finding the sequence domain(s) responsible for tomato necrosis and unraveling the molecular mechanism(s) by which this disease is induced in the CARNA 5 trilateral relationship with CMV and the host plant.

B. Cucumber Mosaic Viral Satellites and Other Forms of Disease Modulation
1. Disease Attenuation
   Although disease attenuation as a result of CARNA 5 replication had been implied in the first report on this satellite[16] dramatic visual evidence of such an effect was provided in 1979, when a tomato necrosis-inducing CARNA 5 was shown to attenuate CMV disease in a number of other plant species.[24] This Dr. Jekyll and Mr. Hyde-type effect is illustrated in Figure 2 for tabasco pepper (*Capsicum frutescens* L.) and tomato, respectively, where the upper row of plants was infected with the genomic RNAs of CMV alone, whereas the lower row of plants was infected with the genomic RNAs plus CARNA 5. It was also shown that CARNA 5 progeny produced in the pepper plants was qualitatively as necrogenic (by tomato necrosis assay) as the CARNA 5 progeny obtained from the tomato plants, although no sequence comparisons were carried out.[24] Shortly thereafter, a similar attenuating effect was reported on the part of a CMV satellite isolated in Australia.[25] The fact that this latter disease attenuation was in tomato showed that there existed "strains" of CARNA 5 capable of exerting different disease-modulating properties in the same host species.
   CARNA 5-mediated attenuation of CMV disease has since been experimentally confirmed in several laboratories.[26-28] A biochemical mechanism underlying this attenuation effect is not difficult to envision, provided the validity of one basic assumption, namely, that CARNA 5 and CMV RNA replication involves some common factor(s) (for instance, a viral-coded replicase) for which the two types of RNA compete. In such a case, successful competition by CARNA 5 would be expected to result in disease attenuation due to suppression of viral

multiplication.[29] The time course of the relative rates of viral RNA vs. CARNA 5 synthesis in CMV/CARNA 5-infected tobacco plants has been determined and seems to be consistent with this putative competitive replication mechanism.[30]

a. CARNA 5 as Potential Biological Control Agent Against CMV Infection

Competitive replication of CARNA 5 and CMV RNA has also become the basis for experimental, as well as real-world, attempts at biological control of CMV infection by means of its satellites. The first report of successful protection of tomato plants preinfected with a CMV strain containing a nonnecrogenic CARNA 5 against challenge infection with another strain of CMV was in 1982.[31] Although the emphasis in these experiments was on protection achieved against the necrosis-inducing properties of the challenge strain (implying competition between CARNA 5 species, a phenomenon also frequently observed in our laboratory), one experiment showed protection against challenge by a CMV strain devoid of CARNA 5. A very similar report was recently published by a Japanese group.[18] This work was more comprehensive in that appropriate controls demonstrated that conventional cross-protection effects (among CMV strains), although present, were superseded by satellite protection. Figure 3, taken from this work, illustrates the type of satellite protection achieved for tomato plants. Successful satellite-mediated control of CMV in the field has been reported for pepper,[32,32a] and is presently also being carried out for tomato.

The use of CMV and its satellites in preventive inoculation of crop plants on a large scale has some danger in that minute amounts of necrogenic CARNA 5 variant(s) already present with the virus may quickly dominate the satellite population upon passage in certain host plants,[33,34] and, therefore, rigorous testing for the presence of necrogenic variants is essential. In addition, there are the reservations usually voiced with respect to the use of any attenuated virus forms for cross protection.[7] Some of these reservations may now have been overcome, in principle, since recently it has been shown possible to insert cDNA copies of CARNA 5 into the genome of tobacco plants with an expression vector based on the binary Ti plasmid system of *Agrobacterium tumefaciens*. Regenerated transformed plants were shown to produce transcripts containing the CARNA 5 sequence. These transcripts apparently were recognized, processed, and subsequently replicated by the virus-induced replication mechanism following challenge infection with a satellite-free CMV.[35]

If this type of transformation could be extended to some of the 774 other plant species that are known to be susceptible to CMV infection, we may have accomplished by genetic engineering what could also be the unique design of nature to contain an otherwise perhaps uncontrolled spread of plant viruses such as CMV.

b. Natural Occurrence of CARNA 5 and Possible Host Origin

Nothing is known about the origin of cucumoviral or other satellites, but some data are beginning to emerge with regard to their occurrence in nature. In two widely separated geographical locations, large accumulations of dsCARNA 5 have been found to occur in conjunction with natural CMV infections of *Nicotiana glauca*.[36,37] Such accumulations of dsCARNA 5 have previously been shown to represent evidence for CMV/CARNA 5 replicative competition that has entered the steady state of a low-level persistent infection.[29,30,38,39] Other *Nicotiana* species are known for their propensity to favor the replication of CARNA 5 over its helper CMV.[16,25-27,30,33,40] A popular view of the origin of viral satellites is that they could have arisen from the genetic apparatus of host plants,[41-45] although other hypotheses also exist.[7] However, for none of the known viral satellites has convincing experimental evidence been produced in favor of any of these putative origins. To firmly validate the plant-origin hypothesis, not only should positive hybridization signals be produced between the DNA or RNA of certain plants species and appropriate radioactive satellite probes, but also the same plant species should respond to challenge infection with RNA transcripts of full-length cloned cDNA of the appropriate helper virus with the emergence of satellite progeny. Such tests are within the realm of present-day experimental techniques.

FIGURE 3. Satellite-mediated protection of tomato against challenge infection by cucumber mosaic virus. Upper panel shows plants that were challenge infected with CMV containing a necrogenic CARNA 5; lower panel shows plants challenge infected with CMV alone. Plants on the left were preventively inoculated with CMV containing a nonnecrogenic CARNA 5 and are almost symptomless. Plants on the right were not preventively inoculated and show the symptoms of the CMV strains used for challenge infections. (From Yoshida, K., Goto, T., and Iizuka, N., *Ann. Phytopathol. Soc. Jpn.*, 51, 238, 1985. With permission.)

2. Apparent Exacerbation of Disease Symptoms

In addition to the tomato necrosis syndrome, two other cases have been reported where CMV satellites have caused an apparent exacerbation of symptoms in CMV infections.

a. The 368-Nucleotide Satellite of CMV Strain Y and Tobacco Yellowing

With CMV-Y and a number of other Japanese CMV isolates, satellite association has correlated with strikingly yellow symptoms in tobacco, which was how some of the isolates were actually spotted in the field.[26] Other CMV isolates, devoid of satellite, displayed a green-white mosaic in tobacco, but acquired the yellow symptom when satellite was added to the inoculum. Conversely, the green-white mosaic symptoms could also be made to appear upon experimental removal of the satellite from those CMV isolates that contained it naturally. Each of the CMV/satellite combinations was also reported to induce necrosis in tomato, which suggested that the satellite of CMV-Y (designated Y-CARNA 5 from here on) should resemble the previously described necrogenic CARNA 5.[4] However, subsequent nucleotide sequence determination of Y-CARNA 5[22] showed that in spite of extensive homologies with other known CARNA 5 sequences, the Y-CARNA 5 sequence deviates from them in that it is 30 to 35 nucleotide residues longer.[23] This difference is caused by a complicated set of insertions, deletions, and nucleotide substitutions in a specific domain in the 5' half of the molecule (see Figure 4).[22,23] Whether the bright yellow symptom in tobacco[26] is related to these changes in CARNA 5 structure, or whether it follows from a reduced presence of the virus (and thus represents a form of disease attenuation in this particular virus-host combination) remains to be determined.

Recently, we have confirmed the association of the 368-nucleotide Y-CARNA 5 with CMV-Y, as well as its nucleotide sequence.[34] However, after rigorous gel-purification of Y-CARNA 5 from possible contaminating traces of necrogenic CARNA 5, we were unable to assign it the property of lethal tomato necrosis induction in the presence of CMV-Y or other CMV helper virus strains.

b. Satellite-Induced White Leaf Disease in Tomato

The second case of apparent CARNA 5-induced disease exacerbation, like the first, was also discovered in the field in tomato plants which displayed a striking whitish-green mottle of the leaf laminae and other symptoms uncharacteristic of CMV.[28] This disease has been named tomato white leaf, and CMV (designated the WL strain) has been consistently recovered from such plants. In secondary infections of tomato, some plants displayed a different set of symptoms in which they retained their green color, but where leaf laminae were narrowed and had a distinct downward curl. From the RNA components in virus recovered from plants with the white leaf or with leaf curl symptoms, it was concluded that a satellite (WL-CARNA 5) was responsible for the white leaf syndrome. This was confirmed upon separation of the WL genomic RNAs 1, 2, 3 and WL-CARNA 5 and reinoculation of tomato plants. In contrast, inoculation of tobacco with CMV-WL genomic RNAs alone produced intense chlorosis, whereas in the presence of WL-CARNA 5, essentially complete disease attenuation was observed. Thus, in terms of satellite-induced disease modulation, this is exactly the opposite of the situation with CMV-Y and Y-CARNA 5.[26] With details of the structure of WL-CARNA 5 recently published, its nucleotide sequence seems to bear more resemblence to that of nonnecrogenic rather than necrogenic CARNA 5.[46,23a]

C. Secondary RNA Agents Other than Cucumber Mosaic Viral Satellites and Disease Modulation

Although viral satellites have always been known to interfere with helper virus production, the main biological properties reported prior to CARNA 5-induced tomato necrosis[4] have been the effects of satellites on helper virus local lesion size and numbers.[47] More recent reports of disease-modulating effects have been in conjunction with newly discovered satellites or other types of secondary RNA agents. In this subsection we shall attempt to give brief descriptions of these disease-modulating effects and the agents responsible for them.

1. Secondary RNA Agents Causing Apparent Disease Exacerbation
a. The Satellite of Arabis Mosaic Virus (SArMV)[48]

Relatively important, because of its association with one of the earliest plant diseases

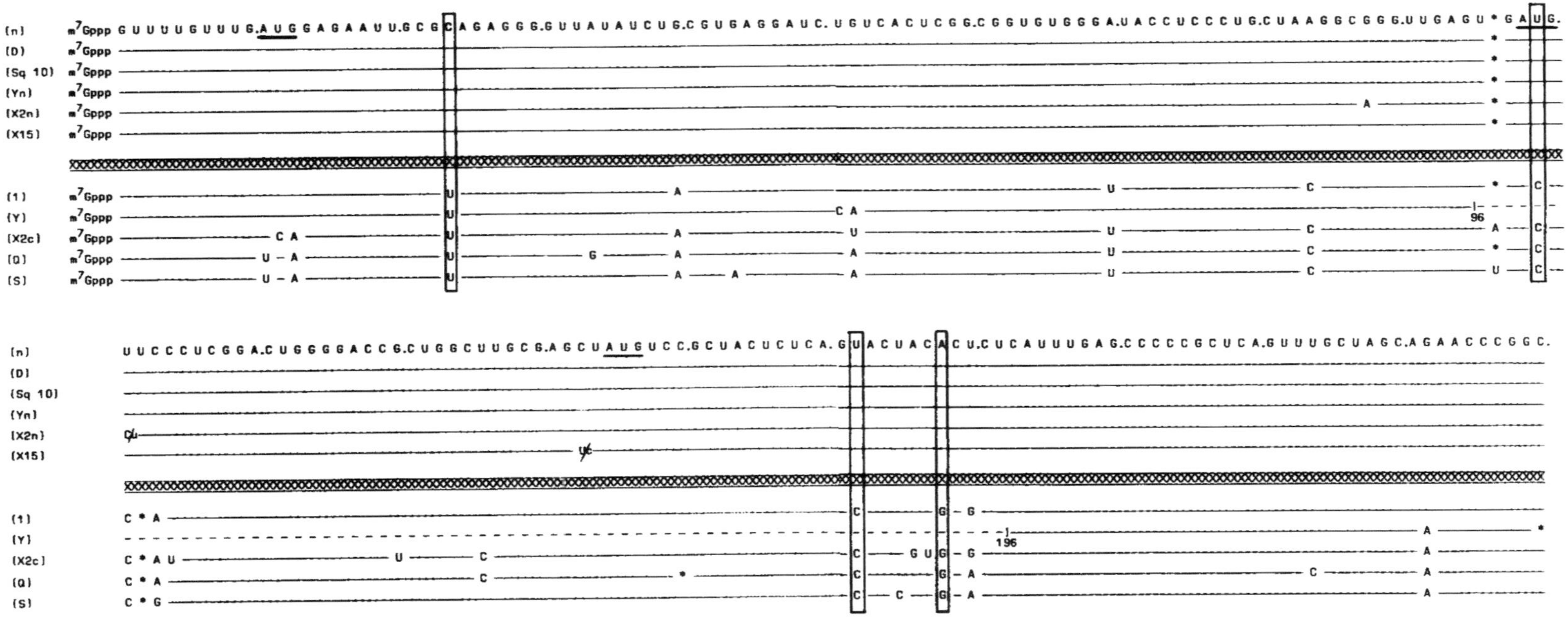

FIGURE 4.   Sequence comparison of cucumber mosaic viral satellites. Heavy cross-hatched line divides nucleotide sequences of necrogenic (above line) and nonnecrogenic (below line) CARNA 5 variants. Variants D, Sq 10, Yn, X2n, X15, X2c were sequenced in our laboratory;[23b] sequences of n,[19] 1,[21] Y,[22] Q,[20] and S[23] were published previously. Changes in the sequences relative to n-CARNA 5 are indicated by a letter (a substitution) or by * (a deletion); a horizontal line signifies no change. The dashed line from position 96 to position 196 of Y-CARNA 5 represents a region of significant rearrangements, insertions, and deletions which is not directly comparable to the other satellites. Nucleotides conserved in all necrogenic variants and altered in the same way in comparable sites in all nonnecrogenic variants are boxed. The AUG initiation codons of putative open reading frames ORF I, ORF IIA, and ORF IIB (see text) are underlined in the n-CARNA 5 sequence.

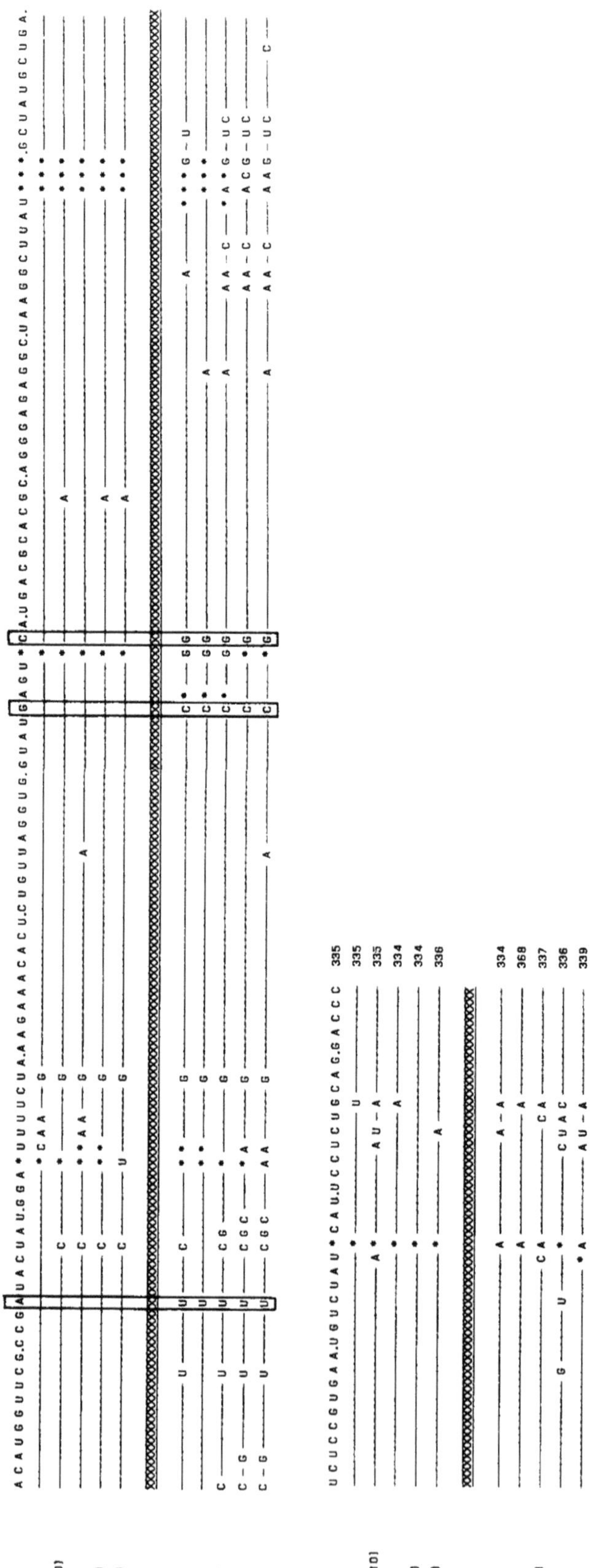

FIGURE 4 (continued)

described for an economically significant crop, is the finding that ArMV isolates from hop plants with nettlehead disease consistently produced unusually severe symptoms of mosaic and distortion in *Chenopodium quinoa*. Progeny virus produced from such infections contained large proportions (80%) of a low-molecular-weight RNA. The genomic RNAs of ArMV alone, after separation from the low-molecular-weight RNA, induced much milder symptoms in *C. quinoa*. With other ArMV isolates, containing different proportions of the suspected satellite, a positive correlation was found between symptom severity in *C. quinoa* and the proportion of low-molecular-weight RNA. In experimental infections of hop plants using ArMV with and without the satellite-like RNA, positive correlation of the presence of the latter with nettlehead symptoms was also found. Furthermore, direct analysis of commercially grown hop plants with nettlehead symptoms suggested the existence of additional low-molecular-weight, satellite-like RNA species. Thus, it seems likely that the satellite-like RNA of ArMV is involved in the symptomatology of at least two plant species, in *C. quinoa* with symptom exacerbation, and as the etiological agent in the nettlehead disease of hop.

Recent work in our laboratory[49] has established that no detectable nucleotide sequence homology exists between the satellite-like RNA and the genomic RNAs or ArMV, and that this small RNA can therefore be classified as a viral satellite. Structural studies, including nucleotide sequence determination, show that the satellite of ArMV (SArMV), when encapsidated, is an approximately 300-nucleotide linear RNA, but that in infected tissues it coexists with a circular form. Its sequence shows partial homology with that of the satellite of tobacco ringspot virus (STobRV, see also Section II.C.2.a) which exceeds significantly the homologies of the 5' terminal domains of the two satellites. Both these domains contain sequences believed to be essential for the self-processing capabilities of these and other viroid-like circular RNAs.[49a]

b. The Chimeric Satellite-DI RNA of Turnip Crinkle Virus (TCV)

In 1981, it was reported that TCV contained a satellite-like RNA capable of increasing significantly the severity of TCV symptoms in turnip plants.[50] This represented the second documented case in which an apparent satellite was shown capable of influencing systemic symptoms of virus disease in plants. In the succeeding 5 years, a considerable amount of additional work has been done with this system.

In TCV infections, several satellite-like small RNAs are found. Three were labeled RNA, C, D, and F by their discoverers, and each is dependent for its replication upon the presence of TCV. RNA C, however, is the agent responsible for the disease-modulating effect in turnips. Its nucleotide sequence has been determined and compared with sequence information available for the genomic TCV RNA and RNAs D and F.[12] This study shows that about 166 nucleotides in the 3' half of the 355-nucleotide RNA C are largely homologous in sequence to those at the 3' end of TCV RNA plus a 15-nucleotide stretch further upstream, whereas in the 5' half of RNA C (189n), a large degree of homology is found with the 5' end of RNAs D (194n) and F (230n). RNAs D and F have no sequence homology with TCV RNA. They apparently are satellites, but do not seem to influence the disease expression of TCV. Thus, RNA C of TCV seems to be a chimera of a satellite and a DI RNA, and is a concrete example of the apparently diverse origins of RNA molecules that are capable of exerting disease-modulating effects in virus-infected plants.

c. The Satellite of Panicum Mosaic Virus (SPMV)[51,51a]

Recent information[52] concerning SPMV indicates that upon its infection of a particular variety of corn (*Zea mays* L. cv. Ohio 28) and pearl millet (*Setaria italica* L. cv. Beauv.) in the presence of the helper virus PMV, characteristically severe symptoms of vein banding and stunting result, whereas infections of PMV alone produce only a mild mosaic.

In molecular properties, SPMV resembles the satellite of tobacco necrosis virus (see Section II.C.3.a). The 17-nm satellite particles can immediately be distinguished in the electron microscope from the 28-nm PMV particles, and they bear no serological relationship to each other.[51] SPMV consists of a 17-kdalton structural protein and a 826-nucleotide RNA, which must therefore encode this structural protein. The nucleotide sequence of SPMV-RNA reveals of the presence of two open reading frames, the first predicting a polypeptide of the size of the above structural protein, and the second a polypeptide of about 6 kdaltons.[51a] However, in vitro translation studies have thus far yielded only one translation product, of the size of the coat protein of SPMV.[51]

d. RNAs 3 and 4 of Beet Necrotic Yellow Vein Virus (BNYVV)

A prime example of viral encapsidated small RNA molecules, incapable of independent replication, possessing disease-modulating characteristics and perhaps other important biological properties, are the RNAs 3 and 4 of BNYVV, a rigid, rod-shaped, fungus-transmitted virus responsible for the rhizomania disease of sugarbeet. Different isolates of the virus from Europe and the U.S. have shown great diversity in length among RNAs 3 and 4, but not with the two longer RNA species 1 and 2. The maximum length encountered for RNA 3 has been 1850 nucleotides, and for RNA 4, 1500 nucleotides; smaller length RNAs 3 and 4 in different isolates are deleted versions of the full-length species.[53] Their lack of extensive nucleotide sequence homology overall (except at the 3' terminus, see below) with RNAs 1 and 2, in addition to the fact that the presence of RNA 3 in the isolates correlates with exacerbated disease symptoms in *C. quinoa*, suggested that RNAs 3 and 4 might be satellites of BYNVV.[54] However, the recent nucleotide sequence determination of RNA 2 revealed extensive sequence homology of the 70-nucleotide stretch at the 3' ends of RNAs 2, 3, and 4, which, in addition to their common terminal features of 5' m^7Gppp caps and 3' poly(A) tails, suggests similar or perhaps even identical replication strategies for the RNAs.[55] All three RNAs have translational open reading frames of significant lengths and have been translated in vitro.[56] While all this is compatible with their being satellites, there are indications that RNAs 3 and 4 could carry certain functions required by the virus in the process of its natural transmission by the soil-borne fungus *Polymyxa betae*. This is suggested by the fact that virus obtained directly from sugarbeet rootlets always contains full-length RNAs 3 and 4, whereas upon mechanical transmission of the virus to the experimental host plants, such as *C. quinoa*, the deleted forms of RNAs 3 and 4 are found (or are totally absent).[57] Thus, RNAs 3 and 4 could be considered genomic RNAs with certain functions essential for viral transmission and spread in natural infections of sugarbeet. Because of selection pressures under conditions of mechanical transmission in *C. quinoa*, where these functions are not needed, these RNAs may undergo progressive deletions. In this deletion process they might acquire properties that enable them to exacerbate certain disease symptoms, and thus resemble satellites.

e. RNA 2, the Circular Satellite of Velvet Tobacco Mottle Virus (VTMoV) and Other Putative Sobemoviral Satellites

An apparent disagreement with regard to the satellite nature of certain small, covalently circular RNA molecules associated with VTMoV and other putative members of the sobemovirus group[7] was resolved recently when a new circular RNA-devoid isolate of VTMoV was shown capable of independently infecting *Nicotiana clevelandii*.[58]

A considerable amount of structural work has been carried out with the RNAs 2 and 3 encapsidated in VTMoV, and three other viruses, solanum nodiflorum mottle virus (SNMV), lucerne transient streak virus (LSTV), and subterranean clover mottle virus (SCMoV), following their initial discovery in the early 1980s. References to the original publications describing this work can be found in a recent review.[7] Some excitement was caused by the fact that the RNA 2 species were found to be covalently circular (RNA 3 being a linear form of RNA 2) and of similar length as viroids (324 to 388 nucleotides), leading to their designation as "viroid-like RNAs" or "virusoids". However, they are different from viroids in that they are incapable of independent replication. While they were initially believed to be indispensable for the replication of the larger coencapsidated RNA 1 (approximately 4500 nucleotides), the satellite nature of the first one of these RNAs[59] and, later, the other, has been shown. The occurrence of oligomeric ds forms of RNA 2 in VTMoV infections has invited suggestions that these satellites, like viroids and the satellite of tobacco ringspot virus (see Section II.c.2.a) are replicated via a rolling circle mechanism.[60]

Recently several experimental studies, using dimeric or permuted monomeric RNA transcripts of cDNA clones of certain sobemoviral satellites, STobRV, and other small RNA molecules, have shown self-cleavage of these molecules at a site surrounded by a specific secondary structure motif determined by certain sequence domains these RNA molecules have in common[49a] (see also Section II.C.1.a). This has further supported the notion that oligomeric forms of viroid-like satellites in infected tissues may represent the primary products of a rolling circle-type transcription process. However, it should be remembered that self-cleavage of satellite RNA has been shown to be reversible in the case of STobRV (see Section II.C.2.a), while it has also been suggested that intermolecular ligation of SNMV RNA 2 monomers could explain the occurrence of its dimeric forms.[61]

2. Secondary RNA Agents Causing Disease Attenuation
a. The Satellite of Tobacco Ringspot Virus (STobRV)

One of the earliest described viral satellites[7,47] and among those best characterized in molecular terms, this 359-nucleotide linear RNA[62] can be found encapsidated in large quantities in the viral coat protein, following combined infections with its helper virus TobRV.[47] Although the interfering effects of STobRV on viral RNA accumulation in such infections have long been known, its attenuating influence in systemic infections of the helper was dramatically demonstrated in a recent report that showed how very severe symptoms of the budblight strain of TobRV in Blackeye 5 cowpeas disappeared almost entirely when different strains of STobRV or an in vitro RNA transcript of cloned STobRV cDNA were coinoculated with the virus.[63]

Oligomeric ds forms of STobRV were the first described for any viral satellite,[47] and they have provided the basis for proposing a rolling circle-type mechanism of replication for STobRV.[64] The subsequent detection of circular forms of ss STobRV,[65] and the reversible self-cleavage capabilities of STobRV oligomers and monomers,[66,66a] have provided additional support in favor of such a type of replication mechanism for STobRV.

b. The Satellite-Like RNA of Chicory Yellow Mottle Virus (CYMV)

The highly characteristic buoyant density pattern of particles of the nepovirus CYMV-T (type strain),[66] and its similarity with the multidense appearance of STobRV,[47] led to early suspicions of the presence of a satellite-like RNA in CYMV-T. This was supported by the fact that such virus preparations contain ample quantities of smaller-size RNAs not needed for infectivity of the two genomic-size RNAs. The smaller, satellite-like RNA fraction is composed of a predominant low-molecular-weight (approximately 0.17×10^6) component, plus lesser quantities of RNA components ranging upward to genomic RNA size.

Comparison of CYMV-T with a more recent isolate from the same field (CYMV-RS), which was found to be devoid of the satellite-like RNAs, showed the two viruses to be closely related strains. However, they differed significantly in symptomatology on *Nicotiana glutinosa*, where the satellite-devoid CYMV-RS causes characteristic ringspot symptoms, whereas CYMV-T infections remain symptomless.[66] This behavior suggest that the presence and replication of satellite-like RNA has an attenuating effect in CYMV infections of *N. glutinosa*.

c. Satellites and DI RNAs Associated with Tomato Bushy Stunt Virus (TBSV)

In 1985, two publications appeared in which symptom-modulating effects of small RNAs found in infections of TBSV were described.[9,44] These reports indicated that two separate types of secondary RNA agent were involved, e.g., satellite(s) without sequence homology to the viral genome, as well as DI RNAs that were derived from the viral RNA.

The TBSV satellite is an RNA, 700 nucleotides in length, which is encapsidated by the helper virus coat protein. Its symptom-modulating effect upon coinoculation with the helper virus is best observed in *Chenopodium amaranticolor*, where the number of local lesions are drastically reduced. In *Nicotiana clevelandii*, an attenuation of systemic symptoms is observed.[44] More recent work has failed to detect satellite-related protein products in protoplast infections or from in vitro translation experiments.[67]

TBSV DI RNAs range 350 to 500 nucleotides in size, but they are not encapsidated in large quantities. Recent work[10] has shown them to possess extensive nucleotide sequence homology with TBSV RNA. The homologous sequences derive from the 3' and 5' ends, as well as from the internal regions of the TBSV genome. In one of these RNAs, of which the nucleotide sequence was determined, these segments are arranged colinearly and form a mosaic-type 396-nucleotide DI RNA. In *N. Clevelandii* infected with TBSV, the presence of the DI RNAs correlates with a reduced viral multiplication and significant attenuation of systemic symptoms.

3. Secondary RNA Agents Without Disease-Modulating Effects

There is some irony in the fact that among the agents discussed in this subsection are some of the viral satellites earliest discovered and, therefore, also best characterized. While falling outside the general scope of this chapter for this apparent lack of biological activity, their properties, most of which have been reviewed extensively elsewhere,[7] will be briefly summarized.

a. The Satellite of Tobacco Necrosis Virus (STNV)

STNV is the first discovered viral satellite.[5] It appears in infections of STNV as spherical particles 17 nm in diameter, which contain the 1239-nucleotide RNA. STNV RNA encodes a 195-amino acid structural protein, 60 copies of which constitute the icosahedral capsid of the satellite, the three-dimensional structure of which has been resolved at 2.5 Å resolution.[68] The nucleotide sequence of STNV RNA, as well as the amino acid sequence of its coat protein, have been experimentally determined. In combined infections, the satellite apparently interferes severely with the replication of the helper virus, but the only reported biological effect is a reduction in the TNV local lesion size and number in French bean plants.[69]

b. The Satellite of Tomato Black Ring Virus (STBRV)

STBRV constitutes a family of related RNA species that can be found in association with different TBRV isolates,[70] the first one of which was identified in 1973. This variant has been completely characterized for its nucleotide sequence; it is 1375 nucleotides long and has a 5' end-linked protein as well as a polyadenylated 3' end.[71] These two features resemble that of the helper virus RNAs, but significant nucleotide sequence homology has not been found. STBRV encodes a protein of 48 kdaltons which has also been identified experimentally both in protoplast and in vitro translation experiments. Unfortunately, the biological function of the satellite-encoded protein is not known, and STBRV coinoculated with TBRV is known only to reduce the number of lesions in *C. amaranticolor*. Recently the nucleotide sequences of four additional satellites of TBRV have been reported.[71a] Their molecular size varied between 1372 and 1376 nucleotides, and although generally sequence related, they segregated into two groups with 90% or more sequence homology each. All four sequences contained a single open reading frame encoding a protein of approximately 48 kdaltons.

c. Peanut Stunt Virus-Associated RNA 5 (PARNA 5)

Discovered soon after the cucumoviral satellite CARNA 5, PARNA 5 is 393 nucleotides long. However, except for the presence of an m⁷Gppp cap, and nearly homologous ten- and eight-nucleotide stretches at the 5′ and 3′ termini of the molecule, respectively, PARNA 5 has no significant sequence homologies with CARNA 5.[45] The two satellites also cannot be supported by the helper viruses of each other. Unlike the different CARNA 5 variants, no significant symptom-modulating effects have been noted thus far in combined infections of PSV and PARNA 5.

d. The Satellite of Tobacco Mosaic Virus (STMV)

Revival of interest in viral satellites has recently culminated in the discovery of a satellite in association with infections of the archtypical TMV. Identified initially as a dsRNA (molecular weight approximately 0.6×10^6) in *N. glauca* naturally infected with TMV U5 and with CMV,[37] it was later found to be an ssRNA encapsidated in its own coat protein (molecular weight approximately 18 kdaltons), forming a 17-nm spherical particle.[72] Aside from its natural helper, TMV U5, STMV can also be supported by TMV U1, but in combined infections with neither one has STMV shown reproducible symptom-modulating effects.

III. MOLECULAR BASIS OF PLANT DISEASE MODULATION BY SECONDARY RNAs

A. Approaches

Since nothing is known at this time as to precisely how plant viruses cause disease, it follows that attempts to understand and explain how secondary RNAs interact with that process are fragmentary and largely speculative in nature. Given that it is necessary (although not always sufficient) for viruses to replicate in order to cause disease, the previously mentioned interference with that process by secondary RNAs most likely has an effect on disease development. However, whether such interference results in the simplest case from direct competition between viral and secondary RNAs that share limited replication machinery (see Section II.B.1), or rather from some more complicated kind of disruption of fine-tuned controls that regulate viral replication and viral gene expression, will have to be determined for each virus-secondary RNA combination. Attempts to demonstrate examples of the former would certainly benefit from the availability of isolated replicase complexes, so that competitive replication could be studied directly in vitro and related functionally to specific structural features of the two molecules. However, such an in vitro approach will most likely be of limited value for the following reasons. First, few such in vitro replicase systems with the required specificity for exogenous plant viral RNAs are currently available, and those that are have no associated secondary RNAs reported to date. Second, available evidence already seems to suggest that while a common replicative mechanism may be operative with some virus/satellite combinations, in other combinations (e.g., TobRV/STobRV) there may be considerable divergence, making the successful purification of a replicase system capable of replicating both molecules in vitro quite unlikely. Third, even the modulation by secondary RNAs of viral replication itself, not to mention the more complex phenomenon of disease modulation, may be effected through interactions with other plant machinery, e.g., the translational apparatus, not present in an isolated replicase complex.

While protoplast systems offer a partial solution to this latter problem, they are often unavailable and by their nature do not reflect the disease reaction of the whole plant. Finally, even those experimental systems that might permit discovery of the theoretically perfect correlation between interference with viral replication and plant disease modulation (attenuation) by secondary RNAs fail to address the larger question of mechanisms of pathogenesis itself. Furthermore, they are unlikely to explain situations of disease exacerbation where the secondary RNA may exert a deleterious effect of its own in addition to that caused by the virus.

While secondary RNAs add an additional level of complexity to an already complex interaction between virus and host, they might also be seen as model systems that, due to their small size and dispensibility to viral replication, offer unique opportunities to explore structure-function relationships in pathogenic RNA molecules. On one hand, in studying replication interference one may learn about the regulatory controls on normal viral replication and, thus, potential disease control strategies. On the other hand, for those molecules that exacerbate disease symptoms, the current-day feasibility of expressing them in transgenic plants in the absence of their helper viruses may offer a long-awaited opportunity to tease apart the trilateral interactions of host, virus, and secondary RNA in the complex process of disease development.

The first step in all these studies must be to determine the structures of the viral and secondary RNAs and then to identify domains involved in their interactions with each other and with the host. Currently available genetic techniques are essential not only to precisely define these domains, but also to attempt to understand their biological activities — i.e., are they *cis*-acting as regions of recognizable structure, or are they *trans*-acting as either RNA enzymes themselves[72a] or through expression of an encoded polypeptide? Since there are now methods available to make biologically active RNA transcripts from satellite cDNA inserts in recombinant plasmids,[63,73-73b] site-specific mutations directed at regions of suspected biological activity[73c] offers special opportunities to resolve such questions. The following sections attempt to explore the limited progress that has been made to date in these directions.

B. Potential Molecular Mechanisms of Disease Modulation in Cucumoviral Satellites
1. Attenuation of Plant Disease

Phenomenologically speaking, disease attenuation effected by the presence of CARNA 5 in at least some hosts is accompanied by a reduction in virus accumulation and specific infectivity,[16,25-27,41] and, therefore, mechanisms by which CARNA 5 might reduce viral RNA synthesis should be addressed. The observation that CMV RNAs and CARNA 5 are replicated in the same particulate fraction from infected plants,[20,74] that they share some limited structural features in both their genomic RNAs [20,75] and their replicative forms,[76] and that CARNA 5 shows varying degrees of specificity for different CMV strains,[25,27,41] together suggest an interaction between viral and satellite RNAs that is either direct or else mediated through some shared component(s) of the replicative machinery.

A biochemical mechanism explaining reduced viral RNA synthesis as a result of the competitive advantage of CARNA 5 over the viral RNAs for shared replicative machinery has been proposed,[29] although possible molecular mechanisms for such a theoretical advantage were not. As discussed elsewhere,[76] one possible regulatory control on the replication of several plant viral RNAs may be related to their aminoacylation and the subsequent binding of elongation factor I of the host to their 3' ends, which in turn might prevent replicase binding.[77] In the case of CMV RNAs and CARNA 5, the inability of the latter to be aminoacylated[20] would allow its escape from such a hypothetical control mechanism and might explain both its high replicative efficiency and the relatively large amount of $(-)$ CARNA 5 in plant tissues.[38] An alternative possibility is that CARNA 5 encodes a polypeptide that associates with the replicative complex and alters its specificity in a way that favors satellite replication. However, the lack of a conserved polypeptide-encoding region of appreciable length within all CARNA 5 variants (see Section III.B.2) casts doubt on such hypothesis.

On the other hand, two possibilites for direct interaction between viral and satellite RNAs that might regulate replication have been noted to date. First, a comparison of the nucleotide sequences at the 5' ends of CMV-Q RNA 1 (and 2) and Q-CARNA 5 (=sat-RNA)[20] has revealed regions of 18 nucleotides that are partially complementary and therefore have the potential to base-pair intermolecularly.[75] However, since attempts to demonstrate interaction of these sequences in vitro have failed,[78] a more likely hypothesis might be that such sequences at the 5' end of Q-CARNA 5, which are therefore partially homologous to those at the 3' ends of the (−) strands of CMV-Q RNAs 1 and 2, might compete for replicase binding and thus reduce nascent (+) strand synthesis of CMV RNAs 1 and 2. This could explain why in virions the proportion of RNAs 1 and 2 seems depressed by the presence of CARNA 5.[16,25-27,41] Second, a direct interaction between a different region of Q-CARNA and 33 adjacent nucleotides within the coat protein gene of CMV-Q RNAs 3 and 4 has been demonstrated in vitro.[78] As noted by the authors, binding by such "anti-sense" regions of the satellite RNA to the coat protein gene could inhibit coat protein synthesis, which subsequently might decrease viral replication or accumulation. Alternatively, such binding might inhibit viral replication directly through hybrid-arrested RNA synthesis. However, whether such binding in vitro has any significance in vivo has been questioned by studies of the interactions between six additional CMV satellite RNAs and additional helper and nonhelper viruses, where a consistent pattern of binding was not observed.[78a]

2. Exacerbation of Viral Plant Disease

While it can sometimes be difficult to distinguish between exacerbation of plant disease caused more or less directly by the presence of satellite vs. a new, sometimes striking symptom appearing due to reduced production of the helper virus (see Section II.B.2.a), in the case of lethal tomato necrosis, there can be little doubt that the disease is mediated by the satellite RNA itself.[4,73] Because CARNA 5 nucleotide sequence variants can be grouped on the basis of whether or not they induce tomato necrosis,[21] over the past several years our laboratory has isolated, assayed, and sequenced several variants for comparison with each other and with previously published CARNA 5 sequences.[23b] This has revealed five regions where sequences conserved in all necrogenic CARNA 5s are altered in the same way in comparable regions of all nonnecrogenic variants (see boxed regions, Figure 4). These regions have been considered in terms of CARNA 5 both as a *cis*-acting molecule and as one exerting its necrogenicity in *trans*, i.e., through a diffusable polypeptide.

The first of these regions occurs at nucleotide number 24 (numbering refers to n-CARNA 5 sequence in Figure 4), where C in all necrogenic molecules is changed to U in nonnecrogenic variants. Because the corresponding change of G to A at position 40 in all but one non-necrogenic variant maintains strong pairing (or weaker U-G pairing in the one exception) between these two positions in a hairpin predicted by secondary structure models of both necrogenic and nonnecrogenic CARNA 5s,[20,23a,78b] the effect of such a C-to-U change on the amino acid encoding potential was considered. The possibility of a necrosis-inducing protein encoded by this region of the satellite was of particular interest, because the amino acid sequence of a putative translational product beginning at the AUG at nucleotides 11 to 13 is conserved in all necrogenic, but not nonnecrogenic, CARNA 5s sequenced to date.[23b] While some nonnecrogenic variants lack this open reading frame (ORF I) due to loss of the initiating AUG (e.g., variants X2c, Q, S), others contain the ORF, but the predicted poly-peptides contain one or more amino acid changes (e.g., the change at number 24 changes an alanine in necrogenic variants to a valine in the nonnecrogenic variants Y and 1). However, site-directed mutagenesis of an expressible cDNA clone of necrogenic D-CARNA 5 resulted in a stably replicating, mutant D-CARNA 5 that was still necrogenic after the AUG-initiating codon of ORF I had been eliminated.[73c] Since this study appears to eliminate the involvement of a protein encoded by ORF I in the induction of tomato necrosis, if the C-to-U change at position 24 has a critical effect on necrogenicity, it most likely involves a tertiary structure alteration.

There are four additional regions of CARNA 5 where consistent differences between necrogenic and nonnecrogenic variants are noteworthy. Changes in all necrogenic variants in the regions around nucleotides 99 and 259 (99 and 257 in Q-CARNA 5)[20] are not compensatory in an obvious way and therefore might alter the secondary structure predicted by interactions between these two regions in the secondary structure model of nonnecrogenic Q-CARNA 5.[20] In a similar manner, changes in a fourth region around nucleotides 152 and 158 (150 and 156 in Q-CARNA 5) in necrogenic variants could lead to a disruption of the short stem between the two loops of region V in the Q-CARNA 5 structure,[20] again, potentially resulting in a substantial structural change. The final alteration, the change of a U in nonnecrogenic variants to an A in necrogenic variants at nucleotide 214 (212 in Q-CARNA 5), would have little effect on the proposed secondary structure,[20] but might affect tertiary structure. On the other hand, the fact that a single U-to-C substitution at nucleotide 134 of the nonnecrogenic S-CARNA 5 seems to cause a conformational change[73c] not predicted by current secondary structure models[20,23a] suggests that CARNA 5 structures may be more complex than presently envisioned. It has been proposed that small changes in nucleotide sequence in plant pathogenic viroids that result in profound biological effects are mediated through secondary structure changes,[79] and while any of the potential structural changes noted for CARNA 5 could affect necrogenicity, no direct evidence of such yet exists.

Alternatively, the change from U at nucleotide 99 to a C in nonnecrogenic variants destroys an AUG and thus a potential ORF (ORF IIA) present only in necrogenic variants.[23b] However, the fact that transcripts necrogenic to tomato can be derived from a variant clone of D-CARNA 5 containing sequences in this region identical to those of the nonnecrogenic WL$_1$-CARNA 5 implies that neither ORF IIA nor the particular sequences in this region are critical to necrogenicity.[73b] Finally, we have noted elsewhere[23] the existence of third ORF (ORF IIB) beginning in the vicinity of nucleotide 135 in all but one[23a] CARNA 5 sequenced to date, which varies in length and in its predicted amino acid sequence due to deletions and resulting frameshifts in the different CARNA 5 variants. While we have demonstrated that a polypeptide encoded by this ORF in S-CARNA 5 is produced in an in vitro translation system,[80] at present there is no direct evidence correlating a protein product from this or any other ORF with necrogenicity.

Recently, a new CARNA 5 purified from a French isolate of CMV has been implicated in a different kind of disease exacerbation — a severe stunting of tomato plants resulting from its multiplication in association with the Ixora strain of CMV.[42] Because the nucleotide sequence of this CARNA 5 is very similar to that of n-CARNA 5,[81] and since it is the only CARNA 5 thus far supported by CMV-Ixora (measured by its presence in progeny virions), mutagenesis studies may be very fruitful in defining domains involved in both disease exacerbation and/or helper virus support.

C. Potential Mechanisms of Disease Modulation by Other Secondary RNAs

While studies on molecular mechanisms of disease modulation by cucumoviral satellites have barely begun, even speculations on potential mechanisms underlying disease modulation by other secondary RNAs are rare. In most systems, structures and active domains of interactants are yet to be determined.

In no system to date has a polypeptide encoded by a secondary RNA been shown to be involved in modification of disease symptoms. However, as mentioned earlier, among 12 isolates of BNYVV which differ in the presence and the lengths of the satellite-like RNAs 3 and 4, a correlation has been noted between the presence of full-length RNA 3 and, thus, the capacity to encode the polypeptide P25, and the appearance of severe chlorotic local lesions in *Chenopodium quinoa*.[54] Whether RNA 3 represents a satellite RNA, and whether P25 represents a protein essential for fungal transmission among sugarbeet roots that exerts a pathogenic effect in the unnatural environment of a *C. quinoa* leaf, remains to be determined. Satellite RNAs associated with the nepoviruses tomato black ring, myrobalan latent ringspot, and strawberry latent ringspot[7] are known to produce polypeptides of about 38 to 48 kdaltons in vivo, but since such satellites have a very limited effect on the symptoms induced by their respective helper viruses, such proteins cannot be implicated in disease modulation. The two satellite viruses whose genomes have been characterized, STNV and SPMV, cause opposite disease-modulating effects, the former decreasing lesion size and the latter increasing disease on corn or millet. While the RNA of each satellite virus has the potential to encode one or more polypeptide(s) in addition to its capsid protein, there is at present no evidence for either the production or function of such proteins in disease modulation.[51a,52,82]

Alternatively, the defective-interfering RNAs of TBSV[9,10] provide an excellent system in which to search for *cis*-acting structural domains that modulate plant disease. Their ability to interfere with normal TBSV replication can be studied using current models from animal virus DI RNAs[83,83a] in an attempt to locate regulatory sequences possibly involved in competition with viral RNAs for RNA polymerase, viral capsid protein, or host ribosomes. In the case of one DI RNA characterized as a colinear deletion mutant of the viral RNA, interference appears to occur at the level of replication rather than encapsidation, although the particular structure of this DI raises the additional possibility of interference at the level of translation by competition for available ribosomes.[10] The existence of other DI RNAs associated with TBSV offers the opportunity to determine whether their ability to attenuate disease is directly correlated with reduced viral replication, since two different DI RNAs that differ in size by only about 40 nucleotides have been observed to have quite different disease-attenuating abilities.[83b] The development of a system capable of quantitating the replication of both viral and DI RNAs should shed some light on this question and would be of further use in attempts to compare and contrast the effects of these DI RNAs with those of a true satellite RNA of TBSV, which also attenuates disease induced by TBSV in *N. Clevelandii*.[44] Finally, further information on domains of secondary RNAs that modulate plant viral disease should come from studies on the complex family of secondary RNAs associated with TCV. The 166 nucleotides present at the 3' terminus of the chimeric RNA C, but missing in satellites D and F, may be responsible for the unique ability of RNA C to severely increase the TCV symptoms in turnip plants. Since this region of RNA C is homologous to the 3' end of TCV RNA, it has been suggested that increased disease severity may be related to an increased dosage of this domain of the viral RNA, although at the moment there is no experimental support for such a hypothesis.[12]

The wide variety of secondary RNAs associated with these definitive and tentative members[84] of the tombusvirus group (satellites, DI RNAs, and a chimeric RNA with characteristics of each), and their varying capacities for disease modulation, offer a rich opportunity for defining structure and function relationships in secondary RNAs. One additional system from which may come clues on molecular mechanisms of disease modulation is that of ArMV and its satellite RNA, the latter implicated as causal agent of the nettlehead disease of hop. The fact that the presence of a satellite is not always correlated with nettlehead disease[48] may indicate that, as in the case of the cucumoviral satellite CARNA 5, there exist nonpathogenic variants of the satellite. If so, comparative studies may be fruitful.

D. Evolution of Secondary RNAs from a Cellular Defense Mechanism

Thoughts on possible origins and the evolution of plant viral satellites have been recently reviewed[7] and will not be repeated here. However, a look at the emerging kaleidoscopic forms of plant viral secondary RNAs in terms of a recent theoretical paper[85] on the evolution of a defective DNA virus from a cellular defense mechanism might be thought-provoking.

Once such an evolutionary scheme is transposed from DNA to RNA, such a model suggests that a piece of "junk" RNA arising in a cell from either host or pathogen origin, due to some combination of intra- and intermolecular recombination and the high mutation frequency of RNA, might interact fortuitously with the replication machinery of invading virus, resulting in both its own replication and also interference with the replication of the virus. Such a molecule would evolve initially toward intracellular stability, resulting in protection against threatening exo- and endonucleases. Accordingly, each of the following stabilizing structural elements is exhibited by one of more of the plant viral secondary RNAs: increased self-pairing, hairpinning, circularization, increased G-C content, 5'-linked proteins, capping of linear RNAs, tRNA-like ends and double-stranded forms.

Given the rapid evolution of RNA genomes, the next most favorable mutations in interfering RNAs might be those resulting in mRNAs that compete with the helper virus at the additional levels of ribosome binding and/or translation, adding a new dimension to the antiviral defense. However, since a randomly arising, presumably functionless protein might prove harmful or even lethal to the cell, subsequent mutations might be those that allowed the protein to bind to either the RNA of the helper virus, further interfering with the viral life cycle, or to its own RNA, in either case effectively removing the protein from the cellular environment, and in the latter case, resulting in a quite stable satellite virus.

Examples of secondary RNAs resembling each of these putative evolutionary stages are readily apparent. The satellite of TobRV might represent an interfering RNA evolved for structural stability alone, the satellite of tomato black ring virus and possibly CARNA 5 might represent molecules evolved to also compete at the added level of translation, and finally, satellite viruses like STNV and SPMV might be considered the most evolutionarily advanced. Exactly how the disease-modulating capacity of secondary RNAs relates to the other forces shaping their evolution should be expected to have as many different answers as there are unique forms of secondary RNAs.

ACKNOWLEDGMENTS

We gratefully acknowledge those colleagues who provided us with information and manuscripts prior to publication: D. Baulcombe, B. Hillman, N. Iizuka, A. Jackson, R. Koenig, J. Kummert, K. Richards, J. Semal, A. Simon, R. Valverde, and D. Zimmern. C.W.C. has been supported by funds provided by U.S. Department of Agriculture Competitive Research Grants 82-CRCR-1-1136 and 85-CRCR-1-1766.

REFERENCES

1. **Vuittenez, A and Putz, C.,** Catastrophe pour les producteurs alsaciens de tomates, *Alsace,* August 20 and 22, 1972.
2. **Putz, C., Kuszala, J., Kuszala, M., and Spindler, C.,** Variation du pouvoir pathogene des isolats du virus de la mosaique du concombre associee a la necrose de la tomate, *Ann. Phytopathol.,* 6, 139, 1974.
3. **Marrou, J. and Duteil, M.,** La necrose de la tomate, *Ann. Phytopathol.,* 6, 155, 1974.
4. **Kaper, J. M. and Waterworth, H. E.,** Cucumber mosaic virus associated RNA 5: causal agent for tomato necrosis, *Science,* 196, 429, 1977.
5. **Kassanis, B.,** Properties and behavior of a virus depending for its multiplication upon another, *J. Gen. Microbiol.,* 27, 477, 1962.
6. **Kaper, J. M. and Tousignant, M. E.,** Viral satellites: parasitic nucleic acids capable of modulating disease expression, *Endeavour New Series,* 8, 194, 1984.
7. **Francki, R. I. B.,** Plant virus satellites, *Annu. Rev. Microbiol.,* 39, 151, 1985.
8. **Holland, J. J., Kennedy, S. I. T., Semler, B. L., Jones, C. L., Roux, L., and Grabau, E. A.,** Defective interfering RNA viruses and the host-cell response, in *Comprehensive Virology,* Vol. 16, Fraenkel-Conrat, H. and Wagner, R. R., Eds., Plenum Press, New York, 1980, 137.

9. **Hillman, B. I., Morris, T. J., and Schlegel, D. E.,** Effects of low-molecular-weight RNA and temperature on tomato bushy virus symptom expression, *Phytopathology,* 75, 361, 1985.

10. **Hillman, B. I., Carrington, J. C., and Morris, T. J.,** A defective interfering RNA that contains a mosaic of a plant virus genome, *Cell,* 51, 427, 1987.

11. **Hoyer, B., Bonino, F., Ponzetto, A., Denniston, K., Nelson, J., Purcell, R., and Gerin, J. L.,** Properties of delta-associated ribonucleic acid, *Progr. Clin. Biol. Res.,* 143, 91, 1983.

11a. **Wang, K. -S., Choo, Q. -L., Weiner, A. J., Ou, J. -H., Najarian, R. C., Thayer, R. M., Mullenbach, G. T., Denniston, K. J., Gerin, J. L., and Houghton, M.,** Structure, sequence and expression of the hepatitis delta (δ) viral genome, *Nature (London),* 323, 508, 1986.

12. **Simon, A. E. and Howell, S. H.,** The virulent satellite RNA of turnip crinkle virus has a major domain homologous to the 3' end of the helper virus genome, *EMBO J.,* 5, 3423, 1986.

13. **Gilbert, W.,** The RNA world, *Nature (London),* 319, 618, 1986.

14. **Bujarski, J. J. and Kaesber, P.,** Genetic recombination between RNA components of multipartite plant virus, *Nature (London),* 321, 528, 1986.

15. **Marrou, J., Duteil, M., Lot, H., and Clerjeau, M.,** La necrose de la tomate: Une grave virose des tomates cultivees en plein champ, *Pepin. Hortic. Maraich.,* 137, 37, 1973.

16. **Kaper, J. M., Tousignant, M. E., and Lot, H.,** A low molecular weight replicating RNA associated with a divided genome plant virus: defective or satellite RNA? *Biochem. Biophys. Res. Commun.,* 72, 1237, 1976.

17. **Quiot, J. B., Leroux, J. P., Labonne, G., and Renoust, M.,** Epidemiologie de la maladie filiforme et de la necrose de la tomate provoquees par le virus de la mosaique du concombre dans le sud-est de la France, *Ann. Phytopathol.,* 11, 393, 1979.

18. **Yoshida, K., Goto, T., and Iizuka, N.,** Attenuated isolates of cucumber mosaic virus produced by satellite RNA and cross protection between attenuated and virulent ones, *Ann. Phytopathol. Soc. Jpn.,* 51, 238, 1985.

19. **Richards, K. E., Jonard, G., Jacquemond, M., and Lot, H.,** Nucleotide sequence of cucumber mosaic virus-associated RNA 5, *Virology,* 89, 395, 1978.

20. **Gordon, K. H. J. and Symons, R. H.,** Satellite RNA of cucumber mosaic virus forms a secondary structure with partial 3'-terminal homology to genomal RNAs, *Nucleic Acids Res.,* 11, 947, 1983.

21. **Collmer, C. W., Tousignant, M. E., and Kaper, J. M.,** Cucumber mosaic virus-associated RNA 5. X. The complete nucleotide sequence of a CARNA 5 incapable of inducing tomato necrosis, *Virology,* 127, 230, 1983.

22. **Hidaka, S., Ishikawa, K., Takanami, Y., Kubo, S., and Miura, K.,** Complete nucleotide sequence of RNA 5 from cucumber mosaic virus (strain Y), *FEBS Lett.,* 174, 38, 1984.

23. **Avila-Rincon, M., Collmer, C. W., and Kaper, J. M.,** *In vitro* translation of cucumoviral satellites. I. Purification and nucleotide sequence of CARNA 5 from cucumber mosaic virus strain S., *Virology* 152, 446, 1986.

23a. **Garcia-Arenal, F., Zaitlin, M., and Palukaitis, P.,** Nucleotide sequence analysis of six satellite RNAs of cucumber mosaic virus: primary sequence and secondary structure alterations do not correlate with differences in pathogenicity, *Virology,* 158, 339, 1987.

23b. **Kaper, J. M., Tousignant, M. E., and Steen, M. T.,** Cucumber mosaic virus associated RNA 5. XI. Comparison of fourteen CARNA 5 sequence variants relates ability to induce tomato necrosis to a conserved sequence, *Virology,* 1988, in press.

24. **Waterworth, H. E., Kaper, J. M., and Tousignant, M. E.,** CARNA 5, the small cucumber mosaic virus-dependent replicating RNA, regulates disease expression, *Science,* 204, 845, 1979.

25. **Mossop, D. W. and Francki, R. I. B.,** Comparative studies on two satellite RNAs of cucumber mosaic virus, *Virology,* 95, 395, 1979.

26. **Takanami, Y.,** A striking change in symptoms on cucumber mosaic virus-infected tobacco plants induced by a satellite RNA, *Virology,* 109, 120, 1981.

27. **Jacquemond, M. and Leroux, J-P.,** L'ARN satellite du virus de la mosaique du concombre. II. Etude de la relation virus-ARN satellite chez divers hotes, *Agronomie,* 2, 55, 1982.

28. **Gonsalves, D., Provvidenti, R., and Edwards, M. C.,** Tomato white leaf: the relation of an apparent satellite RNA and cucumber mosaic virus, *Phytopathology,* 72, 1533, 1982.

29. **Kaper, J. M.,** Rapid synthesis of double-stranded cucumber mosaic virus-associated RNA 5: mechanism controlling viral pathogenesis?, *Biochem. Biophys. Res. Commun.,* 105, 1014, 1982.

30. **Piazzolla, P., Tousignant, M. E., and Kaper, J. M.,** Cucumber mosaic virus-associated RNA 5. IX. The overtaking of viral RNA synthesis by CARNA 5 and dsCARNA 5 in tobacco, *Virology,* 122, 147, 1982.

31. **Jacquemond, M.,** Phenomenes d'interferences entre les deux types d'ARN satellite du virus de la mosaique du concombre. Protection des tomates vis a vis de la necrose letal, *C. R. Acad. Sci.,* 294 (III), 991, 1982.

32. **Tien, P. and Chang, X. H.**, Control of two seed-borne virus diseases in China by the use of protective inoculation, *Seed Sci. Technol.*, 11, 969, 1983.

32a. **Tien, P., Zhang, X., Qiu, B., Qin, B., and Wu, G.**, Satellite RNA for the control of plant diseases caused by cucumber mosaic virus, *Ann. Appl. Biol.*, 111, 143, 1987.

33. **Garcia-Luque, I., Kaper, J. M., Diaz-Ruiz, J. R., and Rubio-Huertos, M.**, Emergence and characterization of satellite RNAs associated with Spanish cucumber mosaic virus isolates, *J. Gen. Virol.*, 65, 539, 1984.

34. **Kaper, J. M., Duriat, A. S., and Tousignant, M. E.**, The 368-nucleotide satellite of cucumber mosaic virus strain Y from Japan does not cause lethal necrosis in tomato, *J. Gen. Virol.*, 67, 2241, 1986.

35. **Baulcombe, D. C., Saunders, G. R., Bevan, M. W., Mayo, M. A., and Harrison, B. D.**, Expression of biologically active viral satellite RNA from the nuclear genome of transformed plants, *Nature (London)*, 321, 446, 1986.

36. **Rosner, A., Bar-Joseph, M., Moscovitz, M., and Mevarech, M.**, Diagnosis of specific viral RNA sequences in plant extracts by hybridization with a polynucleotide kinase-mediated, ^{32}P-labeled, double-stranded RNA probe, *Phytopathology*, 73, 699, 1983.

37. **Valverde, R. A. and Dodds, J. A.**, Evidence for a satellite RNA associated naturally with the U5 strain and experimentally with the U1 strain of tobacco mosaic virus, *J. Gen. Virol.*, 67, 1875, 1986.

38. **Habili, N. and Kaper, J. M.**, Cucumber mosaic virus-associated RNA 5. VII. Double-stranded form accumulation and disease attenuation in tobacco, *Virology*, 112, 250, 1981.

39. **Yang, X., Qin, B., Liang, X., and Tien, P.**, Satellite RNA as a biological control agent of diseases caused by cucumber mosaic virus. III. Effect of satellite RNA on the content of double-stranded viral RNA in CMV-infected tissues, *Acta Microbiol. Sin.*, 26, 120, 1986.

40. **Piazzolla, P., Gallitelli, D., and Savino, V.**, Appearance of satellite RNA (CARNA 5) in six cucumber mosaic virus isolates from the open field, *Phytopathol. Mediterr.*, 21, 32, 1982.

41. **Kaper, J. M. and Tousignant, M. E.**, Cucumber mosaic virus-associated RNA 5. I. Role of host plant and helper strain in determining amount of associated RNA 5 with virions, *Virology*, 80, 186, 1977.

42. **Kaper, J. M. and Waterworth, H. E.**, Cucumoviruses, in *Handbook of Plant Virus Infections and Comparative Diagnosis*, Kurstak, E., Ed., Elsevier/North Holland, Amsterdam, 1981, 257.

43. **Altenbach, S. B. and Howell, S. H.**, Nucleic acid species related to the satellite RNA of turnip crinkle virus in turnip plants and virus particles, *Virology*, 134, 72, 1984.

44. **Gallitelli, D. and Hull, R.**, Characterization of satellite RNAs associated with tomato bushy stunt virus and five other definitive tombusviruses, *J. Gen. Virol.*, 66, 1533, 1985.

45. **Collmer, C. W., Hadidi, A., and Kaper, J. M.**, Nucleotide sequence of the satellite of peanut stunt virus reveals structural homologies with viroids and certain nuclear and mitochondrial introns, *Proc. Natl. Acad. Sci. U.S.A.*, 82, 3110, 1985.

46. **Palukaitis, P. and Zaitlin, M.**, Satellite RNAs of cucumber mosaic virus: characterization of two new satellites, *Virology*, 132, 426, 1984.

47. **Schneider, I. R.**, Defective plant viruses, In *Virology in Agriculture*, Romberger, J. A., Anderson, J. D., and Powell, R. L., Eds., Allanheld, Montclair, New Jersey, 1977, 201.

48. **Davis, D. L. and Clarke, M. F.**, A satellite-like nucleic acid of arabis mosaic virus associated with hop nettlehead disease, *Ann. Appl. Biol.*, 103, 439, 1983.

49. **Kaper, J. M. and Tousignant, M. E.**, unpublished work, 1986.

49a. **Forster, A. C. and Symons, R. H.**, Self-cleavage of plus and minus RNAs of a virusoid and a structural model for the active sites, *Cell*, 49, 211, 1987.

50. **Altenbach, S. B. and Howell, S. H.**, Identification of a satellite RNA associated with turnip crinkle virus, *Virology*, 112, 25, 1981.

51. **Buzen, F. G., Niblett, C. L., Hooper, G. R., Hubbard, J., and Newman, M. A.**, Further characterization of panicum mosaic virus and its associated satellite virus, *Phytopathology*, 74, 313, 1984.

51a. **Masuta, C., Zuidema, D., Hunter, B. G., Heaton, L. A., Sopher, D. S., and Jackson, A. O.**, Analysis of the genome of satellite panicum mosaic virus, *Virology*, 159, 329, 1987.

52. **Jackson, A. O.**, personal communication, 1986.

53. **Bouzoubaa, S., Guilley, H., Jonard, G., Richards, K., and Putz, C.**, Nucleotide sequence analysis of RNA 3 and RNA 4 of beet necrotic yellow vein virus, isolate F2 and G1, *J. Gen. Virol.*, 66, 1553, 1985.

54. **Kuszala, M., Ziegler, V., Bouzoubaa, S., Richards, K., Putz, C., Guilley, H., and Jonard, G.**, Beet necrotic yellow vein virus: different isolates are serologically similar, but differ in RNA composition, *Ann. Appl. Biol.*, 109, 155, 1986.

55. **Bouzoubaa, S., Ziegler, V., Beck, D., Guilley, H., Richards, K., and Jonard, G.**, Nucleotide sequence of beet necrotic yellow vein virus RNA 2, *J. Gen. Virol.*, 67, 1689, 1986.

56. **Richards, K., Jonard, G., Guilley, H., Ziegler, V., and Putz, C.**, *In vitro* translation of beet necrotic yellow vein virus RNA and studies of sequence homology among the RNA species using cloned DNA probes, *J. Gen. Virol.*, 66, 345, 1985.

57. **Koenig, R., Burgermeister, W., Weich, H., Sebald, W., and Kothe, C.,** Uniform RNA patterns of beet necrotic yellow vein virus in sugarbeet roots, but not in leaves from several plant species, *J. Gen. Virol.,* 67, 2043, 1986.

58. **Francki, R. I. B., Grivell, C. J., and Gibb, K. S.,** Isolation of velvet tobacco mottle virus capable of replication with and without a viroid-like RNA, *Virology,* 148, 381, 1986.

59. **Jones, A. T., Mayo, M. A., and Duncan, G. H.,** Satellite-like properties of small circular RNA molecules in particles of lucerne transient streak virus, *J. Gen. Virol.,* 64, 1167, 1983.

60. **Chu, P. W. G., Francki, R. I. B., and Randles, J. W.,** Detection, isolation, and characterization of high molecular weight double-stranded RNAs in plants infected with velvet tobacco mottle virus, *Virology,* 126, 480, 1983.

61. **Zimmern, D.,** Communication at EMBO Workshop on Molecular Plant Virology, Wageningen, Netherlands, July 6 to 10, 1986.

62. **Buzayan, J. M., Gerlach, W. L., Bruening, G., Keese, P., and Gould, A. R.,** Nucleotide sequence of satellite tobacco ringspot virus RNA and its relationship to multimeric forms, *Virology,* 151, 186, 1986.

63. **Gerlach, W. L., Buzayan, J. M., Schneider, I. R., and Bruening, G.,** Satellite tobacco ringspot virus RNA: biological activity of DNA clones and their *in vitro* transcripts, *Virology,* 151, 172, 1986.

64. **Kiefer, M. C., Daubert, S. D., Schneider, I. R., and Bruening, G.,** Multimeric forms of satellite of tobacco ringspot virus RNA, *Virology,* 121, 262, 1982.

65. **Linthorst, H. J. M. and Kaper, J. M.,** Circular satellite-RNA molecules in satellite of tobacco ringspot virus-infected tissue, *Virology,* 137, 206, 1984.

66. **Prody, G. A., Bakos, J. T., Buzayan, J. M., Schneider, I. R., and Bruening, G.,** Autolytic processing of dimeric plant virus satellite RNA, *Science,* 231, 1577, 1986.

66a. **Buzayan, J. M., Gerlach, W. L., and Bruening, G.,** Non-enzymatic cleavage and ligation of RNAs complementary to a plant virus satellite RNA, *Nature,* 323, 349, 1986.

66b. **Piazzolla, P., Vovlas, C., and Rubino, L.,** Symptom regulation induced by chicory yellow mottle virus satellite-like RNA, *J. Phytopathol.,* 115, 124, 1986.

67. **Burgyan, J., Russo, M., and Gallitelli, D.,** Translation of cymbidium ringspot virus RNA in cowpea protoplasts and rabbit reticulocyte lysates, *J. Gen. Virol.,* 67, 1149, 1986.

68. **Jones, T. A. and Liljas, L.,** Structure of satellite tobacco necrosis virus after crystallographic refinement at 2.5 Å resolution, *J. Mol. Biol.,* 177, 735, 1984.

69. **Kassanis, B.,** Portraits of viruses: tobacco necrosis virus and its satellite virus, *Intervirology,* 15, 57, 1981.

70. **Fritsch, C., Koenig, I., Murant, A. F., Raschke, J. H., and Mayo, M. A.,** Comparisons among satellite RNA species from five isolates of tomato black ring virus and one isolate of myrobalan latent ringspot virus, *J. Gen. Virol.,* 65, 289, 1984.

71. **Meyer, M., Hemmer, O., and Fritsch, C.,** Complete nucleotide sequence of a satellite RNA of tomato black ring virus, *J. Gen. Virol.,* 65, 1575, 1984.

71a. **Hemmer, O., Meyer, M., Greif, C., and Fritsch, C.,** Comparison of the nucleotide sequences of five tomato black ring virus satellite RNAs, *J. Gen. Virol.,* 68, 1823, 1987.

72. **Valverde, R. A. and Dodds, J. A.,** Some properties of isometric virus particles which contain the satellite RNA of tobacco mosaic virus, *J. Gen. Virol.,* 68, 965, 1987.

72a. **Uhlenbeck, O. C.,** A small catalytic oligoribonucleotide, *Nature (London),* 328, 596, 1987.

73. **Collmer, C. W. and Kaper, J. M.,** Infectious RNA transcripts from cloned cDNAs of cucumber mosaic viral satellites, *Biochem. Biophys. Res. Commun.,* 135, 290, 1986.

73a. **Simon, A. E., and Howell, S. H.,** Synthesis in vitro of infectious RNA copies of the virulent satellite of turnip crinkle virus *Virology,* 156, 146, 1987.

73b. **Kurath, G., and Palukaitis, P.,** Biological activity of T7 transcripts of a prototype clone and a sequence variant clone of a satellite RNA of cucumber mosaic virus, *Virology,* 159, 199, 1987.

73c. **Collmer, C. W., and Kaper, J. M.,** Site-directed mutagenesis of potential protein-coding regions in expressible cloned cDNAs of cucumber mosaic viral satellites, *Virology,* in press, 1988.

74. **Jaspars, E. M. J., Gill, D. S., and Symons, R. H.,** Viral RNA synthesis by a particulate fraction from cucumber seedlings infected with cucumber mosaic virus, *Virology,* 144, 410, 1985.

75. **Rezaian, M. A., Williams, R. H. V., and Symons, R. H.,** Nucleotide sequence of cucumber mosaic virus RNA 1. Presence of a sequence complementary to part of the viral satellite RNA and homologies with other viral RNAs, *Eur. J. Biochem.,* 150, 331, 1985.

76. **Collmer, C. W., and Kaper, J. M.,** Double-stranded RNAs of cucumber mosaic virus and its satellite contain an unpaired terminal guanosine: implications for replication, *Virology,* 145, 249, 1985.

77. **Ahlquist, P., Dasgupta, R., and Kaesberg, P.,** Nucleotide sequence of the brome mosaic virus genome and its implications for viral replication, *J. Mol. Biol.,* 172, 369, 1984.

78. **Rezaian, M. A. and Symons, R. H.,** Anti-sense regions in satellite RNA of cucumber mosaic virus form stable complexes with the viral coat protein gene, *Nucleic Acids Res.,* 14, 3229, 1986.

78a. **Garcia-Arenal, F., and Palukaitis, P.,** Interaction of CMV satellite RNAs with RNAs of helper and non-helper viruses, Seventh International Congress of Virology, Edmonton, Canada, Abstract No. OP19.5, 1987.

78b. **Pleij, C. W. A.,** personal communication, 1986.

79. **Schnolzer, M., Haas, B., Ramm, K., Hofmann, H., and Sanger, H. L.,** Correlation between structure and pathogenicity of potato spindle tuber viroid (PSTV), *EMBO J.,* 4, 2181, 1985.

80. **Avila-Rincon, M. J., Collmer, C. W., and Kaper, J. M.,** *In vitro* translation of cucumoviral satellites. II. CARNA 5 from cucumber mosaic virus strain S and SP6 transcripts of cloned (S)CARNA 5 cDNA produce electrophoretically comigrating protein products, *Virology,* 152, 455, 1986.

81. **Tousignant, M. E. and Kaper, J. M.,** unpublished data.

82. **Ysebaert, M., van Emmelo, J., and Fiers, W.,** Total nucleotide sequence of a nearly full-size DNA copy of satellite tobacco necrosis virus RNA, *J. Mol. Biol.,* 143, 273, 1980.

83. **Perrault, J.,** Origin and replication of defective interfering particles, *Curr. Top. Microbiol. Immunol.,* 93, 151, 1981.

83a. **Levis, R., Weiss, B. G., Tsiang, M., Huang, H., and Schlesinger, S.,** Deletion mapping of Sindbis virus DI RNAs derived from cDNAs defines the sequences essential for replication and packaging, *Cell,* 44, 137, 1986.

83b. **Hillman, B. I.,** personal communication, 1986.

84. **Gallitelli, D., Hull, R., and Koenig, R.,** Relationships among viruses in the tombusvirus group: nucleic acid hybridization studies, *J. Gen. Virol.,* 66, 1523, 1985.

85. **Fisher, R. E. and Mayor, H. D.,** Evolution of a defective virus from a cellular defense mechanism, *J. Theor. Biol.,* 118, 395, 1986.

Chapter 11

MODULATION OF VIRAL DISEASE PROCESSES BY DEFECTIVE INTERFERING PARTICLES

Alice S. Huang

TABLE OF CONTENTS

I. INTRODUCTION AND HISTORICAL PERSPECTIVE

Defective interfering (DI) virus particles remain as intriguing biological entities, the role(s) of which in natural disease is still elusive. It has been 40 years since the concept of interference by "incomplete virus particles" was suggested[1,2] and 20 years since these particles were purified and identified as distinct viral populations.[3] Much is now known about the molecular biology of these particles,[4-8] but the suggestion that such defective interfering (DI) viral particles play an important role in natural disease[9] has generated complex and somewhat confusing data. Although the hypothesis of a natural role for DI particles is attractive, exactly what this role is, especially in each of the different viral diseases, remains unclear. Nevertheless, there is a body of published material dealing with DI particles in animals, and this material appears to suggest a cohesive theory which is applicable to furthering our current understanding of viral pathogensis and transmission. Also, this theory impacts significantly on viral diagnostics.

This review focuses on evidence derived from both animal and human studies. It attempts to examine selective evidence which supports a role for DI particles in pathogenesis. Two types of in vivo evidence are available. One is from the inoculation of animals with DI particles, and the other is from the characterization of isolates obtained from diseased hosts. This review will not cover molecular characterizations of DI particles nor the mechanisms involved during DI particle interference. Moreover, in vitro experiments using cell cultures will only be cited when they contribute important insights into viral disease processes. In Chapter 5 of this volume, there is a presentation on the genesis and mapping of DI genomes. In addition, as part of this review on DI particles in animals, a cohesive theory will be presented and related to viral pathogenesis, transmission, and diagnostics. How this theory can be further tested is indicated in the section entitled Perspectives (VIII), which contains experimental approaches that need to be taken.

It is an old idea that viral preparations contain intrinsic factors which alter the outcome of viral infections specifically. The earliest studies with virus infection of animals, as exemplified by those of Olitsky and Sabin[10-13] and Traub[14] in the 1930s, suggest a serotype-specific protection afforded virus-inoculated animals. The protection is not dependent on generalized host immune responses.[15] Different preparations of the same virus lead to different disease outcomes. Also, inoculation of the same virus preparation at different sites in the animals alters the course of the disease. The themes of these important observations are much the same in studies to follow. Differences among viral preparations probably relate to passage level and subsequent heterogeneity in the viral population. The effects from different inoculation sites will hopefully become clearer as pathogenesis is discussed in relation to the theoretical considerations.

These early observations, however, were explained away by ideas of tissue-specific immunity;[14-16] the focus on the viral preparations themselves eventually did lead to correlations between disease outcome and the passage history of the virus preparation. von Magnus[17] reported that influenza virus preparations contain particles with heterogeneous sedimentation rates. The usual method of growing influenza viruses is to dilute stock virus 1000-fold before inoculation. This diluted passage results in progeny of high infectivity. Inoculation of undiluted stock virus into embryonated eggs leads to the formation of preparations that are more heterogeneous, less infectious, and interfere with the growth of influenza virus.[1,2,17] Bernkopf,[18] who made similar observations, foresaw that such interfering preparations may have vaccine potential. He wrote in 1950: "A study of the chemical differences between such mature and immature virus may prove extremely interesting if such differences could be correlated with the ability of a particle to multiply and to cause toxic reactions."

The finding that less infectious virus is produced following inoculation with undiluted stock virus is called the "autointerference" phenomenon. Such preparations contain a high hemagglutinin to infectivity ratio and interfere with the growth of high-titered, diluted passage preparations.[17] Also, preparations with low infectivity, but high hemagglutinin content, were found to be nonneurotropic in mouse studies.[16] These phenomena led to the identification of what are now defined as DI particles.[4,5,9]

Because many of these earlier studies were done in embryonated eggs or mice, it can be concluded that *DI particles are produced in animals, a fact that may have escaped some molecular virologists.* On the other hand, although these early studies are suggestive of a role for DI particles in natural infections, they cannot be cited as evidence of such.

II. STUDIES ON PATHOGENESIS WITH DI PARTICLES IN CELL CULTURE

In the past, investigators using cell cultures focused on interference, where the production of infectious standard progeny is reduced by coinfection with DI particles. These studies naturally led investigators to the expectation that DI particles would have a protective role in animals as well. Unfortunately, when such expectations were unrealized, interest in other roles for DI particles waned. Many properties of DI particles are defined in cell culture, uncluttered by the various parameters of host immunologic defenses and clearance mechanisms. Such studies, however elegant, cannot mimic disease processes; the questions concerning modulation of pathogenesis or protection from lethal effects can only be addressed with the complete host. Nevertheless, cell culture studies are still valuable for understanding certain aspects of viral pathogenesis, particularly when differentiated cells are grown in culture or when the effects of DI particles on host functions are being monitored.

Some recent studies of this kind are especially interesting because they relate altered host functions to the presence of DI particles. For example, certain deletion mutants of retroviruses not only interfere with their helper standard virus, but also carry genes which transform.[19] Also, equine herpes virus DI particles reduce the cytopathogenicity of a virus preparation and permit cellular transformation to occur.[20] This reversal of the cytopathic effect induced by standard virus may also explain why maximal interferon induction by Sendai virus requires preparations made by undiluted passages.[21] Sendai DI particles by themselves do not induce interferon synthesis, so that the induction is due to a concerted mechanism involving both standard and DI particles. In a related finding, when purified standard vesicular stomatitis virus (VSV) is used to induce interferon in NIH/3T3 cells, transcription of interferon genes occurs within the first hour of infection, but the subsequent viral inhibition of host macromolecular synthesis prevents any synthesis of interferon in these cells.[22] A less cytopathic strain or the presence of DI particles permits interferon synthesis.[23]

Interferon is particularly interesting because it has a relationship to standard infectious virus that is similar to DI particles; i.e., it depends on standard virus for its induction, and, in return, it inhibits the production of its inducer. Although interferon is generally thought to inhibit overall viral synthesis, a recent report shows its potentiating effect on DI particle production in rabies virus-infected neuroblastoma cells.[24] In this case, the protection afforded cells by interferon against the lytic infection turned it into a persistent one, resulting in DI production and subsequent selection of DI-resistant standard virus variants. Therefore, in cell cultures, dynamic interactions occur between standard virus and DI particles as well as between each of them and interferon; these lead to complex time courses for the inhibition and induction of overall viral growth and interferon synthesis. The induction of interferon is one of many possible host functions that has been clearly identified to affect the interaction between DI and standard virus. Future studies will need to concentrate on other cellular functions that are affected by or affect DI particles.

III. DO COINFECTIONS OCCUR IN THE HOST?

Before discussing animal experimentation, it is necessary to clarify the issue of whether or not infections in animals involve enough virus, so that if DI particles are generated or are present in the virus population, cells can be multiply infected with DI and standard virus. Past difficulties in obtaining viral isolates, as well as their low titers from infected hosts, argue against multiply infected cells. More recently, however, with improved diagnostic methods, as well as the detection of persistent chronic infections,[25] this view is being changed. In natural disease it may be that several particles, and perhaps even different viruses, infect the same host cells. Moreover, the continuous coexistence of defective retroviruses dependent on helper viruses in murine or avian systems, indicate commonly occurring coinfection of cells within these animals.[19]

To determine whether continuous interaction occurs among mixtures of viruses, mice were fed genetically marked reovirus strains[26] or inoculated intranasally with standard and DI VSV.[27] This results in continued interaction among viral particles during the progression of disease, with subsequent reisolation of the original mixtures from the central nervous system of the animals. With reovirus, reassortants were also found. Many reports now exist for the coexistence of different strains of the same virus in a host. Therefore, it is likely that more than one virus particle initiates disease, and that DI particles and standard virus have the chance to interact within the host even though multiplicities are less well defined under these circumstances.

IV. DI PARTICLES IN ANIMALS

A. Model Systems

Most studies on viral pathogenesis in animals involve small mammals inoculated with pathogens isolated from humans or other large animals. Although such studies are artificial, they do have value in determining the multiple problems involved in defining the role played by DI particles in the animal. In this section we will examine some of these studies utilizing RNA viruses.

Reports on West Nile virus infection of inbred mice show a genetic locus which determines DI particle production.[28,29] Mice which increase the synthesis of DI particles are resistant to West Nile virus. Cells taken from these animals retain their support of DI particle production. These and studies with other flaviviruses indicate that interferon and generalized immunity do not determine the final outcome of the disease, although prior induction of immunity or interferon reduces the initial replication of input virus.[30,31] Also, inoculation of mice with DI particle-containing preparations several hours before superinfection, or at the same time as infection, with a lethal dose of standard Semliki Forest virus leads to increased survival of the mice.[32,33] However, as more viral preparations are examined, the results with Semliki Forest virus in mice are less reproducible.[34] Because the differences depend on passage history, and because DI particles of Semliki Forest virus cannot be easily purified from their standard virus, such variations can be expected.

When DI particles of influenza virus are examined using intracerebral inoculation of 7-week-old Swiss mice, and with mortality as the end point, protection is afforded by DI particles.[35,36] Intranasal inoculations or use of younger Swiss mice (3 weeks) do not lead to protection by DI particles. Again, the amounts of interferon do not correlate with protection.[36] With a persistent influenza virus infection in S57BL mice established by DI particle-containing preparations, virus can be isolated for up to 45 days from the lungs and from 2 to 8 months from the spleen.[37] Such isolates contain heterogeneous populations of influenza virus, including defective virions. These studies are in contrast to acute lethal infections in the same host which terminate within 10 days. In a more recent study with influenza virus,

Dimmock and colleagues[38] demonstrate a protective effect of DI particles in mice, despite their inability to detect any inhibition of virus growth, reduction in interferon titers, or increase in antibody response to influenza antigens. They suggest that the lethal effects of influenza virus are due to immunopathology caused by cytotoxic T lymphocytes, rather than to viral growth per se, and that in some way, DI-rich preparations inhibit the immunopathology. Although DI sequences are not found in these mice, the possible differential interaction of DI particles and standard virus with the lymphoreticular system provides intriguing possibilities for altering the outcome of disease.

With measles virus, passage history of certain variants also affects the outcome of infections in hamsters. Undiluted passage preparations which show "autointerference" titration curves, are more neurotropic.[39] Disease outcome in hamsters or mice with measles virus also varies depending on mouse strain, age of the mouse, and amount of the virus inoculum.[40] Such variabilities make it difficult to conclude that DI particles play any role in this disease. However, as will be seen later, isolates from humans suggest that DI particles of measles virus may be important.

Rabies virus DI particles have been grown in a variety of animal cell cultures and then titrated in animals with very different results. Baby hamster kidney cells do not support "autointerference"; however, neuroblastoma cells, particularly the C1300 strain, support the growth of high titers of virus and also synthesis of DI particles.[41] Undiluted passages of the CVS strain of rabies virus grown on these neuroblastoma cells have reduced mortality for suckling mice, compared to diluted passages of the same virus.[42] Moreover, direct inoculation of mice, i.e., with large amounts of DI particles, protects mice from lethal challenge. Again, however, different viral preparations made on different species of cells do not show direct correlation between DI particles and protection of mice.[42,43] These results point out the difficulties of crossing species barriers and attempting to make correlations between DI particles produced in cells of one species and its biological effects measured in animals of another species. Another complication with rabies virus is the finding of preparations of temperature-sensitive mutants which titrate in mice as if "autointerference" occurs.[43] Although the authors conclude that this phenomenon is due to random reversion to virulence at the higher concentrations, the decrease in mortality vs. increasing concentration of the inoculated mutant CVS-*ts-2* suggests that DI particles may be involved.

Holland and co-workers initiated much of the current interest in pathogenesis studies with DI particles. Using VSV, they showed that intracerebral inoculation of weanling mice with large amounts of DI particles protects mice from lethal challenge.[44,45] However, there is no direct relation between the protective effects of DI particles and its concentration. Because of the large amount necessary, DI particles could be playing an immunologic role.[46] However, the short time course (5 to 7 days) for the development of VSV encephalitis in mice, as well as the finding that UV-inactivated DI particles did not have a protective effect, indicate that the immunologic role, if it takes part, is a minor one.[45] Similar experiments with VSV in hamsters show a protective effect of DI particles against lethal challenge when the ratio of DI to standard virus is greater than 100,000 to 1.[47] Although interferon can similarly protect, its serum levels in hamsters after viral infection do not correlate with protection.[48] When adult mice are inoculated intranasally with VSV DI particles, a very small relative dose increased survival, whereas greater amounts increased mortality.[27,49] These contradictory results led Cave et al.[27] to postulate the following theory: the effects of DI particles in animals are predictable based on the overlapping cyclic multiplication patterns of standard virus and DI particles previously shown to occur in cell culture.[50] This theory will be elaborated upon later in this review.

Studies in nude mice are of particular interest because they demonstrate the complexities that face investigators trying to examine virulence and DI particles in vivo. BHK cells persistently infected with VSV and its DI particles are generally destroyed by natural killer cells in nude mice. However, VSV variants can be selected, which have altered glycoproteins and form BHK cell tumors that are resistant to natural killer cells.[50a] Such evolution of viruses in the presence of DI particles is not an uncommon occurrence in nature.

B. Natural Hosts

For answering the focal question of this review, studies on animal viruses in their natural hosts are more relevant. There are many examples of inoculations of viruses into their natural hosts, but few prove DI particles are modulators of disease.

With reovirus reassortants of type 1 and type 3 in mice, the L2 gene is responsible for viral growth, shedding, and transmission.[51] Although this gene also plays a role in regulating DI particle production, the role of DI particles in lowering the transmissability of reovirus type 3 is not at all clear.[52] Earlier studies done with reovirus DI particles in rats show a definite protective effect, causing the acute, lethal infection to become chronic.[53] Reisolation of standard reovirus from these infections show the standard virus to be unaltered and genetically identical to the initiating virus. These interesting results need confirmation.

Among all of the naturally occurring viral diseases of mice, lymphocytic choriomeningitis (LCM) has been the most extensively studied. The disease is caused by the LCM virus, with symptoms which are determined both by the age of the animal as well as by the virus strain. DI particles of LCM virus have been thoroughly studied, and they are thought to have some properties distinct from DI particles of other virus groups.[54] The LCM virus generates DI particles in cell culture monolayers and in mixed infections produces a bull's eye plaque, indicative of cycles of standard virus replication.[55] Neuroblastoma cell lines infected with DI containing preparations of LCMV develop persistent infections where luxury functions, but not essential functions, are compromised. Unfortunately in the mouse, such effects of DI particles are marginal.[56]

Lytic LCM viruses isolated from persistently infected baby mice resist DI particles generated in the animal and are most resistant to their own DI particles when tested in cell culture.[57] It is difficult to directly detect subgenomic DI RNA in mice persistently infected with LCM virus.[58] Moreover, persistent infections with LCM virus are complicated by the generation of standard virus variants which are capable of suppressing cytotoxic T lymphocytes that are specific for LCM virus-infected cells.[59] Such variants may be related to a previously described phenomenon where LCM virus in the presence of its DI particles reduces the amount of viral antigen on the surface of infected cells, thus presumably altering antigen presentation to lymphocytes.[60] These variants grow well in adult mice, cause persistence, and elicit only antibody responses. Another variant, made by genetic reassortment between two avirulent strains of LCM virus, becomes virulent and causes liver necrosis.[61] The mechanism appears complex and is dependent on the production of one log higher virus titer, in conjunction with interferon, resulting in liver damage. Whether LCM virus DI particles differentially affect lymphocytic functions or determine the effects of interferon on liver cell survival is open to question. Such roles for DI particles in mice may be impossible or hard to prove because of the difficulties in detecting their presence in vivo. With this particular virus system, much can be gained by focusing studies on one virus variant known to produce DI particles, and looking at the immunopathopathology both in vivo and in reconstructed in vitro cell systems to see whether DI particles affect these functions in any way.

In 2-day-old Lewis rats, however, LCM virus appears to cause a more reproducible acute disease. Concurrent inoculation with standard LCM virus and excess DI particles shows a definitive inhibition of disease progression with decreased antigen production by day 3.[61a]

Retroviruses contain different deletions which result in either defective or competent replication, but an analysis of these viruses in animals is complicated by the fact that helper viruses are often present in the host.[19] Nevertheless, focusing on these viruses in relation to defective genomes and helper functions will be important in explaining many of the variable disease processes caused by these viruses. In particular, lentiviruses which are lytic or persistent in sheep show different restriction patterns, including nonequimolar fragments from certain cloned variants.[62] Whether these are indicative of DI particles remains to be proven. In feline leukemia, presence of the virus FeLV and its deleted variants can be correlated to different stages of the disease. When the disease is chronic, cats develop antibodies to the viral core antigens and the virus population is dominated by variants. These differ from the standard virus by the loss of a few kilobases of genetic information and remain unintegrated.[63] Such DI particle-like deletions appear to dramatically alter the outcome of disease.

V. SEEKING DI PARTICLES IN NATURAL VIRAL ISOLATES

A. Characterization of Isolates

Perhaps the best proof of a role in pathogenesis for DI particles in natural disease is to isolate DI particles directly from infected hosts and to correlate these isolations with avirulence or persistence. Although such attempts have been made, the results are just beginning to be understood.

Slow human viral infections, such as subacute sclerosing panencephalitis (SSPE) caused by measles virus, or progressive multifocal leukoencephalopathy (PML) caused by JC virus, are candidate diseases where DI particles may play an important role. However, attempts at isolating DI particles directly from patients yield measles virus with only small amounts of extragenetic information[64] or JC virus with very small insertions or deletions.[65,66] Such small differences are not what would be expected from studies on DI particles of laboratory strains of these two viruses. Nevertheless, these results are valuable when they are considered in light of persistent infections in cell cultures. As was mentioned previously with rabies virus,[24] persistently infected cell cultures tend to select for standard viruses which are no longer inhibited by the DI particles which initiated the persistent infection. Therefore, when patients with SSPE or PML show disease symptoms, it is likely that a fully infectious virus is present in the absence of DI particles. It would be interesting to test whether such isolates, with their small alterations, are resistant to prototypical homologous DI particles. Similarly, when virulent strains are isolated from chronic infections and they appear avirulent,[67,68] their susceptibility to DI particles should be tested.

Along these lines, virulent VSV isolates from the 1982 to 1983 epizootic in the western U.S. were tested for their susceptibility to laboratory-generated DI particles. These isolates are more resistant to various VSV DI particles and they generate DI particles more slowly.[69] Therefore, virus isolates that are more virulent may also show a differential relation to prototypical DI particles. It should not be surprising that standard virus isolates obtained after prolonged infection have multiple alterations in addition to their resistance to inhibition by DI particles.[70-72]

A particularly interesting result was obtained with avian influenza viruses by Webster and colleagues.[73-75] Between April and October of 1983, the avirulent H5N2 strain of influenza virus isolated from chickens in Pennsylvania became virulent. This correlates with the loss of subgenomic RNAs indicative of DI particles. The virulent strain can be attenuated by undiluted passaging with a subsequent increase in hemagglutinin (HA) titers relative to infectivity. Although major changes are not detected by neutralizing monoclonal antibodies, sequence determination of the nucleic acid shows alterations which correlate with enhanced cleavability of the HA gene product from the virulent strain. Thus, both subgenomic RNAs and decreased HA activation may play a role in the conversion of this avian influenza virus to avirulence.

B. Sequential Isolates

Another important approach in examining isolates for DI particles is the use of sequential isolates from the same patient. Sequential isolates have been obtained from immunosuppressed children with rotavirus diarrhea.[76,77] Examination of the rotavirus double-stranded RNA segments in gels indicate changing patterns and sometimes reversion to previous patterns. Although such results can be interpreted to mean different rotavirus strains predominating in time, or the generation of new reassortants, further analysis of the RNA segments may suggest that deletions and rearrangements, indicative of the presence of DI particles, are occurring among these isolates. Similarly, sequential isolations of cytomegalovirus (CMV) from one individual show, upon analysis of the viral DNA, loss of large restriction endonuclease fragments and increases in smaller fragments.[78] Although these isolations were not correlated with disease progression, other isolates of CMV immunosuppress antigen-restricted cytotoxic T-lymphocyte induction.[79] "High passage" or diluted passage laboratory strains have the opposite effect. Although these findings are suggestive, a direct correlation with DI particles has not been made.

A recent study with sequential isolates of hepatitis A virus[80] points to findings that have been well recognized by workers in viral diagnostic laboratories, i.e., detection of different viral products will sometimes give divergent results. In the hepatitis A virus study, fecal excretion yielded hepatitis A virus RNA by cDNA hybridization 56.7% of the time, compared to antigen positivity by radioimmune assay (RIA) in only 25% of the time. These variances are explained by either differences in specificities or sensitivities of the individual diagnostic tests. However, when sequential samples from the same patient are examined, the shedding of virus is intermittent, suggesting a cyclic pattern. A definite cyclic pattern of shedding is found for chimpanzees experimentally inoculated with hepatitis A virus.[81] Unfortunately the role of DI particles in these studies was not evaluated.

C. Guidelines for Examining Isolates

From these efforts in virus isolations, several approaches need highlighting, particularly in regard to determining if DI particles have a role in natural disease. If possible, direct examination of the isolates should be made without an amplification step in cell culture. If viral isolates need to be amplified in cell culture, then one must take into account species and tissue differences in regard to DI particle production and interference. To examine heterogeneity among very few genomes, the nucleic acids of the viral isolates should be cloned directly from the specimens and analyzed. If examination of sequential isolates directly gives varying nucleic acid patterns, then standard virus should be grown out by clonal plaque isolations and compared one to another. If the nucleic acid patterns of these cloned standard viruses appear identical, then the sequential isolates with their varying patterns would be indicative of genomic rearrangements and the presence of DI particles. Lastly, although difficult to do, it becomes imperative to correlate the virulence of the viral isolates to the status of the host immune system and to the status of the host interferon response. Also, in vitro assays with the viral isolates or their infected cells should be measured for their effects on antigen-specific prekiller T lymphocytes and other immune functions.

Because so many host functions complicate the interpretation of results obtained on DI particles, it may be very difficult to design experiments with clear-cut results indicating the exact role played by DI particles in natural disease. Therefore, it is likely that more convincing proof will come from simpler eucaryotic systems, such as yeast, where double-stranded, RNA-containing, virus-like particles and their defective deletion mutants relate to killer or sensitive host strains, respectively.[82]

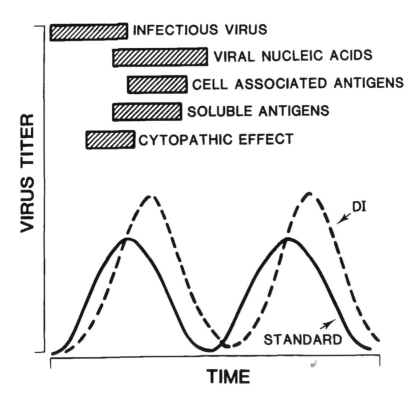

FIGURE 1. Cyclic pattern of production for standard virus and DI particles. Taken from data published on VSV.[50,87]

VI. THEORETICAL CONSIDERATIONS

A. Cyclic Production

Several recent findings suggest that the continuous overlapping cyclic pattern of production described for a population of standard virus and DI particles should be reassessed,[27] particularly in relation to pathogenesis. Evidence to support such a cyclic model comes from serial undiluted passaging of VSV preparations in Chinese hamster ovary cells.[50] As shown in Figure 1, there is alternating production and inhibition of standard virus in time, followed by a similar pattern for DI particles. The amplitude of the cycling may vary for each one, depending on the inhibitory strength of DI particles. Similar cyclic patterns have been reported for many other viral systems; among the best documented examples are herpes simplex virus,[83,84] LCM virus,[85] and Sindbis virus.[86] Such ideal patterns are only obtained when all other parameters affecting virus growth are controlled. Addition of antibody to such in vitro systems suppress both standard virus and DI particles equally.[87] It would be unusual to expect such exact cyclic patterns in vivo. Nonetheless, if the cycling pattern were approximated in vivo, it is worth considering what such a pattern might mean during natural infections.

Consistent with these ideas concerning cyclical production during pathogenesis and transmission are well known observations in disease and viral diagnosis. Evans[88] mentioned the "iceberg phenomenon" for viral infections. It is common to have many more seroconversions and subclinical or mild infections than acute disease. Although much of the subclinical host response can be credited to the host immune system, especially if there has been a previous exposure, differential responses to virus infection in a virgin population may be affected by the infecting virus population. Therefore, as in cell culture, the host response may be dependent on the quality, as well as the quantity, of the virus population used in the inoculum.[50] Also, diagnostic virology laboratories note only 30 to 50% positive isolations or identifications from all incoming samples, even when there is a full-blown viral epidemic with an unambiguous clinical syndrome. This success rate may be the result of intermittent, cyclic shedding. Difficulties in obtaining positive viral isolates may also be due to population heterogeneity where excess DI particles inhibit standard virus growth and prevent the immediate cytopathic effects needed to detect virus. Such DI particles, however, would not affect enzyme-linked immunosorbent assays (ELISA) or nucleic acid hybridization assays, but would contribute to those measurements. In vitro results with continuous passaging of virus in cell cultures support these interpretations. Figure 1 summarizes the time course when it is easy to detect infectious virus as measured by their cytopathic effect. At other times, viral nucleic acids or cell-associated antigens may be easier to detect.

B. Consequences for Pathogenesis and Transmission

Such cycling would explain why there is no direct dose response curve relating survival to the input of DI particles. Since the determination of survival or mortality is so far removed from the initial input virus, it is possible that the stage where the cyclic pattern of viral interaction has reached at a certain time, and not the input population, determines the outcome of infections. For example, an input population with a ratio of high DI particles to low standard virus could produce a large amount of standard virus at some interior location after a few cycles of replication. Therefore, it would be important to determine he cyclic pattern so that one might predict the outcome of a disease merely from the input virus. Also, if an infected individual sheds virus and this population changes during the course of the disease, the quality of the virus transmitted to others would not remain constant during the full infectious period. It is possible that some combinations of DI particles and standard virus would lead to seroconversion and subclinical disease, whereas other combinations would cause acute disease and mortality. If the shed virus were intermittent and followed a known overlapping cyclic pattern of production for standard virus and DI particles, it might be possible to predict the quality of virus that is being shed. Thus, transmission might be timed after the onset of disease symptoms so that there would be only seroconversion among the recipients.

VII. SUMMARY

Despite the incomplete data on roles for DI particles in viral pathogenesis, it is hard to escape the conclusion that DI particles are important in pathogenesis. The major focus of DI work, which has been on the inhibition of viral growth, has led investigators to expect a direct correlation between the amount of DI particles in an inoculum and the degree of attenuation of the disease process. The interactions are much more complicated; they involve a dynamic fluctuating interaction between standard virus and its variants, including DI particles. They also involve the interactions of each of these virus population with nonspecific and specific host defenses. Natural killer cells, interferons, and lymphocytes are all implicated to respond differentially to the presence of DI particles in a viral population. The detailed mechanism of these interactions needs considerable elucidation. Understanding these complex interactions may have significant consequences for controlling viral diseases.

VIII. PERSPECTIVES

Research on DI particles is likely to move into three very different directions. To begin with, the molecular basis of interference, including the exact nucleic acid sequence and proteins affecting them, will continue to be a major focus. These studies will reveal a great deal about replication strategies of different viral nucleic acids and their specific regulatory mechanisms. They will also help to relate essential nucleic acid sequences that bind to structural proteins. Work of this kind is largely underway.

A second direction will be experiments designed to understand the role of DI particles in viral pathogenesis. Much of this has been presented here. It is necessary to highlight the need for rigorous characterization of sequential isolates from patients shedding virus and correlating their properties with disease states. As part of these studies, the growth characteristics of DI particles in relevant differentiated cells will need to be known. Also, in vitro mixed cell culture studies may contribute to our understanding of the complex interactions among DI particles, their standard virus, and different host cells.

The third direction of research will be the manipulation of DI particles so that they may be used to intervene during ongoing disease or as prophylactic agents. Earlier studies of this kind now appear naive. Newer directions are likely to focus on nucleic acid sequences, particularly the regulatory and binding sequences, that are important for interference by DI particles. Engineering these nucleic acid sequences may provide a means for constitutively protecting cells from viruses.

Despite the complications of studying pathogenesis, DI particles provide a natural method for modulating viral infections. Understanding this phenomenon may become important when conventional vaccines fail to protect the host from life-threatening viral diseases.

ACKNOWLEDGMENTS

This work is supported by grant AI16625/20896. I am grateful to many of my colleagues who sent preprints and reprints, especially to Nigel Dimmock for a pre-print of his review on the same subject[89] and Sondra Schlesinger for sharing her ideas and helpful criticisms. I am also grateful to Barbara Connolly for manuscript preparation.

REFERENCES

1. **von Magnus, P.,** Studies on interference in experimental influenza. I. Biological observations, *Arch. Kemi, Mineral. Geol.*, 24(8), 1, 1947.
2. **Gard, S. and von Magnus, P.,** Studies on interference in experimental influenza. II. Purification and centrifugation experiments, *Arch. Kemi Mineral. Geol.*, 24(7), 1, 1947.
3. **Huang, A. S., Greenawalt, J. W., and Wagner, R. R.,** Defective T particles of vesicular stomatitis virus. I. Preparation, morphology, and some biologic properties, *Virology*, 30, 161, 1966.
4. **Huang, A. S.,** Defective interfering viruses, *Rev. Microbiol.*, 27, 101, 1973.
5. **Huang, A. S. and Baltimore, D.,** Defective interfering animal viruses, in *Comprehensive Virology*, Vol. 10, Fraenkel-Conrat, H. and Wagner, R. R., Eds., Plenum Press, New York, 1977, 73.
6. **Kang, C. Y.,** Interference induced by defective interfering particles, in *Rhabdoviruses*, Vol. 2, Bishop, D. H. L., Ed., CRC Press, Boca Raton, Florida, 1980, 201.
7. **Holland, J. J., Kennedy, S. I. T., Semler, B. L., Jones, C. L., Roux, L., and Grabau, E. A.,** Defective interfering RNA viruses and the host-cell response, in *Comprehensive Virology*, Vol. 16, Fraenkel-Conrat, H. and Wagner, R. R., Eds., Plenum Press, New York, 1980, 137.
8. **Perrault, J.,** Origin and replication of defective interfering particles, *Microbiol. Immunol.*, 93, 151, 1981.
9. **Huang, A. S. and Baltimore, D.,** Defective viral particles and viral disease processes, *Nature (London)*, 226, 325, 1970.
10. **Sabin, A. B. and Olitsky, P. K.,** Influence of host factors in the neuroinvasiveness of vesicular stomatitis virus. I. Effect of age on the invasion of the brain by virus instilled in the nose, *J. Exp. Med.*, 66, 15, 1937.
11. **Sabin, A. B. and Olitsky, P. K.,** Influence of host factors in the neuroinvasiveness of vesicular stomatitis virus. II. Effect of age on the invasion of the peripheral and central nervous systems by virus injected into the leg muscles or the eye, *J. Exp. Med.*, 66, 35, 1937.

12. **Sabin, A. B. and Olitsky, P. K.,** Influence of host factors on neuroinvasiveness of vesicular stomatitis virus. III. Effect of age and pathway of infection on the character and localization of lesions in the central nervous system, *J. Exp. Med.,* 67, 201, 1938.

13. **Sabin, A. B. and Olitsky, P. K.,** Influence of host factors on neuroinvasiveness of vesicular stomatitis virus. IV. Variations on neuroinvasiveness in different species, *J. Exp. Med.,* 67, 229, 1938.

14. **Traub, E.,** Factors influencing the persistence of choriomeningitis virus in the blood of mice after clinical recovery, *J. Exp. Med.,* 68, 229, 1938.

15. **Olitsky, P. K., Sabin, A. B., and Cox, H. R.,,** An acquired resistance of growing animals to certain neurotropic viruses in the absence of humoral antibodies or previous exposure to infection, *J. Exp. Med.,* 64, 723, 1936.

16. **Schlesinger, R. W.,** Incomplete growth cycle of influenza virus in mouse brain, *J. Proc. Soc. Exp. Biol. Med.,* 74, 541, 1950.

17. **von Magnus, P.,** Propagation of the PR-8 strain of influenza A virus in chick embryos. III. Properties of the incomplete virus produced in serial passages of undiluted virus, *Acta Pathol. Microbiol. Scand.,* 29, 157, 1951.

18. **Bernkopf, H.,** Study of infectivity and hemagglutination of influenza virus in deembryonated eggs, *J. Immunol.,* 65, 571, 1950.

19. **Weiss, R., Teich, N., Varmus, H., and Coffin, J.,** Eds. *RNA Tumor Viruses,* 2nd ed., Cold Spring Harbor Laboratory, Cold Spring Harbor, New York, 1982.

20. **Baumann, R. P., Dauenhauer, S. A., Caughman, G. B., Staczek, J., and O'Callaghan, D. J.,** Structure and genetic complexity of the genomes of herpesvirus defective interfering particles associated with oncogenic transformation and persistent infection, *J. Virol.,* 50, 13, 1984.

21. **Johnston, M. D.,** The characteristics required for Sendai virus preparation to induce high levels of interferon in lymphoblastoid cells, *J. Gen. Virol.,* 56, 175, 1981.

22. **Zullo, J. N., Cochran, B. H., Huang, A. S., and Stiles, C. D.,** Platelet-derived growth factor and double-stranded ribonucleic acids stimulate expression of the same genes in 3T3 cells, *Cell,* 43, 793, 1985.

23. **Marcus, P. I. and Sekellick, M. J.,** Defective interfering particles with covalently linked [+]RNA induce interferon, *Nature (London),* 266, 815, 1977.

24. **Honda, Y. A., Kawai, A., and Matsumoto, S.,** Persistent infection of rabies virus (HEP-Flury strain) in human neuroblastoma cells capable of producing interferon, *J. Gen. Virol.,* 66, 957, 1985.

25. **Youngner, J. S. and Preble, O. T.,** Viral persistence: evolution of viral populations, in *Comprehensive Virology,* Vol. 16, Fraenkel-Conrat, H. and Wagner, R. R., Eds., Plenum Press, New York, 1980.

26. **Wenske, E. A., Chanock, S. J., and Fields, B. N.,** Genetic reassortment of mammalian reoviruses in mice, *J. Virol.,* 56, 613, 1985.

27. **Cave, D. R., Hendrickson, F. M., and Huang, A. S.,** Defective interfering virus particles modulate virulence, *J. Virol.,* 55, 366, 1985.

28. **Darnell, M. B. and Koprowski, H.,** Genetically determined resistance of infections with group B arboviruses. II. Increased production of interfering particles in cell cultures from resistant mice, *J. Infect. Dis.,* 129, 248, 1974.

29. **Brinton, M. A.,** Analysis of extracellular West Nile virus particles produced by cell cultures from genetically resistant and susceptible mice indicates enhanced amplification of defective interfering particles by resistant cultures, *J. Virol.,* 46, 860, 1983.

30. **Mims, C. A.,** Rift Valley fever virus in mice. IV. Incomplete virus; its production and properties, *Br. J. Exp. Pathol.,* 37, 129, 1956.

31. **Bradish, C. J. and Titmuss, D.,** The effects of interferon and double-stranded RNA upon the virus-host interactions: studies with togavirus strains in mice, *J. Gen. Virol.,* 53, 21, 1981.

32. **Dimmock, N. J. and Kennedy, S. I. T.,** Prevention of death in Semliki Forest virus-infected mice by administration of defective interfering Semliki Forest virus, *J. Gen. Virol.,* 39, 231, 1978.

33. **Crouch, C. F., Mackenzie, A., and Dimmock, N. J.,** The effect of defective interfering Semliki Forest virus on the histopathology of infection with virulent Semliki Forest virus in mice, *J. Infect. Dis.,* 146, 411, 1982.

34. **Barrett, A. D., Guest, A. R., MacKenzie, A., and Dimmock, N. J.,** Protection of mice infected with a lethal dose of Semliki Forest virus by defective interfering virus: modulation of virus multiplication, *J. Gen. Virol.,* 65, 1909, 1984.

35. **Holland, J. J. and Doyle, M.,** Attempts to detect homologous autointerference in vivo with influenza virus and vesicular stomatitis virus, *Infect. Immun.,* 7, 526, 1973.

36. **Gamboa, E. T., Harter, D. H., Duffy, P. E., and Hsu, K. C.,** Murine influenza virus encephalomyelitis. III. Effect of defective interfering particles, *Acta Neuropathol.,* 34, 157, 1975.

37. **Frolov, A. F., Scherbinskaya, A. M., and Sklyanskaya, E. I.,** Properties of influenza virus isolated from mice at different stages of experimental infection, *Vopr. Virusol.,* 5, 544, 1981.

38. **Dimmock, N. J., Beck, S., and McLain, L.,** Protection of mice from lethal influenza by DI virus: evidence that DI virus modulates the immune response and not virus multiplication, *J. Gen. Virol.,* 67, 839, 1986.

39. **Janda, Z., Norrby, E., and Marusyk, H.,** Neurotropism of measles virus variants in hamsters, *J. Infect. Dis.,* 124, 553, 1971.

40. **Rammohan, K. W., McFarlin, D. E., and McFarland, H. F.,** Chronic measles encephalitis in mice, *J. Infect. Dis.,* 142, 608, 1980.

41. **Clark, H. F.,** Rabies serogroup viruses in neuroblastoma cells: propagation, autointerference, and apparently random back-mutation of attenuated viruses to the virulent state, *Infect. Immun.,* 27, 1012, 1980.

42. **Clark, H. F., Parks, N. F., and Wunner, W. H.,** Defective interfering particles of fixed rabies viruses: lack of correlation with attenuation or autointerference in mice, *J. Gen. Virol.,* 52, 245, 1981.

43. **Clark, H. F. and Koprowski, H.,** Isolation of temperature-sensitive conditional lethal mutants of "fixed" rabies virus, *J. Virol.,* 7, 295, 1971.

44. **Doyle, M. and Holland, J. J.,** Prophylaxis and immunization of mice by use of virus-free defective T particles which protect against intracerebral infection by vesicular stomatitis virus, *Proc. Natl. Acad. Sci. U.S.A.,* 70, 2105, 1973.

45. **Jones, C. L. and Holland, J. J.,** Requirements for DI particle prophylaxis against vesicular stomatitis virus infection *in vivo, J. Gen. Virol.,* 49, 215, 1980.

46. **Crick, J. and Brown, F.,** In vivo interference in vesicular stomatitis virus infection, *Infect. Immun.,* 15, 354, 1977.

47. **Fultz, P. N., Shadduck, J. A., Kang, C. Y., and Streilein, J. W.,** On the mechanism of DI particle protection against lethal VSV infection in hamsters, in *The Replication of Negative Strand Viruses,* Bishop, D. H. L. and Compans, R. W., Eds., Elsevier/North-Holland, New York, 1981, 893.

48. **Fultz, P. N., Shadduck, J. A., Kang, C. Y., and Streilein, J. W.,** Mediators of protection against lethal systemic vesicular stomatitis virus infection in hamsters: defective interfering particles, polyinosinate-polycytidylate, and interferon, *Infect. Immun.,* 37, 679, 1982.

49. **Cave, D. R., Hagen, F. S., Palma, E. L., and Huang, A. S.,** Detection of the RNA of vesicular stomatitis virus and its defective interfering particles in individual mouse brains, *J. Virol.,* 50, 86, 1984.

50. **Palma, E. L. and Huang, A. S.,** Cyclic production of vesicular stomatitis virus caused by defective interfering particles, *J. Infect. Dis.,* 129, 402, 1974.

50a. **VandePol, S. B. and Holland, J. J.,** Evolution of vesicular stomatitis virus in athymic nude mice: mutations associated with natural killer cell selection, *J. Gen. Virol.,* 67, 441, 1986.

51. **Keroack, M. and Fields, B. N.,** Viral shedding and transmission between hosts determined by reovirus L2 gene, *Science,* 232, 1635, 1986.

52. **Brown, E. G., Nibert, M. L., and Fields, B. N.,** The L2 gene of reovirus serotype 3 controls the capacity to interfere, accumulate deletions, and establish persistent infection, in *Double-Stranded RNA Viruses,* Bishop, D. H. L. and Compans, R. W., Eds., Elsevier/North-Holland, New York, 1983, 275.

53. **Spandidos, D. A. and Graham, A. F.,** Generation of defective virus after infection of newborn rats with reovirus, *J. Virol.,* 20, 234, 1976.

54. **Lehmann-Grube, F., Peralta, L. M., Bruns, M., and Lohler, J.,** Persistent infection of mice with lymphocytic choriomeningitis virus, in *Comprehensive Virology,* Vol. 18, Fraenkel-Conrat, H. and Wagner, R. R., Eds., Plenum Press, New York, 1983, 43.

55. **Welsh, R. M. and Pfau, C. J.,** Determinants of lymphocytic choriomeningitis interference, *J. Gen. Virol.,* 14, 177, 1972.

56. **Oldstone, M. B. A., Welsh, R. M., and Joseph, B. S.,** Pathogenic mechanisms of tissue injury in persistent viral infections, *Ann. N.Y. Acad. Sci.,* 256, 65, 1975.

57. **Jacobson, S. and Pfau, C. J.,** Viral pathogenesis and resistance to defective interfering particles, *Nature (London),* 283, 311, 1980.

58. **Francis, S. J., Singh, M. K., Oldstone, M. B. A., and Southern, P. J.,** Analysis of LCMV gene expression in acutely and persistently infected mice, *Arenavirus Symp.,* 1985.

59. **Ahmed, R., Salmi, A., Butler, L. D., Chiller, J. M., and Oldstone, M. B. A.,** Selection of genetic variants of LCMV in spleens of persistently infected mice, *J. Exp. Med.,* 160, 521, 1984.

60. **Welsh, R. M. and Oldstone, M. B. A.,** DI particles suppress expression of cell surface ag's, *J. Exp. Med.,* 145, 1449, 1977.

61. **Riviere, Y. and Oldstone, M. B. A.,** Genetic reassortants of lymphocytic choriomeningitis virus: unexpected disease and mechanism of pathogenesis, *J. Virol.,* 59, 363, 1986.

61a. **Welsh, R. M., Lampert, P. W. and Oldstone, M. B. A.,** Prevention of virus-induced cerebellar disease by defective interfering lymphocytic choriomeningitis virus, *J. Infect. Dis.,* 136, 391, 1977.

62. **Querat, G., Barban, V., Sauze, N., Filippi, P., Vigne, R., Rosso, P., and Vitu, C.,** Highly lytic and persistent lentiviruses naturally present in sheep with progressive pneumonia are genetically distinct, *J. Virol.,* 52, 672, 1984.

63. **Mullins, J. I., Chen, C. S., and Hoover, E. A.,** Disease-specific and tissue-specific production of unintergrated feline leukaemia virus, *Nature (London),* 319, 333, 1986.

64. **Hall, W. W. and terMeulen, V.,** RNA homology between subacute sclerosing panencephalitis and measles virus, *Nature (London),* 264, 474, 1976.

65. **Grinnell, B. W., Martin, J. D., Padgett, B. L., and Walker, D. L.,** Is progressive multifocal leukoencephalopathy a chronic disease because of defective interfering particles or temperature-sensitive mutants of JC virus? *J. Virol.,* 43, 1143, 1982.

66. **Rentier-Delrue, F. A., Lubiniecki, F. A., and Howley, P. M.,** Analysis of JC virus DNA purified directly from human PML brain, *J. Virol.,* 38, 761, 1981.

67. **Mirchamsy, H., Bahrami, S., Shafyr, A., Shahrabady, M. S., Kamaly, M., Ahourai, P., Razavi, J., Nazari, P., Derakhshan, I., Lotfi, J., and Abassioun, K.,** Isolation and characterization of a defective measles virus from brain biopsies of three patients in Iran with subacute sclerosing panencephalitis, *Intervirology,* 9, 106, 1978.

68. **Tobler, L. H. and Imagawa, D. T.,** Mechanism of persistence with canine distemper virus: difference between a laboratory strain and an isolate, *Intervirology,* 21, 77, 1984.

69. **Huang, A. S., Wu, T. Y., Yilma, T., and Lanman, G.,** Characterization of virulent isolates of vesicular stomatitis virus in relation to interference by defective particles, *Microb. Pathogen.,* 1, 206, 1986.

70. **Holland, J. J., Grabau, E. A., Jones, C. L., and Semler, B. L.,** Evolution of multiple genome mutations during long-term persistent infections by vesicular stomatitis virus, *Cell,* 16, 495, 1979.

71. **Horodyski, F. M., Nichol, S. T., Spindler, K. R., and Holland, J. J.,** Properties of DI particle resistant mutants of vesicular stomatitis virus isolated from persistent infections and from undiluted passages, *Cell,* 33, 801, 1983.

72. **Wu, C. A., Harper, L., and Ben-Porat, T.,** Molecular basis for interference by defective interfering particles of pseudorabies virus with replication of standard virus, *J. Virol.,* 59, 308, 1986.

73. **Bean, W. J., Kawaoka, Y., Wood, J. M., Pearson, J. E., and Webster, R. G.,** Characterization of virulent and avirulent A/chicken/Pennsylvania/83 influenza A viruses: potential role of defective interfering RNAs in nature, *J. Virol.,* 54, 151, 1985.

74. **Kawaoka, Y., and Webster, R. G.,** Evolution of the A/chicken/Pennsylvania/83 (H5N2) influenza virus, *Virology,* 146, 130, 1985.

75. **Webster, R. G., Kawaoka, Y., and Bean, W. J.,** Molecular changes in A/chicken/Pennsylvania/83 (H5N2) influenza virus associated with acquisition of virulence, *Virology,* 149, 165, 1986.

76. **Spencer, E. G., Avendano, L. F., and Garcia, B. I.,** Analysis of human rotavirus mixed electropherotypes, *Infect. Immun.,* 39, 569, 1983.

77. **Pedley, S., Hundley, F., Chrystie, I., McCrae, M. A., and Desselberger, U.,** The genome of rotavirus isolated from chronically infected immunodeficient children, *J. Gen. Virol.,* 65, 1141, 1984.

78. **McFarlane, E. S. and Koment, R. W.,** Use of restriction endonuclease digestion to analyze strains of human cytomegalovirus isolated concurrently from an immunocompetent heterosexual man, *J. Infect. Dis.,* 154, 167, 1986.

79. **Schrier, R. D. and Oldstone, M. B. A.,** Recent clinical isolates of cytomegalovirus suppress human cytomegalovirus specific human leukocyte antigen restricted cytotoxic T-lymphocyte activity, *J. Virol.,* 59, 127, 1986.

80. **Tassopoulos, N. C., Papaevangelou, G. J., Ticehurst, J. R., and Purcell, R. H.,** Fecal excretion of Greek strains of hepatitis A virus in patients with hepatitis A and in experimentally infected chimpanzees, *J. Infect. Dis.,* 154, 231, 1986.

81. **Bradley, D. W., Gravelle, C. R., Cook, E. H., Fields, R. M., and Maynard, J. E.,** Cyclic excretion of hepatitis A virus in experimentally infected chimpanzees: biophysical characterization of the associated HAV particles, *J. Exp. Med.,* 1, 133, 1977.

82. **Bruenn, J. and Kane, W.,** Relatedness of the double-stranded RNAs present in yeast virus-like particles, *J. Virol.,* 26, 762, 1978.

83. **Stegmann, B., Zentgraf, H., Ott, A., and Schroder, C. H.,** Synthesis and packaging of herpes simplex virus DNA in the course of virus passages at high multiplicity, *Intervirology,* 10, 228, 1978.

84. **Murray, B. K., Biswal, N., Bookout, J. B., Lanford, R. E., Courtney, R. J., and Melnick, J. L.,** Cyclic appearance of defective interfering particles of herpes simplex virus and the concomitant accumulation of early polypeptide VP175, *Intervirology,* 5, 173, 1975.

85. **Staneck, L. D., Trowbridge, R. S., Welsh, R. M., Wright, E. A., and Pfau, C. J.,** Arenaviruses: cellular response to long term in vitro infection with parana and lymphocytic choriomeningitis virus, *Infect. Immun.,* 6, 444, 1972.

86. **Inglot, A. D., Albin, M., and Chudzio, T.,** Persistent infection of mouse cells with Sindbis virus: role of virulence of strains, autointerfering particles , and interferon, *J. Gen. Virol.,* 20, 105, 1973.

87. **Sinarachatanant, P. and Huang, A. S.,** Effects of temperature and antibody on the cyclic growth of vesicular stomatitis virus, *J. Virol.,* 21, 161, 1977.

88. **Evans, A. S.,** Epidemiological concepts and methods, in *Viral Infections of Humans,* 2nd ed., Plenum Press, New York, 1982, 2.

89. **Dimmock, N. J. and Barrett, A. D.,** Defective viruses in diseases, *Curr. Top. Microbiol.,* 128, 55, 1986.

Role of Genome Variation in Disease

Chapter 12

SEQUENCE SPACE AND QUASISPECIES DISTRIBUTION

Manfred Eigen and Christof K. Biebricher

TABLE OF CONTENTS

I. INTRODUCTION: RNA — THE PRIMAL SOURCE OF GENETIC INFORMATION

Self-reproduction, the basis of evolutionary adaptation, actually serves two purposes: first, it ensures conservation of information, despite the steady chemical decomposition of its structural carrier.[1,2] Second, it provides by its autocatalytic nature a mechanism of competitive growth and selection. Although proteins, owing to their more straightforward chemical synthesis mechanisms, might have appeared first on our planet,[3,4] functional adaptation and optimization had to await the arrival of a self-replicative system for information storage.[2,5] There are several strong indications that it was RNA rather than DNA which provided the first information carriers in early evolution:

1.   Ribose forms more easily under potential prebiotic conditions than deoxyribose does (aldol condensation).[4]
2.   The nucleophilicity of the 3'-OH of deoxyribose is weaker than the vicinal 2'-3'-hydroxyls of ribose. The efficiency of phosphodiester formation for any nonenzymic polynucleotide synthesis is thus higher for ribonucleotides.[4]
3.   The biosynthesis of deoxyribonucleotides proceeds via reduction of ribonucleoside diphosphates.
4.   Replication of DNA requires RNA primers[6] or — as Sol Spiegelman once phrased it — ''DNA up to the present day hasn't yet learned to reproduce itself without the help of RNA.''
5.   The ribo base-pairs are more stable than their deoxyribo analogues.[7] Internal folding is further stabilized by additional hydrogen bonds, often involving the 2'-hydroxyl. The larger tendency of RNA to undergo internal folding processes stabilizes single-stranded forms, while DNA prefers the double-helical form of the complementary strands. In the single-stranded forms, RNA offers a richer repertoire of tertiary structures than an analogue DNA strand would provide. As far is known today, all functional nucleic acids are of the ribo-type (cf. ribosomal RNA, tRNA, ''ribozymes'').[8]

All these properties made RNA the more favorable candidate for early self-organization of a genetic memory.[5]

However, RNA is more labile to alkaline hydrolysis than is DNA.[4] Hence, DNA offered selective advantage for storage of information, whenever genome sizes grew large and whenever error correction (utilizing double-stranded forms for reference) became compulsory.

RNA viruses — though latecomers in evolution, requiring the preexistence or concomitant evolution of their hosts — reflect many of the properties mentioned and hence represent suitable model systems for experimental studies of molecular evolution. Such experiments and their theoretical interpretation have brought about an important modification of our comprehension of natural selection and its consequences for evolutionary optimization. In particular, they have revised our picture of the wild type, which reveals itself as a subtly structured mutant distribution centered around one or several master sequences.[1] Such a distribution of sequences, if stationary, is called a quasispecies.[2] It usually has a defined consensus sequence, identical with or close to the master sequence, which may be represented, however, by only a small fraction of the ensemble.

Populations of RNA viruses are of a quasispecies nature.[9] Many of their properties can be understood only on this basis. It is therefore appropriate to conclude this volume on ''RNA variability'' with a discussion of the quasispecies concept which, moreover, is of significance in the evolution of any system of replicators.

II. RNA AS A DARWINIAN SYSTEM

A. The Meaning of "Natural Selection"

Evolution in a Darwinian system is based on "natural selection". What does it mean? Is it a law revealing some regularity of behavior? Or does it only describe some tendency or trend, typical for living matter? Or is all that this principle expresses just the tautology of a "survival of the survivor"?

Natural selection can be shown to be a direct physical consequence of reproduction, be it vegetative (self or complementary) or sexual. The principle hence discloses an "if-then" relation. This can be lucidly demonstrated for some simple models; it holds, however, just as well for a wider class of replication mechanisms of the type discussed in Chapter 1 of Volume I, and it guides rapid evolution of RNA genetic elements in situations that are relevant to viral pathogenesis, as discussed in other chapters of this volume.

In absence of saturation effects, replication was found to observe a simple autocatalytic rate law of the form (cf. Volume I, Chapter 1):

$$dc_i(t)/dt = A_i c_i(t) \tag{1}$$

where $c_i(t)$ is the (time dependent) concentration of a self-replicating molecule "i", $dc_i(t)/dt$ the replication rate, and A_i an overall replication rate coefficient, possibly depending on time-dependent factors and certainly containing a substrate (i.e., nucleoside triphosphate) concentration function (which may be buffered in order to yield constant rate coefficients A_i).

Equation 1 is of an inherently nonlinear form, primarily due to the substrate concentration function contained in A_i. If, on the other hand, one buffers all substrate concentrations to constant values, thereby obtaining a constant rate coefficient A_i, one should include a (possibly time-dependent) flow term which regulates the monomeric substrate concentration to a constant value and removes excess polymeric products.

The sequence i is only one of many sequences k present ($k = 1,2,...n$). Introducing relative concentration variables $x_i(t) = c_i(t)/\Sigma_k c_k(t)$ one obtains from Equation 1

$$dx_i(t)/dt = \{A_i - \overline{A}(t)\} x_i(t) \tag{2}$$

where $\overline{A}(t) = \Sigma_k A_k x_k(t)$ represents the weighted average of all replication rate coefficients at time t.

The inherently nonlinear form of Equation 2 describes a selection process. All sequences i with rate coefficients A_i smaller than the average $\overline{A}(t)$ yield a negative sign for their rates and therefore will disappear, while all sequences i having rate coefficients larger than $\overline{A}(t)$ will grow and thereby increase the value of $\overline{A}(t)$ until the average matches the maximum rate coefficient A_{max}. Hence, we have an extremum principle

$$\overline{A}(t) \to A_{max} : x_{max} \to 1$$

$$x_{i \neq max} \to 0 \tag{3}$$

All relative concentrations $x_i(t)$ will vanish except x_{max} (i.e., the relative population number belonging to A_{max}) which reaches one. This self-ordering process (Equation 3) we call "natural selection". It is an immediate consequence of self-replication as expressed by the simple autocatalytic rate law: $dc_i(t)/dt = A_i c_i(t)$.

Of course, this model is too simplistic as to be of practical value for describing any real situation where natural selection occurs. (Its main shortcoming is the neglect of mutations.)

However, it serves excellently to demonstrate the meaning of "natural selection", which is not of a trivial tautological nature. Equation 3 states that survival ($x_m \rightarrow 1$; $x_{i \neq m} \rightarrow 0$) is a particular form of presence that is correlated with a maximum reproduction rate. The particular form of the selection equation is the consequence of the inherently autocatalytic rate law and the absence of chemical equilibrium. At equilibrium, survival would mean always coexistence of all partners, the concentrations of which depend on free enthalpies of reaction $c_i/c_j = \exp[-\Delta G^0_{ij}/RT]$. At equilibrium, even any autocatalytic system must approach this distribution.

However, Darwinian behavior means more than mere selection of preexisting replicators. A more precise description must also include the formation of mutants.

B. Mutation and Natural Selection

One way of introducing the mutation rate is to superimpose on Equation 1 an additional term. In classical population theory, mutants were not identified individually and, therefore, the added term was just some unspecified expression modifying the wild-type rate law only in a minor fashion.[10] As a consequence, an interpretation of Darwinian behavior has become common, that reflects this presumed, unspecific, unbiased nature of mutation. Selection — apart from neutral drift[11] — has been associated with maximum fitness. Mutations, however, were thought to appear stochastically, regardless of their fitness.

The difficulties which such an interpretation will encounter — even in the evolution of self-replicating sequences of very limited length (i.e., single genes) — becomes immediately obvious if we specify mutations more quantitatively. In Figure 1, assuming sequences of length ν and a uniform (Poissonian) error model, the probabilities or relative frequencies of mutant production are shown as a function of the Hamming distance d_{ik} between the template sequence i and the replica k that is produced by erroneous copying. The definitions are as follows:

ν —	The sequence length, i.e., the number of nucleotides contained in the particular sequence i.
d_{ij} —	The Hamming distance between two sequences i and j, i.e., the number of positions at which these two sequences differ.
κ —	The number of symbol classes, i.e., four nucleotides (A, U, G, C).
q —	The fidelity, i.e., the probability of inserting the correct nucleotide for any position. Correspondingly $(1-q)$ is the error rate per nucleotide.
$Q_d = \binom{\nu}{d} q^{\nu-d}(1-q)^d$ —	The relative frequency of inserting $(\nu-d)$ correct and (d) incorrect nucleotides into a sequence with the chain length ν.
$N_d = \binom{\nu}{d} (\kappa-1)^d$ —	The number of different copies in the error class with Hamming distance d. Dividing Q_d by N_d yields that below.
$Q_{ij} = q^\nu p^{d_{ij}}$ with $p = (q^{-1}-1)/(\kappa-1)$ —	The relative frequency of producing mutant j by miscopying sequence i, if the Hamming distance between i and j is d_{ij}.

For simplicity, we assumed a uniform error rate. Rate equations can also be written for site-specific error rates by considering each sequence position individually. For RNA genomes, the extent of differences in miscopying individual positions is not known (cf. Section VII and Chapter 7, Volume II). Likewise, error forms referring to insertion or deletions may be added individually if they occur individually. (Nonreproducible effects of this kind may be of importance in evolutionary adaptation; they do not contribute to the population structure at stationary selection, cf. Section VII.E.)

As is seen from Figure 1 for four sequences with length $\nu = 30, 300, 3000,$ and 30,000,

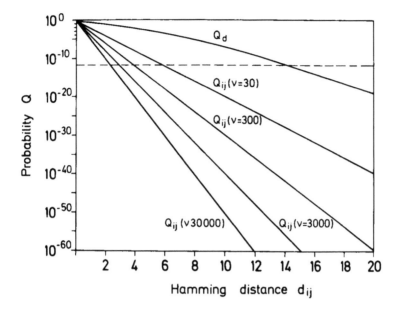

FIGURE 1.   Probabilities of producing error copies in sequences comprising v nucleotides. Q_d refers to the production of any (i.e., unspecified) error copy having a Hamming distance d from the template, while Q_{ij} refers to the production of a specified individual copy i that is formed by miscopying template j, d_{ij} being the Hamming distance between i and j. v is the sequence complexity. The error rate $(1 - \bar{q})$ is assumed to be v^{-1}. In the range considered ($v > 20$), Q_d then turns out to be nearly independent of v, while Q_{ij} at given d_{ij} decreases strongly with increasing sequence complexity v, reflecting the increasing size of the error class N_d. The values show that in a population of size typical for laboratory conditions, about 10^{12}, (dashed line) one may expect on average still just one 14-error copy (largely independent of v), while the largest Hamming distance of error classes with almost completely represented mutant spectra are expected to be six for $v = 30$, four for $v = 300$, three for $v = 3000$, and two for $v = 30,000$. The expectances are based on the binomial expressions quoted in the text and neglect any selective bias. (After Eigen, M., *Chem. Scr.*, 26B, 13, 1986. With permission.)

Q_d is almost independent of sequence length (which is not true anymore if the sequence length becomes very short), while Q_{ij} depends strongly on sequence length and drops to very low values for any appreciable Hamming distance. The case $v = 300$ is representative for a (short) gene, $v = 3000$ to 30,000 covers the range of most RNA viruses. Typical laboratory populations of RNA molecules or viruses are in the range of 10^{12} individual particles (dashed line). As is seen, for $v = 30$, all six-error mutants, for $v = 300$, all four-error mutants, and for $v = 3000$, all three-error mutants are to be expected (almost) deterministically. However, if for $v = 300$ a particular ten-error mutant is required, the population size has to grow to at least 10^{20}, for $v = 3000$, it has to increase even to above 10^{30} before its appearance gets highly probable. The sizes of natural populations of RNA viruses and their relation to the biological relevance of RNA variance have been reviewed.[9]

Since the fitness landscape — like any landscape on earth — is to be expected to be rugged rather than smooth or even monotonic, and since mutational jumps of such and larger lengths are to be expected in order not to get stuck on any minor fitness hill, evolution in the common interpretation of Darwinian models, i.e., supposing unbiased and undirected stochastic mutational jumps, would become hopelessly slow.[12]

III. HOW TO MAP SERIAL INFORMATION: THE CONCEPT OF SEQUENCE SPACE

In order to take mutations into account, one has to construct a space in which all mutual kinship distances are correctly represented. The concept of such a space has been introduced to information theory by Hamming.[13] A corresponding point space for the four-digit alphabet of nucleic acids has been suggested by Rechenberg.[14] How such a space may be constructed in a recursive way is shown in Figure 2 using the simpler model of binary sequence. A plane would just suffice to represent sequences consisting of two positions only. Increasing the sequence length to three positions requires a third coordinate. The three one-error mutants then can be arranged at unit distance from the origin (the reference sequence), the three two-error sequences at distance two from the origin defining three planes which connect at the corner diagonal to the origin, assigned to the three-error mutant completing the unit cube. It is obvious now how the construction principle is to be iterated for sequences including ν positions, although our capacity of imagination is exhausted if the number of dimensions surpasses 3. Nevertheless, the example $\nu = 6$ in Figure 2 provides us with some intuitive answers to the question: what are the major differences between low- and high-dimensional spaces? Three features essentially distinguish the high-dimensional point space:

1. The storage capacity greatly increases with increasing dimension ν. If κ is the number of digit classes, the storage capacity is κ^ν. With $\kappa = 4$, a 180-dimensional space (representing a nucleic acid sequence comprising 180 positions) would be sufficient to map the whole universe (viewed as a sphere with a diameter of about 10 billion light years) with a resolution in the cubic Ångstrom range. (The volume of the universe is about $4^{180} = 10^{108}$ [Å]^3.)

2. Despite the large capacity, the longest direct (i.e., detour-free) distances cannot exceed the number of dimensions. In other words, one is never far apart to reach any target if guided by a directional bias or some gradient. Otherwise one gets lost in the many-dimensional network, just as one would in the vast depths of the universe. To come back to the above example: if we map the universe cubic-Ångstrom by cubic-Ångstrom in the 180-dimensional space, the largest direct distance is 180 Å rather than some ten billions of light years.

3. Finally, the large connectivity among all points of the ν-dimensional hypercube is intuitively obvious from Figure 2. If in such a space mutational jumps over a distance d are possible, one may reach from any point immediately $\binom{\nu}{d}$ neighbors. With $\nu = 180$ and d = 5 (Figure 1) this means about 10^9 different points, through which evolution could proceed.

The concept may be generalized as to account for a system of sequences with varying lengths. Since all coordinates are equivalent, one may add any number corresponding to the variation in sequence length. By introducing blank space symbols one may thus account also for insertions or deletions.

We now ask, what kinds of gradients are available to provide a directional orientation in such a space. If the evolutionary process were entirely of a random-walk nature (involving a random fitness distribution), on average, all points in sequence space would have to be checked before any definite could be reached. In such a case no advantage of a low-distance and high-dimensional space over a long-distance and low-dimensional space could be claimed.

IV. THE FITNESS TOPOGRAPHY OF SEQUENCE SPACE

The distribution of altitude on earth is neither monotonic nor is it entirely random. Altitudes usually change smoothly, and mountains cluster in connected regions. According to

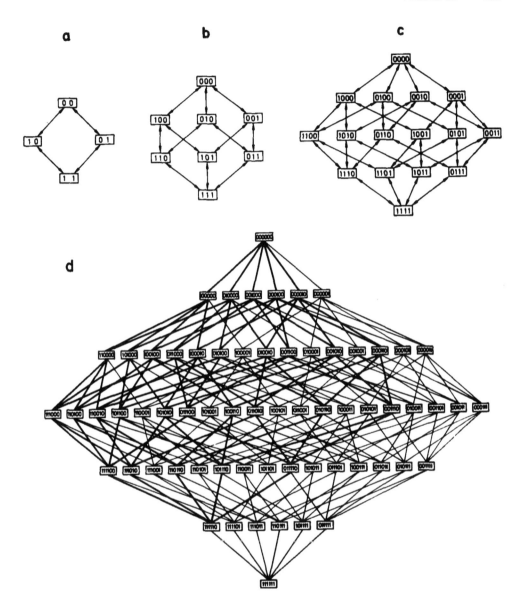

FIGURE 2. Sequence space. The correct representation of kinship distances among mutants requires a ν-dimensional point space. Constructions are exemplified for binary sequences of lengths $\nu = 2$ (a), 3 (b), 4 (c) and 6 (d). If four-digit classes (e.g., nucleotides) were involved, to each point in binary sequence space a ν-dimensional binary subspace is attributed. This representation was chosen to provide an intuitive feeling for three important properties of sequence space: (1) the enormous capacity or volume (2^ν or 4^ν states, respectively), (2) the comparably small distances (not exceeding ν for any direct connection line), and (3) the large connectivity among states ($\binom{\nu}{d}$ neighbors may be directly reached through a mutational jump over the Hamming distance d). (From Eigen, M., *Ber. Bunsenges. Phys. Chem.*, 89, 658, 1985. With permission.)

Mandelbrot[15] the kind of randomness one may expect is a distribution of a self-similar nature. He coined the term "Brownian landscape". The meaning of this term is obvious from Figure 3. Consider a Brownian process such as the cumulative results of tossing a coin. The ordinate records in the upward direction the cumulative wins of "head" and in the downward direction those of "tail", while the abscissa represents the number of trials. If all points are connected, the curve looks like a one-dimensional cut through a mountainous landscape (Figure 3). The

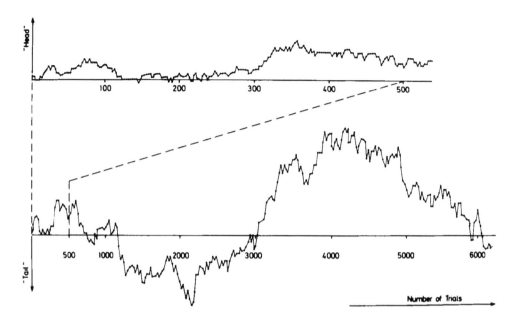

FIGURE 3. A Brownian landscape. The notion of a Brownian landscape was introduced by Mandelbrot[15] in order to represent altitude distribution on earth. A similar concept may be applied to the value topography in (ν-dimensional) sequence space. The Brownian relief in this picture is produced through (simulated) coin tossing by counting cumulative successes for head (upward) and for tail (downward). A typical property of the Brownian relief — reminiscent of contour lines in a mountainous landscape — is the self-similarity of the curve for different magnifications of the abscissa, expressed by a fractal dimension larger than that of an one, but smaller than that of a two-dimensional space. (Modified from Mandelbrot, B. B., *The Fractal Geometry of Nature*, W. H. Freeman, New York, 1983.)

self-similarity of the distribution becomes obvious if we look at the curve at different magnifications of the abscissa. The curve fills part of a two-dimensional plane. The mathematician defines a "fractal dimension" which for these curves turns out to be 1.5. Similarly, Mandelbrot constructed Brownian landscapes on two-dimensional planes. They indeed do look like true mountainous landscapes, where the fractal dimension 2.5 reminds one of typical badland formations, while the smoother forms, as obtained with dimension 2.1, resemble mountains like the Alps or the Rocky Mountains.

Mandelbrot considers the problem of a rain drop falling on a mountainous island and its path to the ocean. If the landscape were a height distribution on a one-dimensional line, as represented by the projection in Figure 4, the droplet would run into the next "cup", as Mandelbrot calls it, and stay there until the cup is filled. Only if all cups (with self-similar shapes) have been filled, a rain drop falling at any point of this landscape could reach the ocean along the indicated path $B_H^*(x)$, a terrace avoiding all ascending sections of the true (one-dimensional) profile $B_H(x)$. Mandelbrot shows that an extension to higher-dimensional landscapes lowers the value of $B_H^*(x_1,x_2...x_M)$ relative to $B_H(x_1,x_2...x_M)$. We may visualize this from looking at a cup in two dimensions. The lowest points of the rims of the cups almost certainly will be lower than the maximum positions in the one-dimensional cut. Since the water escapes through the lowest points of the rims, the water level in nearly all cups will be lower than in the one-dimensional representation. Adding more dimensions will cause the water level to drop further until for an infinite-dimension space every droplet falling on such a landscape will reach the ocean, i.e., along a connective path $B_H^*(x_1,x_2...x_M)$ will approach $B_H(x_1,x_2...x_M)$, a result which had been derived before by Levy.[16]

The evolutionary uphill motion in the multidimensional fitness landscape appears to be a

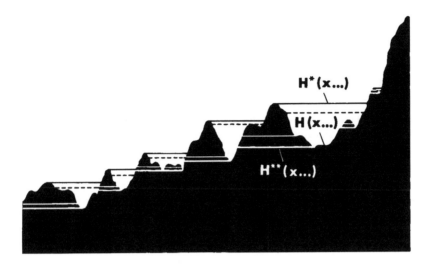

FIGURE 4. Mandelbrot's rain droplet problem. The problem is concerned with the path of a rain droplet falling onto a mountainous island down to the ocean. In a one-dimensional relief, all cups (defined as loci between two successive maxima of the contour line $H(x)$) have to be completely filled before a rain drop can move along the terrace represented by the function $H^*(x)$. If a second dimension is added, the water may escape through outlets in the rim of the cup, which would only accidentally correspond to the height of the maximum position in the one-dimensional case, but usually are lower and hence allow for a water level below that represented in the one-dimensional case. If the number of dimensions M increases further, the water level will decrease accordingly until for $M \to \infty$ there will always be a connective pathway for which $H^*(x_1, x_2 \ldots x_M) = H(x_1, x_2 \ldots x_M)$. The hill climbing procedure of evolutionary optimization may be represented in a complementary way. However, in the evolutionary route, the system always seeks the locally highest points in the value topography, defining a function $H^{**}(x_1, x_2 \ldots x_M)$ for which a similar argument holds as for $H^*(x_1, x_2 \ldots x_M)$. (H, H^*, and H^{**} are abbreviations for Mandelbrot's Brownian functions B_H, B_H^*, B_H^{**}, referring to a fractal dimension H^{-1}.) (From Eigen, M., *Chem. Scr.*, 26B, 13, 1986. With permission.)

quite analogous case. The role of cups is taken over by gendarmes and the riverbed through which water reaches the ocean now is represented by a ridge along which the highest fitness point is to be approached. The problem is not to get stuck at some isolated gendarme. The difference from the former case is the following. Rain water, because of gravity, always seeks to reach the locally lowest points. Evolutionary motion, in contrast, approaches the locally highest fitness point. Hence, it does not follow the terrace along which water will move (Figure 4), but rather along a second terrace that would be formed by grinding off the peaks between successive wells, defining a function $B_H^{**}(x_1 \ldots x_M)$. This function, upon increasing the dimensionality, changes in a similar way with respect to $B_H(x_1 \ldots x_M)$ as does $B_H^*(x_1 \ldots x_M)$. The analogy becomes apparent if Figure 4 is turned upside down. Wells then become gendarmes and vice versa. Moreover, if — as for the problem of evolution — we are dealing with a point space and motion through discrete jumps, there might be — as Mandelbrot conjectured — a finite *critical dimension M* for which a connected route can always be found. This critical dimension must depend on how finely grained the discrete landscape is and on the size of possible mutational jumps. Values of M realized in problems of evolution may easily exceed 10^{10}.

Each point in the sequence space, depending on the evolutionary problem posed, now can be assigned a fitness value. We may, so to speak, assign a color to each point in sequence space which characterizes its fitness value, thereby constituting a fitness topography. The evolutionary process then can be described by a jump route for which fitness gradients provide a directional bias. That such a picture indeed corresponds to reality is suggested by a number of experimental facts.

1. Sequence comparison of different isolates of the same virus[17-21] indicates that the fitness landscape indeed is quite connective and smooth. Most low-error mutants produced by site-directed mutagenesis (in exon regions) were found to produce translation products with functional fitnesses not much different from that of the wild type.[22] There are, of course, critical positions where the correct nucleotide is indispensable for function.[23] Likewise, there are steep slopes or gorges in any mountain landscape on earth.
2. The existence of phylogenetic trees suggests the existence of long-range gradients as well as far protruding connective ridges between (nearly) neutral mutants. The fact that phylogenetic divergence often tolerates homologies less than 50% (e.g., levivir-idae)[24,25] suggests that the sequence space is pervaded by broad and far-extending connective mountain regions.
3. The known existence of unrelated sequences with almost identical functional destination and efficiency (examples: proteases, capsid proteins of viruses, lysozymes,[26] etc.) suggests a degenerate fitness landscape in which different fitness peaks of similar height exist that give the appearance of having been reached in a deterministic manner.
4. Evolutionary experiments carried out with RNA sequences produced *de novo* by Qβ replicases (Volume I, Chapter 1), which adapt rapidly to optimal performance in a given environment, endorse the above conclusions (Section VII.D).

How do we get from a fitness topography to a population topography, or how is fitness expressed through population numbers? Since evolution proceeds through the population of mutant states, this is the central issue of Darwinian theory, and a quantitative answer should be provided.

V. HOW ARE POPULATION NUMBERS RELATED TO FITNESS VALUES?

Survival expresses itself in terms of population numbers, which are extensive parameters. Fitness, on the other hand, is intensive and should be expressed by rate or force parameters. It is the dynamical balance between buildup and removal or decomposition that determines the extensive structure of the population.

Since the mutant space of a nucleic acid sequence of length ν includes 4^ν states, we are now looking into a phase space of dimension 4^ν where the axes refer to the relative population numbers x_i. For each of these states we may write down a deterministic rate equation like Equation 2, taking into consideration both precise and erroneous reproduction:[1,2]

$$dx_i(t)/dt = (W_{ii} - \bar{E}(t))\, x_i(t) + \sum_{k \neq i} W_{ik}x_k(t) \qquad (4)$$

The meaning of the parameters is apparent from Figure 5. $\bar{E}(t)$ is the average excess productivity, i.e., $\bar{E}(t) = \sum_k A_k x_k - \sum_k D_k x_k$. Equation 4 differs from Equation 2 by the sum term describing the formation of sequence i through miscopying sequence k, separated by a Hamming distance d_{ik}.

The off-diagonal coefficients W_{ik} result from the product of a vector of reproduction rate coefficients A_i introduced with Equation 1 and 2 and the error matrix Q_{ik} introduced in Section II. The diagonal coefficient is composed of two rate parameters, i.e., for exact reproduction (A_i) and for decomposition (D_i):

$$W_{ii} = A_i \bar{q}^\nu - D_i \qquad (5)$$

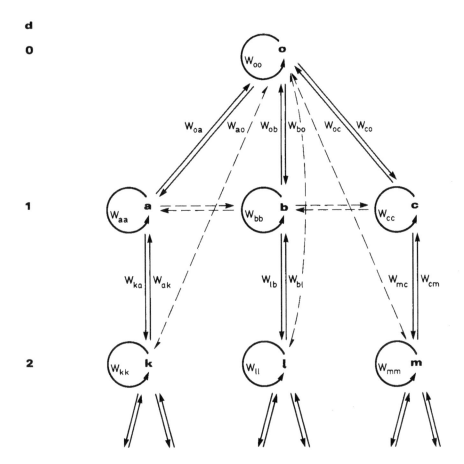

FIGURE 5. Reaction scheme of a quasispecies. The index o is assigned to the master copy, W_{oo} being the rate coefficient for correct (excess) production of the master copy. Likewise, W_{ii} refers to the correct (excess) production of mutant copy i. The off-diagonal rate coefficients refer to mutation rates. W_{ij} represents the production of mutant i by miscopying template j. The rate coefficients are always associated with relative concentration terms x_j that refer to the second of the two subscripts (in agreement with conventional matrix notation). The Hamming distances in this scheme are relative to the master copy. It is important to note that diagonal coefficients are by order of magnitude $(\kappa - 1)/(q^{-1} - 1)$ larger than off-diagonal coefficients that refer to one-error miscopying, and that these, in turn, are correspondingly larger than the coefficients for two-error miscopying and so on. The error matrix may be ordered recursively in a similar way, as was demonstrated for the buildup of the sequence space (Figure 2), such that the antidiagonal represents the mutation terms with extreme errors. In this way, explicit solutions could be obtained for certain rate coefficient schemes in the uniform error model.[33]

The system of rate equations refers to a simple model describing mutagenous replication. According to kinetic studies carried out with various RNA-replicators (described in Chapter 1, Volume I) it is sufficient to account for a large variety of realistic situations including enzyme-catalyzed and (complementary) template-instructed production. The pairs of rate equations referring to a plus and minus strand involve a positive and a negative eigenvalue. The latter one describes relaxation into a fixed ratio of plus and minus strand concentration, while the positive eigenvalue is equivalent to the growth term in Equation 4 referring to the sum of plus and minus strand concentration. Wherever superimposed effects modify the straightforward exponential growth resulting from Equation 4, they do it in an accountable way without principally invalidating the results and conclusions presented in this paragraph.

The inherently nonlinear system of rate equations (4), originally solved by perturbation theory,[1] has been shown to be exactly soluble. With the help of the transformation,[27,28] $z_i = x_i \int_0^t \exp\{-\overline{E}(\tau)d\tau\}$, they can be transformed into a (quasi-) linear system of equations. (Again assuming buffered monomeric substrates and controlled enzyme level.) After transforming back into relative population numbers the system (Equation 4) assumes the form

$$dy_i(t)/dt = \{\lambda_i - \overline{\lambda}(t)\}\, y_i(t) \tag{6}$$

where y_i now is a population variable associated with the eigenvalue λ_i, which refers to a whole clan and which comprises contributions from all mutant terms weighed according to their kinship relations. The average of all eigenvalues $\overline{\lambda}(t)$ equals the average excess production rate $\overline{E}(t)$. From this form it is easily seen that again an extremum relation holds

$$\overline{\lambda}(t) \to \lambda_{max} \tag{7}$$

The mutant distribution with maximum eigenvalue appears to be the target of selection. We call it a "quasispecies". The essential difference with respect to the extremum principle (Equation 3) — caused by the mutation terms — is a shift of emphasis. Instead of specifying individual sequences, one is looking now at an equivalent number of ways of dividing them up into clans, each comprising all sequences weighed according to their kinships within the clan. The target of selection then is a particular way of kinship order dominated by one master sequence (or a degenerate set). The master itself may constitute only a minor fraction of the whole "wild-type" distribution (Figure 6b). Despite this fact, the wild type usually exhibits a defined sequence which is a consensus sequence that is not necessarily identical with that of the master, nor does the master sequence have to be the one characterized by the maximum diagonal coefficient W_{mm}, because of the mutant contributions. It will be seen that the details of mutant population are decisive for the evolutionary route of optimization and that not only neutral, but also near-neutral mutants are of utmost importance.

One immediate consequence of the extremum value relation, i.e., the existence of maximum eigenvalue, is the existence of a threshold relation. It correlates the mean error rate $(1 - \overline{q})$ with a length of the sequence ν_{max} that can be reproducibly maintained by selection.[1,2]

$$\nu_{max} < \ln\sigma_o/(1 - \overline{q}) \tag{8}$$

In this relation, the logarithm of a function σ_o appears, which we call the *selectivity* of the master:

$$\sigma_o = A_o/(D_o + \overline{A}_{k \neq o} - \overline{D}_{k \neq o}) \tag{9}$$

where the subscript o refers to the master and $k \neq o$ to its mutant distribution, for which the averages are taken, i.e., $\overline{A}_{k \neq o} = \Sigma_{k \neq o}A_k\overline{x}_k/\Sigma_{k \neq o}\overline{x}_k$. (Note that in this case, $\Sigma_{k \neq o}\overline{x}_k = 1 - \overline{x}_o$.) The master selectivity refers to an excess of reproductive power relative to decomposition and to mean excess production of mutant competitors. If decomposition terms are negligible, as is often the case (removal by dilution), σ_o becomes $A_o/\overline{A}_{k \neq o}$ and thereby is simply related to the average of the reciprocal individual selectivities A_i/A_o *selectivities* through $\sigma_o^{-1} = \Sigma_{k \neq o} A_k x_k/A_o$. We shall see that individual selectivities may vary between one and infinity, introducing a much more subtle individualization of mutant population numbers than resulting from individual error rates. The error threshold relation (Equation 9) requires not only the presence of a superior master with $\sigma_o > 1$, but in addition, $\sigma_o > \overline{q}^{-\nu} \approx \exp\{\nu(1 - \overline{q})\}$. Therefore, if the error rate is given (through some replication mechanism), the limitation of the sequence length depends on the selectivity obtainable. Violation of the threshold relation results in an error catastrophe, i.e., an accumulation of

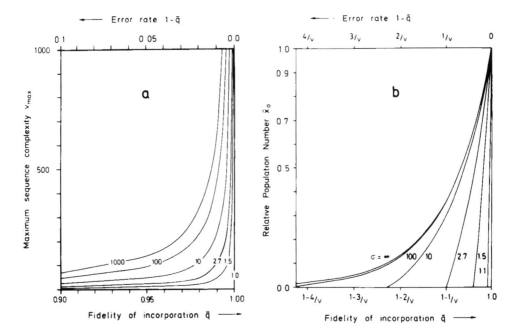

FIGURE 6. Error threshold and wild type composition. The error threshold relation has two important implications. First, it limits the sequence complexity ν that can be stably selected, (i.e., the amount of information that can be reproducibly accumulated) because for any given error rate $(1 - q)$ per nucleotide the fraction of correct copies strongly decreases with increasing ν as $q^\nu = \exp\{-\nu(1 - q)\}$. Second, near the error threshold, the number of master copies becomes strongly reduced. Near the error threshold the quasispecies is evolutionarily most versatile because it produces a wide variety of mutants without destabilizing the master sequence. Both dependences, i.e., the maximum tolerated information content ν_{max} and the fraction of master copies x_o as function of the mean error rate $(1 - \bar{q})$, shown in (a) and (b), are dependent on σ_o, the selectivity of the wild type. Only if σ_o reaches infinity does the information content become unrestricted. When the σ_o values are in a range between 2 and 10, their logarithms do not deviate much from 1. However, there are cases where σ_o may be close to unity, resulting in very small values for $\ln\sigma_o$. At larger error rates the fraction of master copies gets very small, even for large values of σ_o. Near the (apparent) intersection of the curves with the abscissa ($\bar{x}_o = 0$) the master sequence becomes unstable. The deterministic x_o values then become exceedingly small, even though they never vanish, as simulated by the results of second order perturbation theory which represent good approximations down to (invisibly in the figure) small values of x_o. (After Eigen, M., *Chem. Scr.*, 26B, 13, 1986. With permission.)

errors in successive replication rounds until the information is entirely lost. Figure 6a depicts the correlation between $\nu_{max} \cdot (1 - \bar{q})$, and σ_o, and Figure 6b shows how the master is populated correspondingly. While the relation between ν_{max} and reciprocal error rate $(1 - \bar{q})$ is straightforward, the term $\ln\sigma_o$ introduces an influence reflecting the fine structure of the mutant distribution. Explicit expressions have been obtained for certain assumed continuous distributions using renormalization procedures.[29] If many neutral mutants are present, σ_o might come close to one requiring an increased fidelity of reproduction to maintain the information. Vice versa, an absence of neutral or nearly neutral mutants (which usually are the ones that can easily be detected) is not per se indicative of a low error rate. If, on the other hand, a part of the information is dispensable under the prevailing conditions, the pertinent σ_i values approach unity and the fidelity of reproduction does not suffice to conserve the information, resulting in rapid degeneration of the dispensable part of the genome.[30,31] Moreover, parts of the genome not contributing to the information, e.g., "selfish DNA" sequences,[32] must be neglected for the maximum information content ν_{max}.

In order to obtain explicit solutions and thereby determine the population fine structure of the mutant spectrum, the system of rate equations (4) has to be solved in its explicit form. This could be achieved for various model cases by ordering the rate equations in a recursive way, similar to the recursive reconstruction of sequence space, as illustrated in Figure 2. It turned out that the product of the rate parameter vector and the (uniform) error matrix can be written in a form that is well known to mathematicians (hyperoctahedral group).[33] Another important insight was the realization of a close analogy between the quasispecies model and the two-dimensional Ising model used in physics to describe cooperative phenomena such as ferromagnetism.[34]

Rather than going further into theoretical details, let us present some numerical examples which will shed light on the remarkable properties of the quasispecies model, and which will let the Darwinian concept appear in an entirely new light.

VI. THE QUASISPECIES CONCEPT: EXAMPLES AND CONCLUSIONS

The existence of a threshold relation for selection suggests a new interpretation of the phenomenon of natural selection, namely, as a kind of condensation phenomenon, the condensation or localization of a sequence distribution in a limited area in sequence space. Let us demonstrate the clues of the new view with three numerical examples (Figures 7 to 9).[35] In Figure 7, the relative population numbers (x_d) are plotted as functions of the average symbol copying fidelity $\bar{q}$ (error rate: $1 - \bar{q}$) for the following model system: master sequence (binary digits, $\nu = 50$), uniform mutant distribution with degenerate rate parameters ($W_{ii} = 0.1W_{oo}$). The limiting case $(1 - \bar{q}) = 0$ corresponds to the trivial case of survival of the wild type as treated in Section II.A. It is seen that with increasing error rate the different error classes build up while the wild-type or master sequence decays strongly towards the error threshold, which occurs at $(1 - \bar{q}) = \ln10/50 = 0.046$ or $\bar{q} = 0.954$. The logarithmic plot in Figure 7 demonstrates the drastic drop of master sequence population at the threshold, quickly approaching uniform population of all error types. The sum of all copies for $\nu = 50$ is at maximum for $d = 25$. (There is only one master vs. $\binom{50}{25} \approx 10^{14}$ 25-error copies). Below, but close to the error threshold, the conditions for evolution are optimal in that the wild type is stable and a maximum number of mutants is present in the distribution. Such a behavior suggests annealing procedures, i.e., exceeding the error threshold for short periods of time in order to speed up evolution under controlled laboratory conditions, analogous to zone melting in crystallization. Hence, error rate indeed somehow plays the role of temperature in a melting or evaporation process. "Melting" or "evaporation", however, is to be interpreted as total loss of information, i.e., it is the delocalization or "thinning out" of the distribution in sequence space. The analogy to the melting of spin orientation at the Curie temperature (as described by the two-dimensional Ising model) becomes immediately apparent.[34]

The change of the classical Darwinian concept becomes even more obvious if we consider mutant distributions with individual rate parameters, especially distributions including neutral or nearly neutral mutants. Consider the example shown in Figure 8. Here, two master sequences differ in fitness by 10%. However, the inferior sequence is assumed to be surrounded by 50 closely related mutants that are somewhat better adapted than those surrounding the absolutely fittest. Near the error threshold, the inferior mutant outgrows the fitter one by virtue of its better mutant environment. This simulation clearly demonstrates that the target of selection is not the single species, but rather the distribution of the quasispecies as a whole.

Another typical case of individual value distribution[12] is represented in Figure 9. A binary sequence of 100 positions with an adapted uniform error rate $(1 - \bar{q})$ of 10^{-2} is assumed. In the upper part of the figure, we depict a particular ridge in the fitness value landscape for mutants with increasing Hamming distance. The lower part of the figure shows the expectation values for the population numbers of those mutants (filled symbols). The open symbols in this logarithmic plot represent the Poissonian expectance values, i.e., the pop-

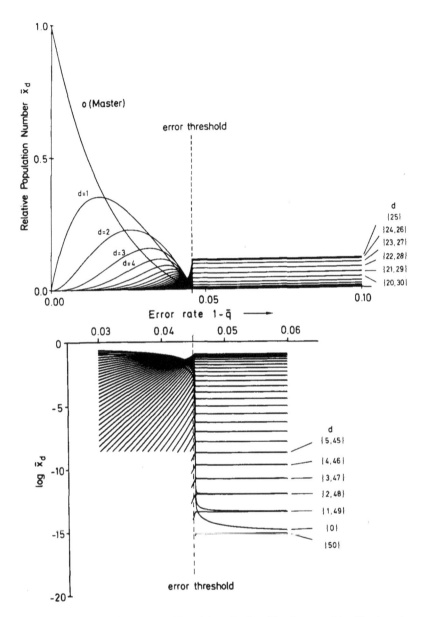

FIGURE 7. The physical nature of Darwinian selection. The phase transition-like character of natural selection, governed by the error threshold, is demonstrated in this computer simulation by Schuster and Swetina.[35] Assumed is a sequence distribution of complexity v = 50 and a selectivity σ_o of 10, all mutants having degenerate selection values (W_{ii} = 0.1 $\times$ W_{oo}). Mutant classes with increasing Hamming distance d from master, build up with increasing (mean) error rates (1 − q). Their relative population numbers $\bar{x}_d$ are obtained by summing over all $\bar{x}_i$ values of their members. The following features are observed: (1) only if $(1 - \bar{q}) \to 0$, does the master o survive exclusively; (2) near, but below the error threshold (at $(1 - \bar{q})$ = ln10/50 = 0.046), the master drops to a very low relative population number (cf. the logarithmic plot in the lower part) without loss of its information; (3) only by surpassing the error threshold, the information of the wild type is lost: all mutants (including the master) then reach the same expectation value 2^{-v}. Since the number of individual sequences with a Hamming distance 25 is highest, i.e., $\binom{50}{25}$, $\bar{x}_{25}$ dominates. Selection hence may be viewed as a kind of condensation of information in sequence space; exceeding the threshold means that the information volatilizes through an error catastrophe, the error threshold being an analogue of a phase transition temperature. (After Schuster, P., *Chem. Scr.*, 26B, 27, 1986. With permission.)

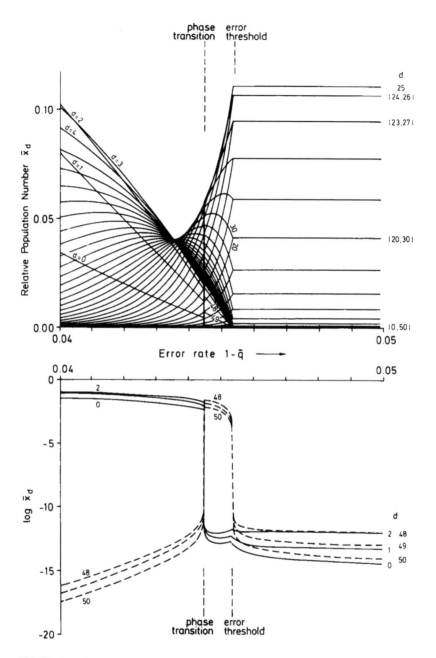

FIGURE 8.   The target of natural selection. The second example demonstrates that it is the quasispecies distribution as a whole that is rated for the selection. The data are, again, from Schuster and Swetina. In the simulation, a binary sequence of complexity 50 and degenerate mutant selective values of $W_{ii} = 0.1 \times W_{oo}$ are assumed again. In the value landscape a second peak only slightly lower than W_{oo} is assumed at the Hamming distance 50, having a selective value of $0.9 \times w_{oo}$. The second maximum, however, is assumed to be surrounded by 50 one-error mutants, all having selective values of $W_{dd} = 0.5 \times W_{oo}$ (for d = 49).

Two phase transitions are seen. At low mutation rates $(1 - q)$ the sequence o is the winner of the competition. At increasing error rate a threshold is surpassed where the slightly inferior sequence "50" is strongly favored by virtue of its better mutant environment. The logarithmic plot (lower part) emphasizes the sharp transition and the strong selection. Further increases of the error rate lead to surpassing a second (and final) error threshold where the selection forces do not suffice anymore to stabilize any master. (Courtesy of Drs. Schuster, P. and Swetina, J., University of Vienna.)

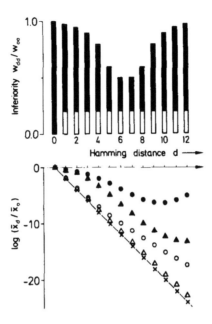

FIGURE 9. The importance of mutant bias in evolution. The final simulation demonstrates the role of ridges in the selectivity value landscape in directing the route of evolution. A binary sequence of the complexity 100 is assumed. In the upper part of the picture two value profiles are shown as functions of Hamming distances from the reference master o. Two selective value maxima (the second peak at d = 12 has a relative selectivity σ_o^{-1} of 0.98) are connected by three different landscapes: (1) a lowland, where all intermediate mutants have an inferiority of 0, (2) an altiplane, where all mutants have σ_o^{-1} values of 0.2 (open bars), and (3) a smooth mountain ridge with a monotonous decrease of the σ_o^{-1} values to 0.5 at d = {6,7}, followed by a rise to 0.98 at d = 12 (Filled bars). Adding a sequence at d = 13 that is neutral with the master ($\sigma_o^{-1} = 1.0$) would produce a symmetric distribution between the two maxima. For each profile, two limiting scenarios have been calculated. Triangles: elevated σ_o^{-1} values (W_{ii}/W_{oo}) are found for only one particular route comprising 12 subsequent defined mutation steps. The σ_o^{-1} values of all other mutants are zero. Circles: all mutants with the same Hamming distance to the nearest maximum have degenerate σ_o^{-1} values (W_{dd}/W_{oo}), i.e., all direct routes are possible because the succession of mutation steps is arbitrary. (Consult Figure 2d in order to correlate with the routes. Note that the routes for d = 12 comprise only a minute fraction of the possible connections in the 100-dimensional space, i.e., $2^{12}/2^{100} = 10^{-26.5}$.) The solid curve in the lower part of the picture represents the relative mutant populations according to a Poissonian distribution ($W_{dd}/W_{oo} = 0$). The open and filled symbols refer to the relative selectivity value distributions of the altiplane and smooth ridge profiles, respectively. The following features are seen: the relative populations around a peak in the selection value landscape that arises abruptly from a lowland drop rapidly with increasing Hamming distance from the master. The probability of reaching another peak with a Hamming distance of 12 is very low for typical laboratory populations. In the case where a sharp crest with low selection values is attached to the peak, the probability of finding a 12-error mutant is only slightly higher. If the peak is surrounded by a low plateau (between d = 0 and d = 12), this probability rises by nearly seven orders of magnitude. The effect of higher selectivity values is drastic: just one sharp crest with the profile depicted above brings the probability of occurrence of a 12-error mutant up to such an order of magnitude that the population necessary to find one strand can be realized readily in the laboratory. If all routes leading away from the summit have the same profile, a large multiplicity of 12-error mutants will occur (deterministically) in a laboratory population. The bias of selectivity guiding the evolutionary path along ridges in the selectivity value landscape is clearly demonstrated. Note also the cumulative effect of connected high value regions in the ν-dimensional landscape.

ulation numbers expected for low fitness values (i.e., mutants with $W_{ii} << W_{oo}$). The cumulative effect of a connected value landscape is clearly demonstrated. The 12-error mutant (probability of appearance 10^{-6}) then would appear deterministically in a micro-population (e.g., 10^9 particles), while it would never be found (probability 10^{-24}) in absence of the value ridge, i.e., in mutant regions of low fitness. The process of evolutionary optimization therefore can speed up by many orders of magnitude, depending on the fine structure of mutant fitness, which maps very sensitively into the fine structure of mutant population. By this biased guidance, a mutant in an environment of high values is favored to appear by many orders of magnitude more frequently than a corresponding mutant in an environment of low fitness. The multidimensional structure of sequence space largely enhances this effect with the consequence that optimization is a target-directed process despite the inherent stochastic nature of the mutational event.

These examples show in particular the importance of neutral or nearly neutral mutants which distort the smooth Poissonian landscape into a topography of bizarre structures. In deterministic theory there is no ambiguity in deciding how closely a mutant has to resemble the wild type in order to be called "neutral". All mutants are rated by their relative fitness according to $W_{ii}/(W_{oo} - W_{ii})$, and the cumulative enhancement of population numbers in mutant environments of high fitness can be taken into account in a quantitative way (Figure 9). On the other hand, since the mutant environment contributes to the value rating, and since small fitness differences $W_{oo} - W_{ii}$ map very sensitively into population numbers, one scarcely finds true degeneracies. Even if mutants are (nearly) neutral with respect to W_{oo}, their environments can hardly be sufficiently identical that they yield degenerate population numbers. Hence, there is usually a defined consensus sequence of the wild type that is found in sequence analysis.

Neutral theory had been introduced as inherently stochastic in nature.[11] What we have presented in this review, in contrast, is a deterministic model. It is clear that stochastic effects are present also in the deterministic model. Mutants, for instance, appear stochastically at the periphery of the wild type distribution. Such effects, too, have been considered recently in random-walk models.[36,37] Where are the limits of application? Stochastic models hold for relatively limited population numbers and large genome sizes. Mutants — even with low Hamming distances — in such a case appear with low expectance and are not populated deterministically. Actually, such models have been developed in population genetics. Consider a human genome with its three billion nucleotides of which a substantial part appears to be genetically significant. Local populations through which neutral mutants may spread can hardly be of a size adequate to warrant deterministic considerations. RNA or DNA molecule distributions, viruses, or even autonomous microorganisms easily appear in numbers up to 10^{12} and more, while their genomes are small enough to allow for large populations of low-error mutants (Figure 1). The important result of theory is that mutants with a high degree of fitness still appear deterministically at large Hamming distances from the master sequence (Figure 9), whereas mutants with low fitness soon become negligible. Because of the connectiveness of fitness landscapes, however, the mutants with high fitness are the ones that determine the evolutionary route, the uncertainty of which is considerably reduced as compared to any simple trial and error model. This, of course, does not exclude or reduce other chance effects, e.g., due to short- or long-term fluctuations of environment, super-imposed upon the historical process of evolution.[38,39]

VII. EXPERIMENTAL STUDIES OF QUASISPECIES

A. Model and Reality

The purpose of a physical theory is not so much to describe how processes occur in nature in every detail, but rather to understand why certain regularities can be observed. In this respect theory is no *alternative* to experiment, but rather *supplement*, enabling us to interpret correctly experimental results as well as to suggest further meaningful experiments to be carried out. There are numerous examples in molecular biology where a superficial interpretation led to a miscomprehension of the nature of the process, and reference to Darwinian behavior is among the most frequently used superficial methods of interpretation.

A rigorous treatment of the quasispecies model has shown us what kind of regularities are to be expected in selection and evolution of (vegetative) replicator systems. In some cases they even tell us where probabilitistic gaps exist or where a certain outcome will turn out to be inevitable. Theory, however, should not be applied to any experimental situation without carefully checking whether the assumptions and simplifications made really do apply.

Application of the quasispecies concept to bacterial genetics seems straightforward. Bacteria have a clearly defined and a highly reproducible growth kinetics. Already 40 years ago, the stochastic nature of mutations and the error propagation resulting from growth of mutants were recognized[40-42] and interpreted in a quantitative way.[43] On the other hand, analysis was restricted to a few mutations which could be detected as genetic markers. However, one should be aware of the low error rates of bacteria adapted to their genome size. Evolutionary rates for a single gene (about 3000 nucleotides) are thus — for laboratory time scales — relatively slow, and polymorphism of expression products is expected to be rather limited except when caused by some recombination mechanism such as transposition. For the same reason, direct screening of small parts of the genome should not uncover appreciable sequence heterogeneities.

Applications of theory to RNA replication and comparison with experimental results have been discussed in Chapter 1, Volume I. The fact that replication processes are usually catalyzed by replicases, that templates are copied in a complementary manner, and that all these reactions are of a multiple step nature, involving initiation, elongation (including pause sites), replica release, template release, or reactivation and various kinds of inhibition, does not cause any serious problems of interpretation. In general, the promises of the quasispecies model do apply and, where limitations are observed, they are obvious on the basis of the mechanisms involved. Dissymmetries in the kinetic parameters of plus and minus strands have been considered. The behavior found guarantees a concomitant selection evaluation of both strands.

An application to the mechanisms of viral infection requires some precautions. The mechanism of replication inside the host cell is not of the simple autocatalytic nature as was assumed in the model system. In fact, RNA viruses must contribute the essential part of their replication machinery by expression of their information, and the various ways this is realized are described in Volumes I and II. In contrast, DNA viruses — for which the quasispecies principles equally apply — were able to modify and supplement an already existing sophisticated replication and repair apparatus. Additional selection pressure is thus introduced by the processes of translation and packaging. Here we have to distinguish replication rounds from infection cycles. During an infection cycle, many replication rounds take place. All replicators within the host cell share a common replication and translation machinery; hence, selection pressure from expression of that machinery appears to be quite reduced within the infection cycle. Upon burst with a typical size of some thousand virus particles, only a relatively small fraction turns out to be infectious. Furthermore, the selection pressure depends sensitively on the conditions, and careful studies require experimental devices to keep conditions constant, e.g., a special flow reactor.[44]

B. Determination of Mutation Rates

There is ample experimental evidence for the genetic variance of organisms, particularly the strikingly high one of RNA viruses.[9,45] However, due to the lack of generally accepted definitions, sequence data are often differently interpreted. The main source of mutations is the inaccuracy of the replication process, and the mutation rate or frequency $1 - q_i$ is thus defined as a fraction of misincorporations at position i. Often *mutation* frequencies are not clearly distinguished from *mutant* frequencies, i.e., the proportion of a certain mutant appearing in the population. The latter are often also called mutation rates, but the two meanings are not identical and must be distinguished from one another.

An early approach to measuring the mutation rate was carried out by Benzer in a classical experiment[46,47] which became instrumental in elucidating genome structure and organization. Benzer chose for analysis the rII gene of bacteriophage T4, which is dispensable under a certain set of conditions, while being absolutely required under another. Thus, a large number of spontaneously arising or chemically induced rII mutants could be picked and amplified under permissive conditions. Mutants with a measurable rate of reversion to the wild-type phenotype (detected under nonpermissive conditions) were assumed to contain base substitutions which could be mapped by complementation. In the topographic map of the rII region of the T4 genome the frequencies of substitutions at the particular position were indicated. Thus, what has been determined is a *mutant* frequency, not a *mutation* rate. Positions where extraordinarily high mutant frequencies are found have been termed "hot spots", which creates the impression that the mutation rate at these positions is particularly high. This may be true, in particular, if recombination is involved, but the hot spot could also be produced by selection processes. The mutant frequency is a measure of the mutation rate only if two conditions are satisfied:

1.   Under permissive conditions, the mutations must be essentially neutral ($\sigma_o^{-1} \approx 1$), otherwise their concentration ratios are changed during amplification by selection forces.
2.   Under nonpermissive conditions, amplification (including DNA replication) must be completely wiped out ($\sigma_o^{-1} = 0$).

In fact, Benzer excluded detectably "leaky" mutants for his determinations; however, DNA replication (which continues to take place under non-permissive conditions)[48] raises the DNA population of some mutants considerably and can produce a hot spot even though the mutation rate is not markedly enhanced. If, on the other hand, the spontaneous formation of mutants is measured, high mutant frequencies are more likely to be caused by their neutrality rather than by the (definitely existing) variability of mutation rates.

The mutant frequencies, measured in terms of relative population numbers $\bar{x}_i$, assume defined values only in an equilibrated quasispecies distribution. Direct measurement of the mutation rate, on the other hand, must avoid such equilibration with its concomitant error propagation. Mutation rates may be optimally determined by measuring the mutation frequencies of replicas synthesized from a homogeneous population of templates in just one replication round.[49-53] Of course, the mutation frequency for any given mutant is very small and, thus, quite difficult to detect in a direct way. Therefore, no generally applicable method has yet been developed.

An elegant method, described already above, is to measure the reversion rate of a conditionally lethal mutant.[46,47,49-55] The method assumes that the mutation is perfectly neutral under permissive conditions and no selection takes place. Under nonpermissive conditions, the number of revertants can be accurately determined. (Frameshift mutants are mostly lost.) In a strict sense, mutation rates $1-q$ obtained in this way are only valid for the sequence positions tested. However, even though an experimental proof is still lacking, incorporation fidelities in RNA replication probably do not vary from position to position more than one order of magnitude. Hence, errors by assuming average mutation rates $1-\bar{q}$ are probably not large. (There seems no molecular basis for the assumption of site-specific extraordinary misincorporation rates in replication itself; hotspots are rather due to differences in "ironing

Table 1
POLYMERASE ERROR RATES $(1-q)^a$

Subject	Error rate	Method
Nonenzymic	10^{-1}—10^{-2}	I^b
DNA polymerases		
E. coli pol I	3×10^{-4}—1.2×10^{-5}	I
E. coli pol I	$\sim 10^{-6}$	AKQV
E. coli pol II	8×10^{-5}	I
E. coli pol III	6×10^{-5}	I
E. coli pol III	$<10^{-7}$	AKQV
T4 pol	5×10^{-5}	I
T4 pol	$<2 \times 10^{-6}$	AKQV
AMV rev. T.	1.4×10^{-3}—7×10^{-4}	I
AMV rev. T.	3×10^{-3}—6×10^{-5}	AKQV
Eu polα	3×10^{-4}—6×10^{-5}	I
Eu polα	3×10^{-5}	AKQV
Eu polβ	2×10^{-4}—5×10^{-5}	I
Eu polβ	2×10^{-4}	AKQV
Eu polγ	2×10^{-4}	I
Eu polγ	1.4×10^{-4}	AKQV
RNA polymerases		
E. coli rpo	10^{-3}—10^{-5}	I
E. coli rpo	1×10^{-4}	AKQV

Note: Abbreviations: E. coli — *Escherichia coli*, T4 — coliphage T4, AMV — Avian myeloblastosis virus, Eu — eukaryotic, pol — DNA polymerase, rev. T — reverse transcriptase, and rpo — RNA polymerase.

[a] Data are from Reference 49.
[b] Method I: incorporation of mismatched nucleotides into synthetic polymers. Other methods are described in Table 2.

out''[46] base substitutions. Indeed, hotspots in the lac gene were the result of defects in detecting and repairing the substitutions rather than of unusual misincorporation rates.[56] The molecular basis for the hotspots in the rII gene has been extensively discussed;[57,58] recombination processes[59] are also involved.) A simpler approach for measuring the replication accuracy has been the following: a homopolymer is replicated and the misincorporation of a radioactively labeled nucleotide into the replica is determined directly.[60,61] Homopolymers are not physiological templates, however, particularly if the "replication" mechanism required slippage.[6,62] Furthermore, they often contain "wrong" nucleotides as impurities (e.g., deamination products). It is thus not surprising that mutation rates determined by the latter method may be by one or two orders of magnitude higher than those determined by the former method (Table 1).

The mutation rates of polymerases are much higher for RNA-dependent RNA polymerases,[63] DNA-dependent RNA polymerases,[64] and RNA-dependent DNA polymerases[65] than for DNA polymerases,[49-53] mainly because the former lack a proofreading mechanism. All reliable values are found to be of the order of 10^{-4} to 10^{-3}, their predominant products being base transitions. It is conceivable that this is already the lower bound of mutation rates in the absence of proofreading,[65] being dictated by the tautomerization equilibrium of uridylic acid residues[7] and possibly by GU pairing.

Table 2
METHODS EMPLOYED TO DETERMINE MUTATION RATES

Virus Population

I	Pick conditionally lethal clones, amplify under permissive conditions (A)	Pick arbitrary clones, amplify (B)	Passage several times (C)
II	Isolate nucleic acids, replicate for one round (K)	Amplify (label) subclones (L)	(Label), passage (M)
III	Amplify (in- or transfect) under nonpermissive conditions (Q)	Isolate nucleic acids, digest with base-specific nucleases (R)	Retrotranscript, clone into DNA plasmids (S)
IV	Plate, count revertants (V)	Compare fingerprint patterns (W)	Compare sequences (read-off) (X)
IV	Determine specific product formed (Y)	Determine sequence ambiguity in read-off (Z)	Compare sequences of different strains (O)

The different methods of measuring mutation rates using viral populations rather than nucleic acids are listed in Table 2 (cf. Chapter 7, Volume II). They have in common the disadvantage that the relevant selection pressure also includes gene expression and packaging; thus, selection forces inevitably play a decisive role.

On first sight, a direct sequence comparison of the different mutants in the population seems to be most straightforward.[66] However, one cannot sequence single copies of RNA, but must accumulate about 10^9 copies by amplifying subclones. 10^9 particles correspond to about 30 generations or replication rounds. After 30 generations of the wild type — exponential growth assumed — a mutant with a reproduction rate reduced by only 10% will have completed 27 generations and is thus present in about 1/10 of the wild-type concentration. Hence, a mutant, in order to be detected by this method, has not only to be just viable, it must also be almost neutral. Some quasispecies distributions contain a much larger fraction of neutrals than others do; moreover, the variability tolerated at different regions of the same genome (e.g., structural proteins vs. functional proteins) differ considerably. Large differences in mutation rates may be simulated by this fact.

If the subclones used for sequence comparison are selected from a clone that has undergone only a relatively small number of replication rounds, one certainly is not dealing with a distribution representative for equilibrium appearance of the mutants. For practical reasons, usually only a part of the sequence is analyzed which is much below the critical sequence length ν_{max}. The majority of the sequences will be thus identical to the master accompanied by a few mutants.[66] A mutant sequence appearing more than once should be counted as only one mutation event, since it is probably produced by error propagation. The number of different positions that have been found mutated in relation to the number of positions unchanged should be a measure of $1 - q$, again with the reservation that only nearly neutral mutants are observable upon amplification. The values $1 - q$ thus found (e.g., from the data for influenza [66] we calculate a value of 7.5×10^{-5}) represent lower bounds if they are related to the average mutation rate including nonviable and those viable mutants that are clearly nonneutral (e.g., with reduced burst sizes).

Another often-applied method, developed by Domingo et al.,[67] screens the oligonucleotide fingerprint patterns of viral subclones for sequence alterations. Generally, only a fraction of sequence alterations produce discernible pattern changes, but one can correct for this fact. Much more serious is the bias introduced by amplification. Again, only nearly neutral mutations are detected, while lethal mutants or those with strongly reduced burst sizes remain undetected. No numerical correction for the fraction of the mutants lost is possible, since the proportion of sequence alterations seriously affecting virus amplification is unknown.

Table 3
MUTATION RATES OF DIFFERENT
ORGANISMS DETERMINED IN VIVO

Organism	Rate	Method[a]	Ref.
Bacteria			
Escherichia coli	2×10^{-10}	AQV	55
Salmonella	2×10^{-10}	AQV	55
typhimurium			
Neurospora crassa	7×10^{-12}	AQV	55
DNA viruses			
Coliphage λ	2.4×10^{-8}	AQV	55
Coliphage T4	1.7×10^{-8}	AQV	55
RNA viruses			
Coliphage Qβ	3×10^{-4}	DQE[b,c]	69
Coliphage Qβ	1×10^{-3}	BLRW	67
Foot-and-mouth V	$10\%^{d}$	SO	18
Foot-and-mouth V	$1\%^{d}$	MRWO	99
Poliovirus	$<2 \times 10^{-6}$	BLX	66
Polio V	$>100/8000^{d}$	MRWO	20
Polio V	3×10^{-5}	AV	45
Vesicular stomatitis V	5×10^{-5}	AV	96
Vesicular stomatitis V	5×10^{-4}	AKRY	63
Vesicular stomatitis V	3×10^{-4}	AV	97
Entero V	$320/8000^{d}$	MRWO	21
Sindbis V	$<1 \times 10^{-6}$	AV, KZ	83
Influenza V	2.7×10^{-4}	CMSX	72
Influenza V	1.5×10^{-5}	BLX	66
Influenza V	3×10^{-5}	AV	97
Influenza V	$5\%^{d}$	MRWO	17
Sendai V	3×10^{-5}	AV	97

[a] Methods are listed in Table 2.
[b] D = produce mutant by site directed mutagenesis.
[c] E = determine outgrowth kinetics of wild type.
[d] Number of positions changed in sequence.

Considering the many different roles of the viral RNA required for amplification, one may presume that the mutation rates are underestimated by one or two orders of magnitude. This factor may vary considerably from virus to virus and may explain the striking differences in mutation frequencies found for different members of the picornavirus family, even though their replication mechanisms are probably similar (Table 3). The drawback of the fingerprint method is its limited sensitivity. It has thus been used generally to screen the heterogeneity of a virus population after many passages, i.e., after a quasispecies distribution has been established. A derivation of mutation rates from the mutant frequency of a quasispecies, however, would be only possible if the selective values of the mutants were also known. More sensitive methods to detect oligonucleotide pattern changes were developed,[63] but then the advantage of simultaneously screening many positions had to be sacrificed. The measurement of mutation frequencies after a limited amplification was possible, thus avoiding error propagation and extensive selection.

The most reliable determinations of mutation rates are based on the reversion rates of mutants generated artificially by site-specific mutagenesis.[23,69] Since the method involves a determination of the selective values of mutant and wild type, it will be described in the next section.

The vast experimental material available making use of these or similar methods (Table

2) is listed in Table 3. The numerical values described were extracted from the sources without correcting them for mistakes in evaluation.

C. Determination of Selective Values

When the sequence heterogeneity of a viral clone is screened after many passages, error propagation is extensive and a steady state quasispecies has been established where each mutation is represented according to its rate of formation by mutation and its competition success as reflected in its selective value. We introduced earlier the individual selectivities of mutants (in relation to the wild type) by dividing their selective values (W_{ii}) by that of the mutant (W_{oo}). The selective values σ_o^{-1} of virus mutants are certainly complicated functions of replication rates, qualities of expression products, as well as of the rates of packaging to infectious viruses. Effects of infection multiplicity must also be taken into account.[70] Therefore, it does not suffice simply to measure eclipse times and burst sizes of different virus mutants to derive selective values, particularly when different mutants interfere with each other (e.g., defective interfering particles.)[45,71] There are only a few exact experimental determinations of selectivities. Weissmann and collaborators estimated the relative selectivity of a cloned Qβ mutant from the outgrowth kinetics of a revertant wild type. Once the relative selectivity was known, it was also possible to extrapolate back to the appearance of the first wild type revertant in the mutant population and obtain a value for the mutation rate also.[69] Again, this method is only applicable for individual mutants with not-too-small reciprocal selectivity values. Unlike the mutation rates, selectivities may vary over many orders of magnitude and the σ_o^{-1} values may cover the full spectrum between 0 (a lethal mutation) and 1 (a neutral mutation). Quasispecies distributions are likely to contain a vast number of different mutant types with σ_o^{-1} values near 0, each present only in trace amounts, and a small number of different types with σ_o^{-1} values near 1 that are present in high abundance. The later class contains all the mutants that are found when sampling subclones.[67]

The technique of cloning viral RNAs into DNA plasmids avoids the bias of selectivity in the amplification of clones to macroscopic amounts for sequence analysis.[72] It does, of course, involve retrotranscription which also may produce errors and may also have a strong bias against some mutants. However, since only one round of replication is involved and the bias follows criteria completely different from those of RNA expression, it is nonetheless likely to produce valuable results.

D. Mutation and Selection of Self-Replicating RNA Species

By carrying out the RNA replication in vitro, all constraints on the selective value other than RNA replication itself disappear. This is the rationale behind the famous "extracellular Darwinian evolution" experiment introduced by Spiegelman and co-workers.[73] Starting with infectious Qβ RNA, and ensuring virtually indefinite growth conditions by serially diluting the products into fresh incubation mixture containing precursors and enzyme, they ended up with an RNA species with a chain length only a fraction of that of Qβ RNA, but with an approximately 15-fold faster duplication rate. Unfortunately, at the time of the experiment, the technology of analyzing what was going on at the molecular level was not yet available, and thus we cannot interpret the results of these and other early experiments of selecting mutants adapted to certain test tube conditions.[74,75] However, there seems to be no plausible alternative to the interpretation that these "RNA variants" are formed by mutation and selection.[76]

In our laboratory, we have undertaken to analyze mutation and selection in a quantitative way, using the (meanwhile) quite well-understood replication of short-chained RNA templates by Qβ replicase. In the self-replicating RNA model system, the selective bias is always working *cis*, i.e., on the genotype itself; thus, the influence of competitors is limited to sharing the same environment. We have shown in a previous article (Chapter 1, Volume I)

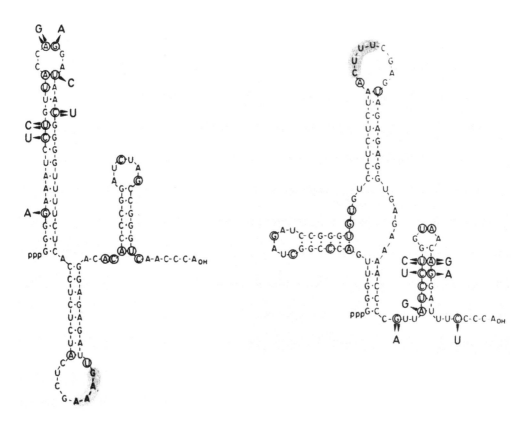

FIGURE 10.   Sequence heterogeneity of nanovariant RNA. The sequence of the most abundant subquasispecies WS1 (x_1 = 0.4) is shown. The base transitions designated with one and two arrows are found in WS2 (x_2 = 0.24), the mutations with two arrows are found also in WS3 (x_3 = 0.07). Base transitions found in relative population numbers >0.05 are circled. The shaded nucleotides were found to be deleted in a minor fraction of the RNA copies. Note that the foldings of plus and minus strands (calculated as the most stable secondary structures) are not antiparallel, because GU base-pairs do not disrupt a double helix as the complementary AC base-pairs do. (After Schaffner, W., Ruegg, K. J., and Weissman, C., *J. Mol. Biol.*, 117, 877, 1977. With permission.)

how to calculate the selective value of different competing RNA species in this situation. Since the chain lengths of these species are at least an order of magnitude below the inverse of the error rate of Qβ replicase, the majority of replication copies can be assumed to be error-free. The W_{ii} values of the quasispecies equation can thus be replaced by the selective values described in Chapter 1 of Volume I, and the relative selectivities can be calculated from the ratio of the selective values of mutant and wild type (W_{ii}/W_{oo}).[77]

The results from sequence analysis indicate a broad quasispecies distribution: fingerprints of all species show an unusually large number of "minor spots", i.e., oligonucleotides present only as small fractions of the stoichiometric amount.[78,79] A nanovariant RNA species was analyzed in more detail and found to contain a variety of base transitions and a deletion (Figure 10).[80] Although an unambiguous interpretation is not possible from the sequence data of the total RNA population, the results indicate the following picture: a rather broad mountain range in the value landscape was found, centered around three peaks at Hamming distances of 3, 4, and 7, and separated by gorges of "missing links" (i.e., mutants with x_i-values too small to show in the analysis). The relative peak heights were found to depend sensitively on slight differences in the growth conditions. A strong correlation between selective value and secondary structure was indicated by the observation that base exchanges between the peaks compensated each other to conserve the same loop and stem structure.

Note that RNA secondary structures of complementary strands are usually not antiparallel since G:U pairs are tolerable in a helix, while A:C pairs disrupt it. Presumably the missing links contain mutants with strong asymmetry in plus-minus synthesis because mutations often affect the structure in only one of the complementary strands. The selective values are indeed functions of the rates of the plus and minus cycles whose rate constants enter in a multiplicative manner (Chapter 1, Volume I).[81] Calculation of the selective values shows furthermore that in the linear growth phase, where the RNA concentration is high, mutants with extensive sequence alterations have a selective advantage because their loss rate by formation of hybrid double strands with the wild type is lower.[77]

Base insertion and base deletion were found to be frequent replication errors in self-replicating RNA species.[82] This is probably a consequence of the replication mechanism of single-stranded RNA. Presumably, the double helix between template and the newly synthesized replica is only a few base-pairs long, and the impact of a slippage error is particularly small where homonucleotide clusters occur. Insertion and deletion mutations have also been detected in viral genomes;[57,58,83,84] however, in cistronic regions they cause frameshifts with usually disastrous consequences for the gene product and are thus usually weeded out. On the other hand, it is conceivable that viruses may use such errors for stretching their information content: the lysis gene of class A phages is read only if a frameshift at a certain position in the coat cistron occurs, which may be caused either by a frame error of the ribosome or by a deletion of a nucleotide in replication. Since expression of the lysis gene is also observed when the gene is cloned into DNA, the former explanation seems more likely.[85]

Subcloning of self-replicating RNA species by amplifying single RNA strands by Qβ replicase,[86] did not result in the production of sequentially different subclones. Instead, a rapid reformation of the original quasispecies distribution was observed. This result is not unexpected: we have seen above (Figure 1) that the probability of finding a particular error mutant is highly dependent on the sequence complexity (which is in this case equal to the template chain length). Thus, the population sizes required for sequencing (10^8 to 10^9 copies) contain certainly all one-nucleotide-error mutants and a substantial proportion of all possible two-error mutations. Rapid drifting can thus occur. For determination of error frequencies, we used the technique of annealing the RNA strands to double strands. The double strands obtained are then treated with single-strand-specific nucleases.[87,88] Imperfections in the double strand due to hybridization of different mutants are cleaved by ribonucleases; the 5'-OH ends formed can be labeled and sequenced (Figure 11). The self-replicating RNAs species analyzed by that method show surprisingly broad quasispecies distributions. Cloning a self-replicating RNA population into DNA (Biebricher and Luce, unpublished), on the other hand, indeed produced clones that have homogeneities within the clone, but showing sequence differences from one clone to another. Screening of a real quasispecies is hence possible, even though quite troublesome.

E. Evolution Experiments with Self-Replicating RNA

Evolution and selection is best understood under conditions of autocatalytic exponential growth with enzyme and substrate in large excess.[76] Under these conditions all RNA mutants grow independently of each other and their selective values are equal to their overall growth rates in the exponential phase.[77] The selective values remain constant and independent of time as long as exponential growth conditions are maintained. The growth of RNA has to be balanced by an influx of nutrient and an outflux of products, experimentally readily realized by serial transfer of an aliquot of the growth mixture into fresh medium containing substrate and enzyme. Spiegelman and co-workers chose these defined conditions for selecting an ethidium bromide-resistant mutant from the well-characterized RNA species MDV-1.[89]

O G   A L A+U U+C L   O   G   A   L A+U U+C L

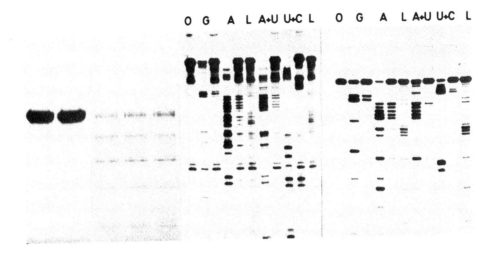

FIGURE 11.   Sequence heterogeneity of minivariant RNA. MNV-11 RNA was annealed to the double-stranded form and treated with RNase A which produces nicks at mispaired positions (left two lanes of electropherogram) that become detectable upon melting to single strands (next lanes). Mispairing is not randomly distributed over all positions and is frequently caused by slippage errors. Insertions or deletions of one nucleotide blur the sequence by producing superposition of the same sequence at one nucleotide displacement (left sequence). Cutting the mispaired nucleotides eliminates the ambiguities in the read-off of the sequence, as shown by the sequencing gels[95] on the right side where a band from the electropherogram was used for sequencing. The "hot spots" — due to mutations that do not affect replication — can be easily mapped on the sequence. 0, untreated; A, G, C, U, nucleotide ladders; L, random hydrolysis ladder.

The experimental design starts with 2×10^6 strands of MDV-1 which is amplified in the presence of ethidium bromide by a factor of about 10^5 and then diluted 10^5-fold into fresh mixture. After 19 such transfers (corresponding to 315 replication rounds or to a 10^{95}-fold amplification), a mutant had been selected which grew faster in the presence of ethidium bromide than wild type did. Its sequence was found to be altered by three base transitions that apparently arose one after the other, rather than simultaneously.

An inherent flaw of the technique used was to start each transfer with a small fraction of the population, an experimental procedure to avoid a rapid increase of the reaction volume to astronomic dimensions. The population dropped thus to numbers as low as 10^6, and the chance of finding one or a few strands of an advantageous mutant formed during the previous transfer in the randomly selected population used for inoculation of the next transfer is very small. Nearly exclusively, those mutants present already in the quasispecies distribution of the previous transfer can be selected since only they have undergone sufficient amplification to survive the decimation. The population size used for the next inoculation certainly did include all one-error mutants, but only a fraction of all possible two-error mutants. Thus, the second point mutation could only appear after selection had raised the population size of the one-point mutations sufficiently. These assumptions are supported by evaluation of the selection kinetics.[89] Since we have seen above that structural conservation rules render it unlikely to find another selective value maximum with single point mutations, methods must be developed to enlarge mutational jumps. Increasing the error rates artifically could help avoiding experimentally infeasibly large populations in order to reach more distant selection peaks. Indeed, the adaptation to the drug achieved in the described experiment was quite weak, whereas self-replicating RNA species synthesized *de novo* in the presence of the drug[90] score much better (Figure 12). An improvement in the size of adaptation can be obtained by (1) working with higher population sizes (e.g., 10^{10} to 10^{12} strands), (2) adapting the mutation rate to the error threshold conditions, and (3) raising intermittently

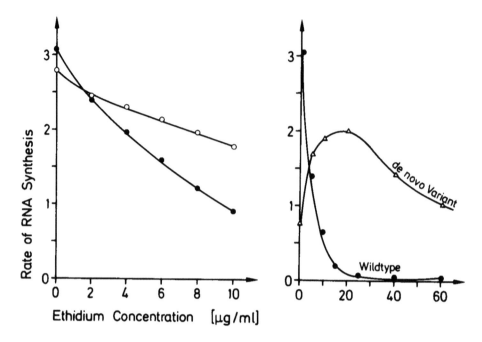

FIGURE 12.   RNA species selected for resistance against ethidium bromide. Replication of MDV-1 is inhibited by ethidium bromide (a) and (b). Replication of MDV-1 for 315 rounds resulted in the selection of a resistant MDV-1 mutant that grew faster in the presence of ethidium bromide than MDV-1 did (a). *De novo* synthesis in the presence of ethidium bromide led to a resistant RNA species not only resistant to quite high concentrations of ethidium bromide, but was even stimulated by moderate concentrations of the drug (b). (After Kramer, F. R., Mills, D. R., Cole, P. E., Nishihara, T., and Spiegelman, S., *J. Mol. Biol.*, 89, 719, 1974 and Sumper, M. and Luce, R., *Proc. Natl. Acad. Sci. U.S.A.*, 72, 162, 1975. With permissions.)

artificially the error rate in order to produce high error copies (''annealing''). Development of automated devices to realize these conditions are in progress.

We have shown that selection in the linear growth phase is much more efficient than in the exponential growth phase.[76,77] The quantitative interpretation is, however, much more difficult: the selective values change with time and are dependent on the concentration of other competitors. Furthermore, during the long time span of the slow linear growth, side reactions occur. We have consistently observed that mutants are selected which have undergone extensive sequence alterations, including insertions of new genetic material. These mutational reactions include random nucleotide additions at the 3'-end of a template (similar to the mechanism of *de novo* synthesis) as well as reactions similar to recombinations (Biebricher and Luce, unpublished). Their mechanism is not yet understood, but is under investigation. These mutations — even though they occur rarely — were found to be far more effective in optimization of RNA species than point mutations.[91] During such a reaction, the quasispecies distribution, which has been established around a local selective value maximum, is replaced by a new one when one of these rare mutations has accidently hit a mountain region belonging to a higher selection maximum. Point mutations lead then to the new maximum around which a new quasispecies distribution is established.

In viral RNA replication recombination-like processes have also been observed;[92,93] different mechanisms, e.g., copy choice during replication or transesterification, have been discussed (see the contributions of Volume II for details). The rarity of those reactions makes their investigation an experimental challenge. However, it is likely such reactions play a major role in the evolution of viruses, since they allow for mutation jumps sufficiently large enough to escape a local fitness maximum.

VIII. SUMMARY

A. Population Structure

1. The target of selection is not a single genotype and its corresponding phenotype, but rather a distribution of sequentially related genotypes which are selectively rated as a whole.
2. The distribution is centered around one or several degenerate master sequences characterized by high (at least locally maximal) efficiency of (excess) reproduction.
3. Close to the error threshold (Equation 8), where evolution proceeds most rapidly, the fraction of individual master sequences (relative to the total population) is rather low.
4. Under the same conditions, i.e., near the error threshold, mutants dominate the population. Mutants of low reproductive efficiency essentially are populated according to their probability of being produced through erroneous copying of the master. For a uniform error model, this probability would be essentially p^d, where d is the Hamming distance between master and mutant, $p = (1-q) / 3q$, and $1-q$ the (uniform) error rate. If all mutants are low in reproductive efficiency (as compared to the master), the total mutant distribution would be Poissonian ($\epsilon^d \exp(-\epsilon)$)/d! where $\epsilon = \nu(1-q)$ is the expectation value of errors in a sequence of ν positions and d the Hamming distance to the master sequence).
5. Mutants of high reproductive power (relative to the master sequence) modify the population structure drastically. They are rated with respect to the master in terms of the form $W_{ii}/(W_{oo} - W_{ii})$ where W_{ii} and W_{oo} are the (accurate) reproduction rates of mutant i and master o. These ratios are hyperbolic with respect to a variation of W_{ii} relative to W_{oo}. In regions of mutants of high reproductive efficiency, the population numbers may be modified by many orders of magnitude due to cumulative effects described in the legend of Figure 9.
6. Since the value landscape is multiply connected, mutants will be highly populated in mountain regions of the fitness landscape, though being sparsely populated in valleys and planes. Since the maximum selective value is to be expected in a mountainous region where the mutant population is high, there is a strong mutational bias towards improvement up to the maximum value, despite the chance nature of the mutation process itself. This guidance effect of nearly neutral mutants is extremely powerful. It changes the picture of a wild type that moves through sequence space by random walk into the picture of the quasispecies with its subtle mutant structure migrating through sequence space in an internally controlled manner and thereby guiding itself to the peaks of a connected value landscape. Due to the difference in the magnitude of diagonal and off-diagonal rate paramets (cf. p^d) the motion is not smooth, but rather a discontinuous alternation of condensation in sequence space (selection) and dislocation (evolutionary advancement).

B. Wild Type Sequence

1. According to the new picture, the wild type is not characterized by a single sequence.
2. Despite 1., the wild type often shows an unambiguous consensus sequence.
3. The consensus sequence often is, but need not be, identical with the master sequence.
4. The master sequence of a selected distribution is not necessarily the sequence distinguished by the highest individual reproductive efficiency. The mutant environment is rated also and may contribute decisively.

C. Error Threshold

1.   The error threshold plays a central role in determining the population structure and its adaptive behavior. It provides a condition for stability of genetic information. It is analogous to a melting temperature. Violating the threshold causes a "melting" of information.

2.   The error threshold condition correlates the mutational error rate with sequence length and functional efficiency. If $\bar{q}$ is the average fidelity of nucleotide reproduction (and, accordingly, $1 - \bar{q}$ the average error rate per nucleotide), then $\bar{q}^{\nu}$ is the probability of correctly reproducing a whole sequence comprising ν positions. A stable master sequence then has to be by a factor σ_o (selectivity of master) more efficient in its reproduction than the average of its mutant spectrum: $\sigma_o\bar{q}^{\nu} > 1$, yielding (for $\bar{q}$ near one) $\nu < \ln\sigma_o/(1 - \bar{q})$. For large sequence length ν,σ may become close to one. As a consequence, we observe for autonomous organisms (pro- and eukaryotes) usually $(1 - \bar{q}) \ll 1/\nu$.

3.   Quasispecies populations adapt to the error threshold, i.e., try to correlate sequence length with error rate, depending on the master selectivity. At this threshold, the evolutionary rate becomes maximal. (Adaptation then means that those distributions which correlate error rate with sequence length evolve most rapidly).

4.   Persistent violation of the error threshold causes deterioration of the quasispecies. i.e., "melting" of the information (error catastrophe). Occasional violations of short duration, on the other hand, may be advantageous in providing faster adaptation through annealing (in analogy to "zone melting").

5.   As a consequence, there is nothing odd with RNA viruses having high error rates as compared to autonomous organisms. Their error rate is just adapted to their limited length. Since this adaptation depends on σ_o (and thus on the value distribution of mutants), different viruses may show quite different population structures, even if they are similar in length.

6.   It is important to distinguish the *mutant* frequency $\bar{x}_i$ from *mutation* rate $(1 - \bar{q})$. The latter is much more uniform than the former. So-called hot spots have been usually derived from mutant frequencies rather than from mutation rates.

D. Peculiarities of RNA Viruses

1.   RNA viruses usually appear with separated plus and minus strands. Selective evaluation is modified correspondingly (cf. Chapter 1, Volume I).

2.   The reproduction of RNA viruses requires viral gene expression to synthesize the replication machinery and, hence, the simple quasispecies model has to be modified accordingly.

3.   One has to distinguish replication rounds from infection cycles. During an infection cycle, many replication rounds take place for which the population of viral RNA in a host cell shares the same machinery. This may explain the low fraction of viable, i.e., infectious virus particles observed for RNA viruses. One may take all these effects into account by suitable adaptations of the model and test them by computer simulation. So far, experimental data strongly support the relevance of the quasispecies model for RNA viruses.

ACKNOWLEDGMENTS

We are indebted to Dr. W. C. Gardiner for critical reading of our manuscript.

GLOSSARY OF SYMBOLS

A_i	Rate coefficient for amplification of template i (including miscopying, cf. Q_i).
$\bar{A}(t) = \Sigma_k A_k x_k$	Weighted average of A-values taken over complete population.
$\bar{A}_{k \neq 0}$	Average of A-values taken over the mutant spectrum (i.e., over the total population minus the master copy $\Sigma_{k \neq 0} A_k x_k / \Sigma_{k \neq 0} x_k$).
B_H	Brownian function (cf. Mandelbrot) of fractal dimension H^{-1}.
$c_i(t)$	Time-dependent concentration of component i.
c_i	Equilibrium concentration of component i.
D_i	Rate coefficient for decomposition of template i.
$\bar{D}(t), \bar{D}_{k \neq 0}$	Cf. $\bar{A}(t)$ and $\bar{A}_{k \neq 0}$ (defined analogously).
d	Hamming distance (number of nucleotides in two aligned sequences that differ in homologous positions).
d_{ij}	Hamming distance between sequences i and j.
$E_i = A_i - D_i$	Rate coefficient for excess production of template i (including miscopying), cf. W_{ii}.
$\bar{E}(t) = \Sigma_k E_k x_k(t)$	Mean excess productivity at time t.
$\bar{E}_{k \neq 0}$	Mean excess productivity of mutants (without master).
$\epsilon = \nu(1 - \bar{q})$	Expectation value of errors in a sequence of length ν and average copying error rate $(1 - \bar{q})$ per nucleotide.
G	Gibbs free energy (free enthalpy).
ΔG_{ik}^o	Standard free enthalpy difference between states i and k.
H	Reciprocal fractal dimension (used in Figure 4 for B_H).
i, j	Indices characterizing single states or sequences.
k	Running index in summations.
κ	Number of digit classes in sequential information (for nucleic acids: $\kappa = 4$).
λ_i	Eigenvalue referring to matrix of rate coefficients.
$\bar{\lambda}(t) = \Sigma_k \lambda_k y_k(t)$	Average eigenvalue ($\bar{E}(t)$).
M	Number of dimensions in Brownian function (Mandelbrot).
N_d	Number of different sequences belonging to an error class d (i.e., all sequences having the Hamming distance d from a reference sequence).
ν	Sequence length (i.e., number of nucleotides comprising the sequence under consideration), also called complexing.
o	Index referring to master sequence (error class d = 0).
$p = (q^{-1} - 1)/(\kappa - 1)$	Probability expression occuring in Q_{ij}.
$Q_d = \binom{\nu}{d} q^{\nu - d}(1 - q)^d$	Probability of producing an unspecified d-error copy in one replication round.
$Q_{ij} = q^\nu p^{d_{ij}}$	Probability of producing a specified error copy i by one replication round of template j.
$Q_{ii} = q^\nu$	Probability of producing a correct copy of template i of length ν in one replication round.
q	Single digit copying fidelity, i.e., the probability of copying correctly a single nucleotide. Correspondingly, $(1 - q)$ is the probability of making an error per nucleotide copied. $(1 - q)$ is usually called "error rate".

$\bar{q}$	Average copying fidelity. If the different positions in a sequence show different copying fidelities, $\bar{q}$ results as a geometric mean according to $\prod_{k=1}^{\nu} q_k = \bar{q}^{\nu}$.
RT	Thermal energy.
$\sigma_o = A_o/(D_o + \bar{A}_{k \neq o} - \bar{D}_{k \neq o})$	Selectivity of master copy relative to its mutant spectrum.
$\sigma_o^{-1} = A_i/A_o$	Reciprocal selectivity of mutant i assuming negligible or degenerate decomposition rate coefficients.
t	Time.
τ	Time (variable in time integral).
$W_{ii} = A_i Q_{ii} - D_i$	Rate coefficient for error-free excess reproduction of template i. Diagonal coefficients appearing in system of rate equations.
$W_{ij} = A_j Q_{ij}$	Rate coefficient for producing copy i by miscopying template j. Off-diagonal coefficients in systems of rate equations.
$x_i(t) = c_i(t)/\Sigma_k c_k(t)$	Relative population variable characterizing the fraction of sequences i in total population.
$y_i(t)$	Transformed relative population variable (containing contributions of all sequences) referring to an eigenvalue λ_i.
$y_o(t)$	Refers to maximum eigenvalue constituting the quasispecies whose final structure is established if $y_o(t)$ has reached 1.
$\bar{x}_i, \bar{y}_i$	Variables referring to established selection equilibrium of master and mutant states.

REFERENCES

1. **Eigen, M.,** Self-organisation of matter and the evolution of biological macromolecules, *Naturwissenschaften,* 58, 465, 1971.
2. **Eigen, M. and Schuster, P.,** *The Hypercycle — A Principle of Natural Self-Organization,* Springer-Verlag, Heidelberg, 1979.
3. **Fox, S. W.,** Metabolic microspheres: origins and evolution, *Naturwissenschaften,* 67, 378, 1980.
4. **Miller, S. L. and Orgel, L. E.,** *The Origins of Life on Earth,* Prentice-Hall, Englewood Cliffs, N.J., 1974.
5. **Eigen, M. and Schuster, P.,** Stages of emerging life: five principles of early organization, *J. Mol. Evol.,* 19, 47, 1982.
6. **Kornberg, A.,** *DNA Replication,* Freeman-Cooper, San Francisco, Calif., 1980.
7. **Saenger, W.,** *Principles of Nucleic Acid Structure,* Springer-Verlag, Heidelberg, 1984.
8. **Cech, T. R. and Bass, B. L.,** Biological catalysis by RNA, *Annu. Rev. Biochem.,* 55, 599, 1986.
9. **Domingo, E., Martina-Salas, E., Sobrino, F., de la Torre, J. C., Portela, A., Ortin, J., Lopez-Galindez, C. Perez-Brena, P., Villanueva N., Najera, R., VandePol, S., Steinhauer, S., DePolo, N., and Holland, J.,** The quasispecies (extremely heterogeneous) nature of viral RNA genome populations: biological relevance — a review, *Gene,* 40, 1, 1986.
10. **Kimura, M. and Ohta, T.,** *Theoretical Aspects of Population Genetics,* Princeton University Press, Princeton, N.J., 1971.
11. **Kimura,** *The Neutral Theory of Molecular Evolution,* Cambridge University Press, London, 1983.
12. **Eigen, M.,** The physics of molecular evolution, *Chem. Scr.,* 26B, 13, 1986.
13. **Hamming, R. W.,** *Coding and Information Theory,* Prentice-Hall, Englewood Cliffs, N.J., 1980.
14. **Rechenberg, I.,** *Evolutionsstrategie,* Problemata Formann-Holzboog, Stuttgart-Bad Cannstatt, 1973.
15. **Mandelbrot, B. B.,** *The Fractal Geomety of Nature,* W. H. Freeman, New York, 1983.
16. **Levy, P.,** Le mouvement brownien fonction d'un ou de plusier paramètres, *Rend. Mat. (Roma),* 22, 24, 1963.
17. **Ortin, J., Nájera, R., Lopez, C., Davila, M., and Domingo, E.,** Genetic variability of Hong Kong (H3 N2) influenza viruses: spontaneous mutations and their location in the viral genome, *Gene,* 11, 319, 1980.

18. **Martinez-Salas, E., Ortin, J., and Domingo, E.,** Sequence of the viral replicase gene from foot-and mouth disease virus C$_1$-Santa Pau (C-S8), *Gene,* 35, 55, 1985.
19. **Kew, O. M., Nottay, B. K., Hatch, M. H., Nakano, J. H., and Obijeski, J. F.,** Multiple genetic changes can occur in the oral Poliovaccines upon replication in humans, *J. Gen. Virol.,* 56, 337, 1981.
20. **Nottay, B. K., Kew, O. M., Hatch, M. H., Heyward, J. T., and Obijeski, J. F.,** Molecular variation of type I vaccine-related and wild Poliviruses during replication in humans, *Virology,* 108, 405, 1981.
21. **Takeda, N., Miyumura, K., Ogino, T., Natori, K., Yamakazi, S., Sakurai, N., Nakazono, N., Ishii, K., and Kono, R.,** Evolution of enterovirus type 79: oligonucleotide mapping analysis of RNA genome, *Virology,* 134, 375, 1984.
22. **Fersht, A. R., Shi, J.-P., Wilkinson, A. J., Blow, D. M., Carter, P., Wayre, M. M. Y., and Winter, G. P.,** Analysis of enzyme structure and activity by protein engineering, *Angew. Chem. Int. Ed. Engl.,* 23, 467, 1984.
23. **Weber, H., Taniguchi, T., Müller, W., Meyer, F., and Weissmann, C.,** Application of site-directed mutagenesis to RNA and DNA genomes, *Cold Spring Harbor Symp. Quant. Biol.,* 43, 669, 1978.
24. **Mekler, P.,** Determination of nucleotide sequences of the bacteriophage Qβ genome: organization and evolution of an RNA virus, Ph.D. thesis, University of Zurich, Switzerland, 1983.
25. **Fiers, W., Contreras, R., Duerinck, F., Haegemann, G., Iserentant, D., Merregaert, J., Min Jou, W., Molemanns, F., Raeymakers, T., Van den Berghe, A., Volckaert, G., and Isebaert, M.,** Complete nucleotide sequence of bacteriophage MS2 RNA: primary and secondary structure of the replicase gene, *Nature (London),* 260, 500, 1976.
26. **Weaver, L. H., Rennell, P., Peteete, A. R., and Matthews, B. W.,** Structure of phage P22 gene 19 lysozyme inferred from its homology with phage T4 lysozyme. Implications for lysozyme evolution, *J. Mol. Biol.,* 184, 739, 1985.
27. **Thompson, C. L. and McBride, J. L.,** On Eigen's theory of the self-organization of matter and the evolution of biological macromolecules, *Math. Biosci.,* 21, 127, 1974.
28. **Jones, B. L., Enns, R. H., and Rangnekar, S. S.,** On the theory of selection of coupled macromolecular systems, *Bull. Math. Biol.,* 38, 15, 1976.
29. **McCaskill, J. S.,** A localization threshold for macromolecular quasispecies from continuously distributed replication rates, *J. Chem. Phys.,* 80 (10), 5194, 1984.
30. **Dykhuizen, D. E. and Hartl, D. L.,** Selection in chemostats, *Microbiol. Rev.,* 47, 150, 1983.
31. **Atwood, K. C., Schneider, L. K., and Ryan, F. J.,** Selective mechanisms in bacteria, *Cold Spring Harbor Symp. Quant. Biol.,* 16, 345, 1951.
32. **Orgel, L. E. and Crick, F. C. H.,** Selfish DNA: the ultimate parasite, *Nature (London),* 284, 604, 1980.
33. **Rumschitzki, D.,** Spectral properties of Eigen's evolution matrices, *J. Math. Biol.,* 24, 667, 1987.
34. **Leuthausser, I.,** An exact correspondence between Eigen's evolution model and a two-dimensional Ising system, *J. Chem. Phys.,* 84, 1884, 1986.
35. **Swetina, J. and Schuster, P.,** Self-replication with error — a model for polynucleotide replication, *Biophys. Chem.,* 16, 329, 1982.
35a. **Schuster, P.,** The physical basis of molecular evolution, *Chem. Scr.,* 26B, 27, 1986.
36. **McCaskill, J. S.,** A stochastic theory of macromolecular evolution, *Biol. Cybernet.,* 50, 63, 1984.
37. **Schuster, P. and Sigmund, K.,** Random selection and the neutral theory — sources of stochasticity in replication, in *Stochastic Phenomena and Chaotic Behaviour in Complex Systems,* Schuster, P., Ed., Springer-Verlag, Berlin, 1984.
38. **Demetrius, L.,** The meaning of selective advantage in macromolecular evolution, *Chem. Scr.,* 26, 1986, in press.
39. **Demetrius, L.,** Self-organization in macromolecular systems: the notion of adaptive value, *Proc. Natl. Acad. Sci. U.S.A.,* 81, 6068, 1984.
40. **Luria, S. E. and Delbrueck, M.,** Mutations of bacteria from virus sensitivity to virus resistance, *Genetics,* 28, 491, 1943.
41. **Lea, D. E. and Coulson, C. A.,** The distribution of the number of mutants in bacterial populations, *J. Genet.,* 49, 264, 1949.
42. **Novick, A. and Szilard, L.,** Experiments with the chemostat on spontaneous mutations of bacteria, *Proc. Natl. Acad. Sci. U.S.A.,* 36, 708, 1950.
43. **Lederberg, J. and Zinder, M.,** Concentration of biochemical mutants of bacteria with penicillin, *J. Am. Chem. Soc.,* 70, 4267, 1948.
44. **Husimi, Y. and Keweloh, H.-C.,** Continuous culture of bacteriophage Qβ using a cellstat with a bubble wall-growth scraper, *Rev. Sci. Instrum.,* 58, 1109, 1987.
45. **Holland, J., Spindler, K., Horodyski, F., Grabau, E., Nichol, S., and Van de Pol, S.,** Rapid evolution of RNA genomes, *Science,* 215, 1577, 1982.
46. **Benzer, S.,** On the topography of the genetic fine structure, *Proc. Natl. Acad. Sci. U.S.A.,* 47, 403, 1961.
47. **Benzer, S.,** On the topology of the genetic fine structure, *Proc. Natl. Acad. Sci. U.S.A.,* 45, 1607, 1959.

48. **Mosig, G., Luder, A., Garcia, G., Dannenberg, R., and Bock, S.,** *In vivo* interactions of genes and proteins in DNA replication and recombinations of phage T4, *Cold Spring Harbor Symp. Quant. Biol.,* 43, 501, 1978.

49. **Loeb, L. A. and Kunkel, T. A.,** Fidelity of DNA synthesis, *Annu. Rev. Biochem.,* 52, 429, 1982.

50. **Loeb, L. A., Weymouth, L. A., Kunkel, T. A., Gopinathan, K. P., Beckman, R. A., and Dube, D. K.,** On the fidelity of DNA replication, *Cold Spring Harbor Symp. Quant. Biol.,* 43, 921, 1978.

51. **Fersht, A. R.,** Fidelity of replication of phage φX174 by DNA polymerase III holoenzyme: spontaneous mutation by misincorporation, *Proc. Natl. Acad. Sci. U.S.A.,* 76, 4946, 1979.

52. **Fersht, A. R. and Knill-Jones, J. W.,** DNA polymerase accuracy and spontaneous mutation rates: frequencies of purine:purine, purine:pyrimidine and pyrimidine:pyrimidine mismatches during DNA replication, *Proc. Natl. Acad. Sci. U.S.A.,* 78, 4251, 1981.

53. **Liu, C. C., Burke, R. L., Hibner, U., Barry, J., and Alberts, B.,** Probing DNA replication mechanisms with the T4 bacteriophage *in vitro* system, *Cold Spring Harbor Symp. Quant. Biol.,* 43, 469, 1978.

54. **Drake, J. W., Allen, E., Forsberg, S. A., Preparata, R.-M., and Greening, E. O.,** Spontaneous mutation, *Nature (London),* 221, 1128, 1969.

55. **Drake, J. W.,** Comparative rates of spontaneous mutation, *Nature (London),* 221, 1132, 1969.

56. **Coulondre, C., Miller, J. H., Farabaugh, P. J., and Gilbert, W.,** Molecular basis of base substitution hotspots in *Escherichia coli, Nature (London),* 274, 775, 1978.

57. **Brenner, S., Barnett, L., Crick, F. H. C., and Orgel, A.,** The theory of mutagenesis, *J. Mol. Biol.,* 3, 121, 1961.

58. **Pribnow, D., Sigurdson, D. C., Gold, L., Singer, B. S., Napoli, C., Brosius, J., Dull, T. J., and Noller, H. F.,** rII cistrons of bacteriophage T4: DNA sequence around the intercistronic divide and positions of genetic landmarks, *J. Mol. Biol.,* 149, 337, 1981.

59. **Mosig, G.,** Recombination in bacteriophage T4, *Adv. Genet.,* 15, 1, 1970.

60. **Hall, Z. W. and Lehman, I. R.,** An *in vitro* transversion by a mutationally altered T4-induced DNA polymerase, *J. Mol. Biol.,* 36, 321, 1968.

61. **Inoue, T. and Orgel, L. E.,** A nonenzymatic RNA polymerase model, *Science,* 219, 859, 1983.

62. **Radding, C. R. and Kornberg, A.,** Enzymatic synthesis of deoxyribonucleic acid: kinetics of primed and *de novo* synthesis of deoxynucleotide polymer, *J. Biol. Chem.,* 237, 2877, 1962.

63. **Steinhauer, D. A. and Holland, J.,** Direct method of quantitation of extreme polymerase error frequencies at selected single base sites in viral RNA, *J. Virol.,* 57, 219, 1986.

64. **Englisch, U., Gauss, D., Freist, W., Englisch, S., Sternbach, H., and von der Haar, F.,** Fehlerhäufigkeit bei der Replikation und Expression der genetischen Information, *Angew. Chem.,* 97, 1033, 1985.

65. **Kunkel, T. A., Eckstein, F., Mildvan, A. S., Koplitz, R. M., and Loeb, L. A.,** Deoxynucleoside(1-thio)triphosphates prevent proofreading during *in vitro* DNA synthesis, *Proc. Natl. Acad. Sci. U.S.A.,* 78, 6734, 1981.

66. **Parvin, J. D., Moscona, A., Pan, W. T., Lieder, J., and Palese, P.,** Measurement of the mutation rates of animal viruses: influenza A virus and poliovirus type 1, *J. Virol.,* 59, 377, 1986.

67. **Domingo, E., Sabo, D., Taniguchi, T., and Weissmann, C.,** Nucleotide sequence heterogeneity of an RNA phage population, *Cell,* 13, 735, 1978.

68. **Domingo, E., Flavell, A., and Weissmann, C.,** *In vitro* site-directed mutagenesis: generations and properties of an infectious extracistronic mutant of bacteriophage Qβ, *Gene,* 1, 3, 1976.

69. **Batschlelet, E., Domingo, E., and Weissmann, C.,** The proportion of revertant and mutant phage in a growing population, as a function of mutation and growth rate, *Gene,* 1, 27, 1976.

70. **Spindler, K. R., Horodyski, F. M., and Holland, J. J.,** High multiplicities of infection favor rapid and random evolution of vesicular stomatitis virus, *Virology,* 119, 96, 1982.

71. **O'Hara, P. J., Horodyski, F. M., Nichol, S. T., and Holland, J. J.,** Vesicular Stomatitus Virus mutants resistant to defective-interfering particles accumulate stable 5′-terminal and fewer 3′-terminal mutations in a stepwise manner, *J. Virol.,* 49, 793, 1984.

72. **Fields, S. and Winter, S.,** Nucleotide sequence heterogeneity and sequence rearrangements in influenza virus cDNA, *Gene,* 15, 207, 1981.

73. **Mills, D. R., Peterson, R. L., and Spiegelman, S.,** An extracellular Darwinian experiment with a self-duplicating nucleic acid molecule, *Proc. Nat. Acad. Sci. U.S.A.,* 58, 217, 1967.

74. **Levisohn, R. and Spiegelman, S.,** Further extracellular Darwinian experiments with replicating RNA molecules: diverse variants isolated under different selective conditions, *Proc. Natl. Acad. Sci. U.S.A.,* 63, 807, 1969.

75. **Saffhill, R., Schneider-Bernloehr, H., Orgel, L. E., and Spiegelman, S.,** *In vitro* selection of bacteriophage Qβ variants resistant to ethidium bromide, *J. Mol. Biol.,* 51, 531, 1970.

76. **Biebricher, C. K.,** Darwinian selection of RNA molecules *in vitro*, in *Evolutionary Biology,* Vol. 16, Hecht, M. K., Walace, B., and Prance, C. T., Eds., Plenum Press, New York, 1983, 1.

77. **Biebricher, C. K., Eigen, M., and Gardiner, W. C.,** Kinetics of RNA replication: competition and selection among self-replicating RNA species, *Biochemistry,* 24, 6550, 1985.

78. **Mills, D. R., Kramer, F. R., and Spiegelman, S.,** Complete nucleotide sequence of a replicating RNA molecule, *Science,* 180, 916, 1973.

79. **Biebricher, C. K., Eigen, M., and Luce, R.,** Product analysis of RNA generated *de novo* by Qβ replicase, *J. Mol. Biol.,* 148, 391, 1981.

80. **Schaffner, W., Ruegg, K. J., and Weissmann, C.,** Nanovariant RNAs: Nucleotide sequence and interaction with bacteriophage Qβ replicase, *J. Mol. Biol.,* 117, 877, 1977.

81. **Biebricher, C. K., Eigen, M., and Gardiner, W. C.,** Kinetics of RNA replication: plus-minus asymmetry and double-strand formation, *Biochemistry,* 23, 3186, 1984.

82. **Biebricher, C. K.,** Replication and selection kinetics of short-chained self-replicating RNA molecules, in *Positive Strand RNA viruses,* Vol. 54, UCLA Symp. Molecular and Cellular Biology, Brinton, M. A., and Rueckert, R., Eds., Alan R. Liss, New York, 1986.

83. **Durbin, R. K. and Stollar, V.,** Sequence analysis of the E2 gene of a hyperglycosylated, host-restricted mutant of Sindbis Virus and estimation of mutation rate from frequency of revertants, *Virology,* 154, 135, 1986.

84. **Darlix, J. L. and Spahr, P. F.,** High spontaneous mutation rate of Rous sarcoma virus demonstrated by direct sequencing of the RNA genome, *Nucleic Acids Res.,* 11, 5953, 1983.

85. **Kastelein, R. A., Remaut, E., Fiers, W., and van Duin, J.,** Lysis gene expression of RNA phage MS2 depends on a frameshift during translation of the overlapping coat protein, *Nature (London),* 295, 35, 1982.

86. **Biebricher, C. K., Eigen, M., and Luce, R.,** Kinetic analysis of template-instructed and *de novo* RNA synthesis by Qβ replicase, *J. Mol. Biol.,* 148, 391, 1981.

87. **Shenk, T. E., Rhodes, C., Rigby, P. W. J., and Berg, P.,** Biochemical method for mapping mutational alternations in DNA with S1 nuclease: the location of deletions and temperature-sensitive mutations in Simian Virus 40, *Proc. Nat. Acad. Sci. U.S.A.,* 72, 989, 1975.

88. **Myers, R. M., Larin, Z., and Maniatis, T.,** Detection of single base substitutions by ribonuclease cleavage at mismatches in RNA:DNA duplexes, *Science,* 230, 1242, 1985.

89. **Kramer, F. R., Mills, D. R., Cole, P. E., Nishihara, T., and Spiegelman, S.,** Evolution *in vitro:* sequence and phenotype of a mutant RNA resistant to ethidium bromide, *J. Mol. Biol.,* 89, 719, 1974.

90. **Sumper, M. and Luce, R.,** Evidence for *de novo* production of self-replicating and environmentally adapted RNA structures by bacteriophage Qβ replicase, *Proc. Natl. Acad. Sci., U.S.A.,* 72, 162, 1975.

91. **Biebricher, C. K., Eigen, M., and Luce, R.,** Template-free RNA synthesis by Qβ replicase, *Nature (London),* 321, 89, 1986.

92. **Yang, F. and Lazzarini, R. A.,** Analysis of the recombination event generating a Vesicular Stomatitis Virus delection defective interfering particle, *J. Virol.,* 45, 766, 1983.

93. **Meier, E., Harmison, G. G., Keene, J. D., and Schubert, M.,** Sites of copy choice replication involved in generation of Vesicular Stomatitus Virus defective-interfering particles RNAs, *J. Virol.,* 51, 515, 1984.

94. **Eigen, M.,** Macromolecular evolution: dynamical ordering in sequence space, *Ber. Bunsenges. Phys. Chem.,* 89, 658, 1985.

95. **Donis-Keller, H.,** Site specific enzymatic cleavage of RNA, *Nucleic Acid Res.,* 7, 179, 1979.

96. **Wunner, W. H. and Pringle, C. R.,** A temperature-sensitive mutant of Vesicular Stomatitis Virus with two abnormal virus proteins, *J. Gen. Virol.,* 23, 97, 1974.

97. **Portner, A., Webster, R. G., and Bean, W. J.,** Similar frequencies of antigenic variants in Sendai, Vesicular Stomatitis, and Influenza A viruses, *Virology,* 104, 235, 1980.

98. **Younger, J. S., Jones, E. U., Kelly, M., and Frielle, D. W.,** Generation and amplification of temperature-sensitive mutants during serial undiluted passages of vesicular stomatitis virus, *Virology,* 108, 87, 1981.

99. **Domingo, E., Davila, M., and Ortin, J.,** Nucleotide sequence heterogeneity of the RNA from a natural population of foot-and-mouth disease virus, *Gene,* 11, 333, 1980.

100. **Schubert, M., Harmison, G. G., and Meier, E.,** Primary structure of the vesicular stomatitis virus polymerase (L) gene: evidence for a high frequency of mutations, *J. Virol.,* 51, 505, 1984.

Index

INDEX

D

N

O

Milton Keynes UK
Ingram Content Group UK Ltd.
UKHW050307111024
449327UK00043B/2239